D1064757

Primary Succession on Land

Primary Succession on Land

SPECIAL PUBLICATION NUMBER 12 OF THE BRITISH ECOLOGICAL SOCIETY

EDITED BY

J. MILES

The Scottish Office Central Research Unit
New St Andrew's House
Edinburgh

D. W. H. WALTON

British Antarctic Survey
Natural Environment Research Council
High Cross, Madingley Road
Cambridge

OXFORD

BLACKWELL SCIENTIFIC PUBLICATIONS

LONDON EDINBURGH BOSTON

MELBOURNE PARIS BERLIN VIENNA

1993

and published for them by
Blackwell Scientific Publications
Editorial Offices:
Osney Mead, Oxford OX2 0EL
25 John Street, London WC1N 2BL
23 Ainslie Place, Edinburgh EH3 6AJ
238 Main Street, Cambridge
Massachusetts 02142, USA
54 University Street, Carlton
Victoria 3053, Australia

Other Editorial Offices:
Librairie Arnette SA
2, rue Casimir-Delavigne
75006 Paris, France

Blackwell Wissenschafts-Verlag GmbH
Meinekestrasse 4
D-1000 Berlin 15
Germany

Blackwell MZV
Feldgasse 13
A-1238 Wien
Austria

First published 1993

Set by Setrite Typesetters, Hong Kong
Printed and bound in Great Britain
at the University Press, Cambridge

DISTRIBUTORS

Marston Book Services Ltd
PO Box 87
Oxford OX2 0DT
(*Orders*: Tel: 0865 791155
Fax: 0865 791927
Telex: 837515)

USA
Blackwell Scientific Publications, Inc.
238 Main Street
Cambridge, MA 02142
(*Orders*: Tel: 800 759–6102
617 876–7000)

Canada
Oxford University Press
70 Wynford Drive
Don Mills
Ontario M3C 1J9
(*Orders*: Tel: 416 441–2941)

Australia
Blackwell Scientific Publications Pty Ltd
54 University Street
Carlton, Victoria 3053
(*Orders*: Tel: 03 347–5552)

A catalogue record for this title
is available from the British Library

ISBN 0–632–03547–1

Library of Congress
Cataloging-in-Publication Data

Primary succession on land/edited by J. Miles,
D. W. H. Walton.
p. cm. — (Special publication number 12 of the British Ecological Society)
Includes bibliographical references and index.
ISBN 0–632–03547–1
1. Ecological succession — Congresses. I. Miles, John, 1941– II. Walton, D. W. H. III. Series: Special publication ... of the British Ecological Society: no. 12.
QH540, P74 1993
574.5 — dc20

Contents

Preface

Like many scientific disciplines, ecology suffers from inexact terminology, sometimes because the objective named is imperfectly understood or because of the development of different historical applications of the same term. Succession is one of those imprecise terms. For present purposes we will define succession as the replacement of one community by another.

In their enthusiasm to develop the concepts of succession, ecologists have provided themselves with an extensive and confusing vocabulary. Despite this we can say that there is general agreement that primary succession still means the same today as when Clements first used it almost 90 years ago in his pioneer efforts to classify vegetation. It is, very simply, the establishment and subsequent development of the first assemblage of species on a previously unvegetated surface. Although this is often a soil surface it can be rock or even a man-made substrate.

Why devote a symposium to it? Surely a process as fundamental as this is already well-understood and documented? Unfortunately, this is not the case, and it is clear from our efforts to organize the meeting that there is still a lack of research in this area. In the latter part of the nineteenth century various German and Italian botanists recognized that the earliest stages of colonization of bare ground were associated with algae, lichens and mosses. This initial stage of a community was not of itself useful to humans and, since it was only visible in extreme sites such as sand-dunes, lava flows or freshly exposed moraines, it attracted relatively little interest. Research into succession concentrated on the more interesting, useful and widespread communities of grassland, herbfield and forest. Since any general description of succession had to make reference to the primary stages such data as did exist were frequently referenced but not built upon. Indeed, the studies at Glacier Bay, Alaska, became for many the classic studies on the nature and direction of primary succession.

Ecologists at large, including those in the British Ecological Society and the Ecological Society of America, have generally shown little interest in and sympathy for the microbial or even the cryptogamic components of ecosystems. Thus it is not surprising that, despite the early recognition of their importance as primary colonizers, little attempt was made to include the microbial community and its effects within the framework of primary succession, although a key role was ascribed to lichens. Yet it is at the microscopic level that any consideration of primary succession must start.

The symposium was an attempt to draw together microbiologists and ecologists, botanists and zoologists, into a more coherent discussion on the earliest stages of community development. In this we have been only partly successful, but in the endeavour have identified both strengths and weaknesses in present research. Of these, probably the most significant is the apparent lack of interest in the processes leading to the establishment of animal communities. More exciting is the way in which ecological opportunities offered by the Mount St Helens eruption were exploited.

The templet model for colonization propounded by Southwood can provide a framework within which to place the processes of primary succession. Fundamental to its successful application is the recognition of what constitutes an acceptable niche and its coincident availability with the arrival of a viable propagule. Defining these basic features has yet to be undertaken. Indeed, as quickly becomes clear from a perusal of current ecology textbooks, most of the present descriptions of colonization are based on observation and not experimentation.

New techniques are allowing both observation and experimentation at the microbial level. The interactions between substrate and colonizer in terms of stabilization and nutrient release are being recognized, while the prospect of global warming and ice sheet recession has suddenly made an understanding of primary succession much more fashionable. There is much to do in documenting and adequately modelling colonization and primary succession. This symposium provides an introduction to some recent initiatives. It was held at the University of Liverpool during 5–7 September 1989.

J. MILES
D. W. H. WALTON

Introduction: Understanding the fundamentals of succession

A. D. BRADSHAW
Department of Environmental and Evolutionary Biology, University of Liverpool, Liverpool L69 3BX, UK

Almost all ecologists have their own favourite bit of succession — an area of land and vegetation where nature has been allowed to lead its life unhindered by the external factors which dominate so much of our land surface. Perhaps the most favourite areas are those where nature is in the process of starting off from scratch, from some completely raw substrate where plants and animals have not existed before. For many ecologists these primary successions may be natural areas, such as salt-marsh or reed swamp or mountain scree. For others they may be unnatural, such as a quarry or a colliery spoil heap or a disused railway line. To many the former may seem better than the latter. In many cases, however, the latter may be more relevant to what has happened in the past in many ordinary landscapes, after cataclysms such as ice ages or earthquakes. Whichever they are, the quality that surely appeals is that these areas demonstrate the power of nature in colonizing something raw and skeletal, in which there has been nothing to help life on its way. What is to be found is there because of its own efforts and capabilities. It owes little or nothing to what was there before.

But primary successions have much more than this to offer and to excite us. They demonstrate the creative role of nature in building up complete and complex ecosystems from simple beginnings — ecosystems that in time have considerable species diversity and a complete range of functions without which much life would be impossible, or at least very limited. They also provide a unique opportunity to see all this happening over time, to correct the idea that life is static and that all is just as it was some time ago. This opportunity may come because there is an obvious age (and therefore time) series — or chronosequence — so that at one time we can see the effects of different amounts of time. But this is often not as convincing as it should be, because of the degree of presumption involved. The evidence which is much more powerful comes as we grow older, and have either a good memory or a camera, for then we can see the actual changes which do take place.

Such evidence inevitably challenges us to think about the processes which must be occurring while we watch, and it is likely that we will begin to speculate and analyse. But in the analysis of any ecological process it is all too easy to be woolly and superficial, or just boringly descriptive. There has been no worse place for this than in the study of ecosystems, where until recently there has been either description or philosophy rather than critical analysis. There has even been a movement to deny that ecosystems exist, or are a useful concept, which implies that there are no emergent properties of communities. If this approach means being analytical and reductionist, there is good sense in it. But if it means overlooking general properties and problems which should be thought about, then this approach is misleading.

What therefore are the major processes involved in primary succession, which must be involved if fully functioning ecosystems are to become established? To identify and understand what they are, there is good reason to be as reductionist as we can. It is imperative that we go back to fundamentals. Each process that we identify may have, perhaps, the power to be the important limiting factor to succession and ecosystem development. Each of them we must certainly consider in this symposium.

Let us start by looking at the problem from the point of view of the plant and animal.

1 *Getting there*. This is the obvious first step. If the plants (and animals) cannot get to the new area, nothing can start. It has two components: the organisms must *be available* in the vicinity, and they must be able to *disperse* effectively. We often think that there are no problems over this. But the difficulties of arrival, and the importance of chance effects, could give a stochastic aspect to primary succession which could persist for a long time.

2 *Establishing*. It is no use if what arrives cannot become established. There are problems of *germination*. It is often thought that lack of dormancy is best in colonizing situations, yet the extreme nature of the starting environment could favour species which remain dormant until conditions are favourable. Of course more species arrive than can survive, so there is an important process of *selection*, in which those species and individuals which are not adapted are eliminated. This involves an important property of *tolerance* to all the components of stress generated by the new environment. It seems difficult to accept the idea that there is one overall stress factor to be overcome in these situations; there will be many different and very specific ecological factors to which the immigrants will need to be adapted.

3 *Growth*. Nothing will happen if the plants and animals cannot grow. This can only be achieved by the *acquisition of nutrients* by the plants. Everything has to come from somewhere, and nutrients are no exception. Yet it is easy to forget this when contemplating an established ecosystem; it is much more obvious in an establishing one. This must lead to a consideration of supplies of the most important nutrients, such as nitrogen. We may find that the whole of ecosystem development is held up by difficulties in the development of the supply of a single nutrient. Growth is also dependent on water being available, so the *ensurance of a water supply* is another crucial matter. While this is obvious, the way in which water supply influences primary succession is not well-understood.

We can also look at primary succession from the viewpoint of the *habitat*.

1 *Initial characteristics*. In a primary succession the starting point is very skeletal material. It may therefore have very poor *physical* features, and its *chemical* characteristics will be totally unlike those of a well-developed soil. In some

situations there may be *toxicity*. All these problems set specific difficulties for the organisms which attempt to colonize, and may bias the early stages of succession considerably. But these features of the habitat will soon begin to change.

2 *Influence of allogenic factors*. Many influences arising from outside the developing community of organisms can rapidly have ameliorating effects. It may be that there is *accumulation* of soil-forming material, or *weathering* of materials already present. There may be *inputs of nutrients* from the air, or *leaching* away of toxic materials. Any of these, and other, allogenic factors could have a critical influence in allowing species to colonize and grow, by relieving a single limiting factor.

3 *Influence of autogenic factors*. The activities of the living organisms can clearly have a very profound effect. *Nutrient accumulation* will be brought about by the foraging and retaining powers of plants, which will become more effective as they become more numerous. Nitrogen fixation will begin, early or late depending on habitat conditions. Much of the nutrients will become stored by *organic matter accumulation* and then become available as *decomposition and cycling* processes develop. If these processes do not occur, there may well be problems for the developing ecosystem. At the same time *soil structure* will develop. The whole habitat is likely to become more favourable to plant and animal growth. What is surprising, however, is how little we really know about all these processes and their effects.

This should be the point at which we can terminate our examination of primary succession. But two further matters must be considered.

1 *Progression*. Because succession is a progressive process, conditions must change from one period to the next. As a result, factors that are important at one time may not be important at another. A factor may have important but fleeting effects, which are difficult to observe without very close monitoring.

2 *Species interaction*. So far the effects of one species or individual on another has been disregarded, because many of the most important aspects of succession are independent of them. But inevitably interactions must play an increasingly important role as the ecosystem develops. There may be beneficial interactions giving rise to *facilitation* effects. But equally there can be negative interactions leading to *inhibition* effects. All this can only add to the complexity of what is occurring.

This discussion has at last, and only at last, come round to consider those processes which have dominated so much of our thinking about succession recently, and treated them rather lightly. This is because the aim has been to suggest that studies of primary succession need to consider more simple and basic problems. In trying to make a living, the primary colonists, which are the beginning of the developing ecosystem, are faced with a series of hurdles. It is very encouraging, therefore, that the studies of primary succession which are being discussed at this meeting are about the different ways in which these hurdles are overcome.

Cryptoendolithic communities from hot and cold deserts: Speculation on microbial colonization and succession

J. R. VESTAL*
Department of Biological Sciences,
University of Cincinnati, Cincinnati, Ohio 45221–0006, USA

SUMMARY

1 The cryptoendolithic microbial community is a complete ecosystem containing primary producers and consumer/decomposers located within the pores of certain types of rocks. It has no predators.
2 This unique habitat provides enough protection for microbes to survive a physical environment characterized by extreme temperatures, aridity, low light and limited space.
3 As conditions for metabolic activity are so limited in both frequency and duration, determining colonization and succession events in this environment is very difficult.
4 Two types of community are recognized: one dominated by lichens and the other by cyanobacteria. The distribution of the communities appears to be related to water availability, the presence of iron oxide and rock strength.
5 It is postulated that within these endolithic communities succession may only occur in the decomposer microbiota, and that the plant community constitutes a climax at establishment.

INTRODUCTION

Biotic succession at the microbial level is different from the ecologist's normal expectations formed by studies of plant succession. The basic processes are similar: initial colonization by pioneer species, followed by successional sequences during which species changes occur due to competition and/or changes in the physical and chemical environment. The differences are in the size of the organisms, the length of time for successional events and the relative instability of microbial communities in relation to environmental change. For example, a complete successional sequence of microbes could occur in a 1 ml sample of milk in a few hours.

In this review, a unique microbial ecosystem will be discussed. It is one which

* J. R. Vestal died August 1992.

occurs in extreme hot and cold deserts and is limited by space, light, moisture, temperature and the chemical milieu. These limitations are so severe that growth is extremely slow and therefore succession must be considered on a time-scale of hundreds to thousands of years. The numerous studies on this microbial ecosystem have mainly involved characterization of the species and the effects of environmental variables on growth and metabolism. It has been difficult actually to study succession due to the slow turnover of the microbial community, but studies are underway to attempt to look at changes in the activity, biomass, metabolic status and community structure in this microbial ecosystem. In order to properly discuss the successional events, the system will be described in terms of its physical and chemical features, and how these affect growth and metabolism. Possible routes for colonization and succession will then be discussed.

Endolithic (within rocks) microbial communities exist inside hard substrata, such as sandstones, quartz, limestones, shale and coal. These endoliths can be subdivided into cryptoendoliths which usually live within the pore spaces of rock, (usually sandstones), chasmoendoliths that inhabit the fissures or cracks of a rock and euendoliths which actively penetrate calcareous substrata (Golubic *et al.* 1981).

The particular microbiota discussed here are considered to be crytoendolithic because the organisms inhabit the interstices of sandstones. The physical and chemical environments of cryptoendoliths are important in allowing as well as controlling microbial growth. In the cryptoendolithic microbiota discussed here, lichens, containing photobionts, and cyanobacteria are the dominant primary producers.

THE POLAR CRYPTOENDOLITHIC ENVIRONMENT

In 1976, Friedmann & Ocampo-Friedmann first described a cyanobacteria-dominated microbial cryptoendolithic community in the Ross Desert (Dry Valley) region of Antarctica at about 78°S. Some sandstones were found to contain a lichen-dominated microbial community characterized by the lichens *Buellia* or *Lecidea* (Friedmann 1982; Hale 1987) while others were dominated by cyanobacteria (Friedmann *et al.* 1989). Examples of cyanobacterial and lichen cryptoendolithic communities are shown in Fig. 1. The lichen-dominated community (Fig. 1A) is characterized by a black layer of lichen 1–3 mm inside the rock. This layer overlays a light coloured layer containing filamentous fungi, yeasts and bacteria. Often, but not always, there is a lower green layer characterized by green algae (Tschermak-Woess & Friedmann 1984). In the cryptoendolithic cyanobacteria community (Fig. 1B), there is only one visibly dark layer under the surface containing the cyanobacteria. There are heterotrophic eubacteria also associated with this layer, but invisible to the naked eye.

This unique ecosystem has been extensively studied and members of the community identified. Other components of the microbiota include a new green algal species, *Hemichloris antarctica* (Tschermak-Woess & Friedmann 1984), yeasts

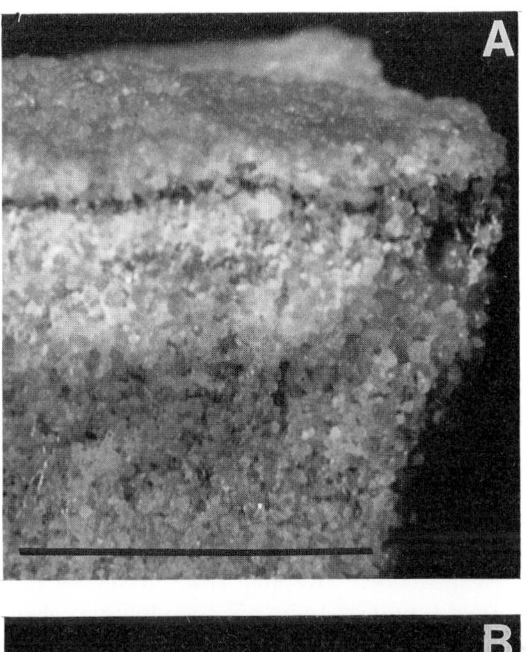

FIG. 1. (A) Cross-section of the cryptoendolithic lichen microbiota from Antarctica. The top dark band is lichen, followed by a white band containing filamentous fungi and other heterotrophic microbes. The grey region below the light band is stained by orange-coloured iron oxides. (B) Cross-section of the cryptoendolithic cyanobacterial microbiota from Antarctica. The top dark band contains the cyanobacteria. Below that band is the rock interior which is not stained by iron oxides. Bar markers are 1 cm.

(Vishniac 1985a), filamentous fungi (Friedmann 1982) and bacteria (Siebert & Hirsch 1988). A new species of bacteria from the Antarctic cryptoendolithic microbiota was found using 5S rRNA sequence analysis (Colwell *et al.* 1989). A detailed description of existing Antarctic cyanobacteria was recently published (Friedmann *et al.* 1989).

Since the discovery of these primitive ecosystems, many studies have focused on this habitat, including geochemistry (Weed 1985; Friedmann & Weed 1987; Johnston & Vestal 1989a, b), nanoclimate (McKay & Friedmann 1985; Friedmann *et al.* 1987), taxonomy (Friedmann 1982; Friedmann *et al.* 1989), nitrogen economy (Friedmann & Kibler 1980), biomass (Tuovila & LaRock 1987; Vestal 1988a), unique freezing properties of the membranes of the lichen-dominated community (Finegold *et al.* 1990) and physiology and carbon metabolism (Vestal 1988b). The functional interactions of these microbes living in such an extreme environment have also been mathematically modelled (Nienow *et al.* 1988a, b). In addition, microbial life in the extremes of polar environments has been recently reviewed (Longton 1988; Vincent 1988; Vincent & Ellis-Evans 1989; Wynn-Williams 1990).

The climate of the Ross Desert, Antarctica, is cold and dry, and there are virtually no signs of exposed life (Kappen *et al.* 1981). In the region where the Beacon sandstone is found colonized with cryptoendolithic lichens or cyanobacteria, the air temperature is consistently below freezing, the mean annual air temperature being −22°C. During the warmest months of the year, December and January, the mean air temperature is *c.* −7°C (Friedmann *et al.* 1987). There is little snow and much of what falls sublimes or is blown away by the frequent katabatic winds. Occasionally, the rocks are warmed by the sun and any remaining snow melts into the pores and serves as the source of water for the microbiota (Friedmann 1978).

The physical environment

The physical factors that impede exposed or epilithic life on the surface of soils or rocks are listed in Table 1. These factors are buffered if the organisms are

TABLE 1. Physical factors which affect the metabolic activity of the Antarctic microbiota, and the advantages of the cryptoendolithic habitat

Factor	Cause	Advantages of the cryptoendolithic habitat
1. Low moisture	Little snow, air humidity <50%	Rock matrix holds moisture for days after snowmelt
2. Low temperature	Air <0°C with frequent fluctuations	Direct solar heating warms rocks and thermal inertia dampens temperature variation
3. Variable radiation	High UV flux and intense visible light	Rock attenuates all radiation particularly UV
4. High wind velocity	Abrasion by strong katabatic winds	Protection by the rock

endolithic and they may operate together to influence the cryptoendolithic nature of these organisms. Most importantly, the rock provides a matrix for holding moisture, generated during the occasional summer snowfalls, to counteract loss to the dry environment (Friedmann *et al.* 1987). After a snowmelt, the moisture level in the rock can remain high (*c.* 80–100% relative humidity, RH) for up to 10–14 days while the outside air RH remains quite low (*c.* 10–30%) (Friedmann *et al.* 1987). Additionally, direct solar heating of the rock can raise the inside temperatures to 5–10°C and the rock provides a relatively thermally stable environment due to its thermal inertia (McKay & Friedmann 1985). Light is another important factor which affects life in the cryptoendolithic community. On the rock surface, the light intensity varies diurnally from *c.* 50–1800 $\mu mol\ m^{-2}\ s^{-1}$ (Friedmann *et al.* 1987). Nienow *et al.* (1988b) showed that light intensity was attenuated about an order of magnitude per millimetre depth within the porous sandstone. Thus, light intensities of 1·0–0·1 $\mu mol\ m^{-2}\ s^{-1}$ would be found in the lower portions of the biotic zone, which is 1–5 mm inside the rock matrix. The rock also provides a mechanical shield against wind abrasion from the strong katabatic winds coming from the polar plateau. Even though the cryptoendolithic microbiota has adapted to living in this 'sheltered' niche, the microbial community is still living under the environmental stresses of low temperature, low moisture, low light intensity and little space for growth throughout most of its existence.

The chemical environment

Chemical factors which affect the Antarctic cryptoendolithic lichen and cyanobacterial microbiota mainly involve the availability of essential elements and pH. In a comparative study of the chemical species in lichen and cyanobacterial communities, Johnston & Vestal (1989a) showed that the chemical environment of the lichen community had higher amounts of the oxides of iron, aluminium and manganese than the cyanobacterial community. In the cyanobacterial community, the dominant inorganic species were water-soluble calcium and magnesium. Potassium and phosphate were readily available in high (0·2–0·4 $mg\ g^{-1}$ rock) amounts in both microbiota. Chloride and sulphate were found in small (50 and 4 μM, respectively) concentrations in the biotic zone of lichen-dominated rocks (C. G. Johnston, pers. comm.) suggesting that the microbiota was not under an abnormally high osmotic stress when the rock interstices were saturated with water. However, as the rock dries, solutes will become more concentrated and could exert a transient osmotic stress, which could affect microbial metabolism for a short time. Water-soluble silica was much more prevalent in the cyanobacterial rocks compared with the lichen-containing rocks. The redox potential (pE) of the cyanobacterial community was about four times lower than in the lichen community (*c.* 8·2). The pH ranged from 5 to 6 in the lichen community and between 8 and 9 in the cyanobacterial community. These differences in pH indicate that the lichen community probably produces acidic organic compounds to maintain the low pH. Johnston & Vestal (1989b) have found rather high amounts of oxalate in water extracts of the lichen community, but much lower amounts in the cyanobacterial community. The cyanobacterial community has a higher pH, probably due to the production of organic amines. Aquatic cyanobacteria commonly prefer a pH in the range of 7–9 (Rippka *et al.* 1981).

Cryptoendolithic lichen community physiology

It was recently shown (Vestal 1988b) that the community photosynthetic metabolism was psychrophilic (Morita 1975); that is, it has a photosynthetic optimum below 20°C and some metabolic activity at 0°C or less. Photosynthetically driven CO_2 fixation was measurable after 12 hours at −8°C and after 4 hours at −5°C. At 0°C, photosynthetic activity was relatively high but only *c.* 40–50% of the optimum rate, which was at 15°C (Vestal 1988b). The period above 0°C for a typical cryptoendolithic community is less than 300 hours per year, with 50–100 hours being more typical (Nienow *et al.* 1988a; Vestal 1989). This suggests that the community rarely has temperatures in its optimum metabolic range of 10–15°C. The community is therefore living under suboptimal temperature conditions, which may place it in a constant state of metabolic stress.

Biomass of the cryptoendolithic lichen community was measured indirectly by lipid phosphate (Vestal 1988a) and adenosine triphosphate (Tuovila & LaRock 1987). Both methods showed *c.* 2·54 g m^{-2} of carbon as viable biomass in the cryptoendolithic community. This compares with 1600 g m^{-2} for a temperate grassland to 45 000 g m^{-2} for a tropical rain forest (Table 2). Using photosynthetic rate measurements (Vestal 1988b), biomass estimates, and assumptions about the length of time temperatures were above −5°C, light intensity was above 10 μmol m^{-2} s^{-1} and moisture was present within the interstices of the rock, it was calculated that the primary production was between 0·11 and 4·4 mg carbon

TABLE 2. Comparison of biomass, production and carbon turnover time of the Antarctic cryptoendolithic microbial ecosystem to other ecosystems around the globe (after Vestal 1989)*

Ecosystem	Biomass (g carbon m^{-2})		Production (g carbon m^{-2} $year^{-1}$)	Turnover (years)
Ross Desert,	2·54	(min.)	0·000108	23 520
Antarctica	2·54	(max.)	0·00441	576
Tropical rain forest	45 000		2200	20·5
Temperate evergreen	35 000		1300	26·9
Temperate deciduous	30 000		1200	25
Temperate grassland	1600		600	2·7
Tundra and Alpine	600		140	4·3
Swamp and marsh	15 000		2000	7·5
Agricultural land	1000		650	1·5
Open ocean	3		125	0·024
Continental shelf	10		360	0·027
Estuaries	1000		1500	0·7

* All non-Antarctic data are from Whittaker (1975) and Antarctic data are from Vestal (1989).

m^{-2} $year^{-1}$ and the carbon turnover was 576–23 500 years (Vestal 1989) (Table 2). This places this cryptoendolithic microbial community as having the lowest primary production and the longest carbon turnover of any ecosystem on Earth (Table 2). The turnover time estimates were based on laboratory and field measurements and many assumptions. Consequently, they must be regarded with caution and as illustrative of the slow growth of these rock communities under natural physical and chemical conditions. It could be that exfoliation of the rock surface (see below) may play an important role in microbial colonization and succession, which could affect the carbon turnover times, making them shorter than the above calculations. It is not clear how such turnover times could be measured directly in the field.

Over long periods of time, the microbiota appears to weaken the structure of the rock by some sort of dissolution. This allows rather large (*c.* 2–5 cm diameter) pieces of rock surface to loosen and eventually be removed by the wind. When this exfoliation occurs, the biotic zone in the rock is exposed, revealing a mosaic of lightly coloured areas leached of iron oxides (Fig. 2). This exfoliative pattern is a common characteristic of lichen-colonized rocks.

In most cyanobacterial rocks, there are usually no iron oxide deposits. However, it has been observed that the rocks may exfoliate more rapidly than lichenized

FIG. 2. A large plate of sandstone containing the cryptoendolithic lichen microbiota from the Ross Desert, Antarctica. Note the mosaic pattern produced by rock exfoliation on the surface of the rock. Scale is noted by the head of the geology pick at the bottom of the picture.

rocks, possibly due to the high pH of these communities, which may cause more rapid dissolution of the silica matrix. The cyanobacteria rocks are quite friable and lack iron oxide, so that exfoliation may occur more at a grain-by-grain level, rather than in large pieces.

In addition to the Antarctic cryptoendolithic communities, there has been a report of cryptoendolithic microbial communities living in sandstones in the high Arctic, at *c.* 80°N (Eichler 1981). It was indicated that there was an exfoliative pattern of weathering of the sandstone surfaces similar to the patterns observed in Antarctic sandstones. Subsequent observations by E. I. Friedmann (pers. comm.) did reveal the presence of the lichenized microbiota in these Arctic rocks, but no detailed studies of these Arctic cryptoendoliths have been made.

HOT DESERT CRYPTOENDOLITHS

Cryptoendolithic microbial communities can also be found in hot deserts such as the Negev, Sinai and in the south-west of the United States (Friedmann & Ocampo-Friedmann 1984). These communities, which are exclusively cyanobacterial, also inhabit the pore spaces of sandstones. It appears that they have adapted to porous sandstones for some of the same reasons as the cold desert cryptoendoliths. Confinement within a fixed space inside the sandstone offers more chances for moisture for growth, and some transient protection from the extremely high temperatures on the surface of the rocks during the daytime. Light availability, which is similar to that in the Antarctic sandstones, restricts the photobionts to the top 1–5 mm inside the rocks. The temperature regime is different in that the rock can cool to 27–28°C during the night and then rise to 46–48°C during the day (Friedmann 1980). It is thought that the source of water for hot desert cryptoendoliths is occasional rain and dew condensation on the rocks at night. This at least increases the relative humidity inside the rocks so that during the morning when light is available, metabolic activity can occur for a short time before the water vapour is evaporated by increasing temperatures. When this occurs, the cells become desiccated and activity ceases until the next diurnal cycle.

The hot desert rocks have not been studied as extensively with regard to their physiology and metabolism, but the effects of water stress on hot desert cyanobacterial cultures have been examined (Potts & Friedmann 1981). In that study, it was demonstrated that a desert *Chroococcidiopsis* species could survive low water potentials (*c.* −20 000 kPa) for a short time (i.e. 24 hours), but was metabolically inert after 72 hours at *c.* −7000 kPa. A marine *Chroococcus* species, however, could tolerate low water potential for at least 72 hours. The nitrogen regimes of cold and hot desert cryptoendoliths have also been studied (Friedmann & Kibler 1980). It appears that the source of combined nitrogen in hot desert cryptoendoliths is the same as for the cold desert, that is, from atmospheric precipitation of nitrates and ammonia. Nitrogen fixation was extremely rare in both hot and cold desert communities.

COLONIZATION AND SUCCESSION IN CRYPTOENDOLITHIC COMMUNITIES

Owing to the limiting environmental parameters just described, the processes of colonization and succession in cryptoendolithic communities are enigmatic. Some speculation is possible. Friedmann & Weed (1987) have shown that in the Antarctic, the Ross Desert sandstone containing the cryptoendolithic microbiota are from 70 000 years to 4 million years old. These sandstones are subject to katabatic winds as well as daily freeze/thaw cycles in the summer, and it seems reasonable to assume that at some point cracks developed either on the surface of the rock, through the rock or around the edges (see Fig. 2). Epilithic (attached surface growing) lichen species from the warmer protected areas in the Ross Desert as well as from the nearby coastal regions of McMurdo Sound have been described (Hale 1987) and are apparently the same as the species found within the rocks. It therefore seems plausible that, over geologic time, some of the lichen phyco- and mycobionts could have become windborne. By chance, they could have become lodged within a crack in the sandstone, and thus were able to grow laterally into the interstices of the rock, 'escaping' to a place where there was space, wind protection, periods of water availability, light and relatively mild temperatures. Lichens are commonly the first macroscopic colonizers on cooled volcanic lava flows (Alexander 1971; del Moral 1993) and survive easily in extremely cold and dry climates (Kappen 1973). As the endolithic lichens grew vegetatively they would completely inhabit the zone a few millimetres below the surface of the rock. Lichen acids as well as fungally produced short chain organic acids, such as oxalate, could be produced and would mobilize the iron oxide precipitates and create the lightly coloured (non-iron stained) biotic zones one observes (Fig. 1a). As growth occurs over time, the microbiota could 'loosen' the cement which holds the sandgrains together and produce the exfoliation pattern characteristic of these lichenized rocks (Fig. 2; Friedmann 1982). The lichen microbiota attached to the exfoliated rock could then serve as potential colonizers for other non-colonized rocks. If the rocks were colonized by aerial transport of lichens in this way, succession in the traditional sense has not taken place. There were no competitors so that the community established was in a 'climax' condition from the beginning. It is hard to conceive of a situation where this community could be replaced by one more suitable to the rock environment. It seems reasonable also to assume that the cyanobacteria could not easily compete with an established lichen community, and vice versa, due to the quite different pH characteristics of each community/rock matrix (Johnston & Vestal 1989b).

The pattern of Antarctic cyanobacterial colonization and succession could have been similar to that of the lichen-colonized rocks in the Ross Desert. The major difference between the two is that cyanobacterial communities are generally associated with rocks that do not contain iron oxide stains, and are in locations which are wetter than lichen cryptoendolithic communities. These occur where windblown snow would accumulate (e.g. cracks, protected ledges, etc.) and provide increased water resources within the rocks. Also, these rocks are generally more protected from direct winds which would blow the snow away. At Battleship Promontory, Ross Desert, cyanobacterial communities are frequently found at the base of sandstone outcrops along the edges of a dolerite pavement. The dolerite has a low albedo, which causes snow to melt rapidly and infiltrate the nearby sandstones. These rocks then become saturated with water, which seems to remain for longer periods than in rocks a few metres higher and more subject to the full force of the wind. These factors may all contribute to the development of a cyanobacterial- rather than a lichen-dominated community.

It can be speculated that similar mechanisms of colonization and succession could account for the presence of cyanobacteria in hot desert rocks.

The only successional events which seem possible in the cryptoendolithic microbiota would have been among the primary consumers/decomposers. These heterotrophic microbes, primarily bacteria, fungi and yeasts, probably arrived as airborne contaminants of snow or attached to dust particles. They would live off the organic excretion products from the lichen or cyanobacteria, as well as from dead cells and their own endogenous reserves. These microbes may have undergone metabolic competition sequences in order to establish a 'climax' decomposer community.

Since cryptoendolithic communities in Antarctica are living under such extreme conditions and have such short metabolically active periods each year (*c.* 50–100 hours of $>-5°C$, light and moisture) (Nienow *et al.* 1988a; Vestal 1989), their growth rates are extremely slow and generation times long. They may not even grow in the sense of increasing biomass, because of space limitations within the rock; they may just be carrying out minimal maintenance metabolism. In order to study succession in such slowly growing communities, long periods of time would be required. Some initial experiments have been set up by the author in the Ross Desert whereby various nutrients and inhibitors have been periodically added to the rock surfaces to percolate into the biotic zones. These chemical treatments include: (a) nitrate, ammonia and phosphate; (b) trace amounts of vitamins and other co-factors; (c) addition of the prokaryotic inhibitors penicillin and streptomycin; (d) addition of the eukaryotic inhibitor cycloheximide; and (e) glucose and acetate. These aqueous solutions have been added two to three times per season for four years (1983/84–1986/87), and it is anticipated that this will be repeated again in the near future. Inverted glass jars were attached to rocks to increase their temperature and thus allow greater metabolism than under ambient conditions. The jars were not completely sealed against the rock surface so that snow could still accumulate as a source of water for the rock microbiota. At some point in the future, the experimental rocks and their controls will be collected and the microbial activity, biomass, metabolic status and community structure will be determined using sensitive lipid analysis techniques (White 1983; Vestal & White 1989). The biochemical differences (if any) will be measured as an assay of microbial community change in response to these allogenic processes. At that time, there may be more information from which actual successional events may be detectable, and the development of these communities can then be more objectively defined.

ACKNOWLEDGMENTS

I wish to thank E. I. Friedmann, M. Meyer, C. P. McKay, M. E. Hale and C. G. Johnston for numerous helpful and stimulating discussions regarding cryptoendoliths. Thanks to D. B. Knaebel and C. G. Johnston for comments on the manuscript. The field work for this discussion was funded by the Division of Polar Programs of the US National Science Foundation (grants DPP 80–17581 and DPP 83–14180) to E. I. Friedmann of Florida State University.

REFERENCES

Alexander, M. (1971). *Microbial Ecology*. John Wiley and Sons, Inc., New York.

Colwell, R.R., MacDonell, M.T. & Swartz, D. (1989). Identification of an Antarctic endolithic microorganism by 5S rRNA sequence analysis. *Systematic and Applied Microbiology*. **11**, 182–6.

del Moral, R. (1993). Mechanisms of primary succession on volcanoes: A view from Mount St Helens. *Primary Succession on Land* (Ed. by J. Miles & D.W.H. Walton), pp. 79–100. Blackwell Scientific Publications, Oxford.

Eichler, H. (1981). Small scale features of high arctic weathering in the Oobloyah Bay region, northern Ellesmere Island, N.W.T., Canada — Genesis and processes. *Heidelberger Geographische Arbeiten*, **69**, 465–85.

Finegold, L.X., Singer, M.A., Federle, T.W. & Vestal, J.R. (1990). Composition and thermal properties of membrane lipids in cryptoendolithic lichen microbiota from Antarctica. *Applied and Environmental Microbiology*, **56**, 1191–4.

Friedmann, E.I. (1978). Melting snow in the dry valleys is a source of water for endolithic microorganisms. *Antarctic Journal of the United States*, **13**, 162–3.

Friedmann, E.I. (1980). Endolithic microbial life in hot and cold deserts. *Origins of Life*, **10**, 223–35.

Friedmann, E.I. (1982). Endolithic microorganisms in the Antarctic cold desert. *Science*, **215**, 1045–53.

Friedmann, E.I. & Kibler, A.P. (1980). Nitrogen economy of endolithic microbial communities in hot and cold deserts. *Microbial Ecology*, **6**, 95–108.

Friedmann, E.I. & Ocampo-Friedmann, R. (1976). Endolithic blue-green algae in the Dry Valleys: Primary producers in the Antarctic desert ecosystems. *Science*, **193**, 1247–9.

Friedmann, E.I. & Ocampo-Friedmann, R. (1984). Endolithic microorganisms in extreme dry environments; Analysis of a lithobiontic microbial habitat. *Current Perspectives on Microbial Ecology* (Ed. by M.J. Klug & C.A. Reddy), pp. 177–185. American Society for Microbiology, Washington.

Friedmann, E.I. & Weed, R. (1987). Microbial trace-fossil formation, biogenous and abiotic weathering in the Antarctic cold desert. *Science*, **236**, 703–5.

Friedmann, E.I., McKay, C.P. & Nienow, J.A. (1987). The cryptoendolithic microbial environment in the Ross Desert, Antarctica: Satellite transmitted continuous nanoclimate data, 1984–1986. *Polar Biology*, **7**, 273–87.

Friedmann, E.I., Hua, M.S. & Ocampo-Friedmann, R. (1989). Cryptoendolithic lichen and cyanobacterial communities of the Ross Desert, Antarctica. *Polarforschung*, **58**, 251–60.

Golubic, S., Friedmann, I. & Schneider, J. (1981). The lithobiontic ecological niche, with special reference to microorganisms. *Journal of Sedimentary Petrology*, **51**, 475–8.

Hale, M.E. Jr (1987). Epilithic lichens in the Beacon sandstone formation, Victoria Land, Antarctica. *Lichenologist*, **19**, 269–87.

Johnston, C.G. & Vestal, J.R. (1989a). Distribution of inorganic species in two Antarctic cryptoendolithic microbial communities. *Geomicrobioloy Journal*, **7**, 137–53.

Johnston, C.G. & Vestal, J.R. (1989b). The inorganic environment of two Antarctic cryptoendolithic microbial communities. *Abstracts Annual Meeting of the American Society for Microbiology*, **89**, 231.

Kappen, L. (1973). Response to extreme environments. *The Lichens*. (Ed. by V. Ahamdjian & M.E. Hale), pp. 310–80. Academic Press, New York.

Kappen, L., Friedmannn, E.I. & Garty, J. (1981). Ecophysiology of lichens in the Dry Valleys of southern Victoria Land, Antarctica: 1. Microclimate of the cryptoendolithic lichen habitat. *Flora*, **171**, 216–35.

Longton, R.E. (1988). *Biology of Polar Bryophytes and Lichens*. Cambridge University Press, Cambridge.

McKay, C.P. & Friedmann, E.D. (1985). The cryptoendolithic microbial environment in the Antarctic cold desert: Temperature variations in nature. *Polar Biology*, **4**, 19–25.

Morita, R.Y. (1978). Psychrophilic bacteria. *Bacteriological Reviews*, **39**, 144–67.

Nienow, J.A., McKay, C.P. & Friedmann, E.I. (1988a). The cryptoendolithic microbial environment in the Ross Desert of Antarctica: mathematical models of the thermal regime. *Microbial Ecology*, **16**, 253–70.

Nienow, J.A., McKay, C.P. & Friedmann, E.I. (1988b). The cryptoendolithic microbial environment in the Ross Desert of Antarctica: light in the photosynthetically active region. *Microbial Ecology*, **16**, 271–89.

Potts, M. & Friedmann, E.I. (1981). Effects of water stress on cryptoendolithic cyanobacteria from hot desert rocks. *Archives of Microbiology*, **130**, 267–71.

Rippka, R., Waterbury, J.B. & Stanier, R.Y. (1981). Isolation and purification of cyanobacteria: some general principles. *The Prokaryotes*. (Ed. by M.P. Starr, H. Stolp, H.G. Truper, A. Balows, Hans G. Schlegel), pp. 212–20. Springer-Verlag, Berlin.

Siebert, J. & Hirsch, P. (1988). Characterization of 15 selected coccal bacteria isolated from Antarctic rock and soil samples from the McMurdo-Dry Valleys (South Victoria Land). *Polar Biology*, **9**, 37–44.

Tschermak-Woess, E. & Friedmann, E.I. (1984). *Hemichloris antarctica*, gen. et spec. nov. (Chlorococcales, Chlorophyta), a cryptoendolithic alga from Antarctica. *Phycologia* **23**, 443–54.

Tuovila, B.J. & LaRock, P.A. (1987). Occurrence and preservation of ATP in Antarctic rocks and its implications in biomass determinations. *Geomicrobiology Journal*, **5**, 105–18.

Vestal, J.R. (1988a). Biomass of the cryptoendolithic microbiota from the Antarctic desert. *Applied and Environmental Microbiology*, **54**, 957–9.

Vestal, J.R. (1988b). Carbon metabolism of the cryptoendolithic microbiota from the Antarctic desert. *Applied and Environmental Microbiology*, **54**, 960–5.

Vestal, J.R. (1989). Primary production of the cryptoendolithic microbiota from the Antarctic desert. *Polarforschung*, **58**, 193–8.

Vestal, J.R. & White, D.C. (1989). Lipid analysis in microbial ecology. *Bioscience*, **39**, 535–41.

Vincent, W.F. (1988). *Microbial Ecosystems of Antarctica*. Cambridge University Press, Cambridge.

Vincent, W.F. & Ellis-Evans, J.C. (eds) (1989). *High Latitude Limnology*. Kluwer, Dordrecht, The Netherlands.

Vishniac, H.S. (1985a). *Cryptococcus friedmannii* a new species of yeast from Antarctica. *Mycologia*, **77**, 149–53.

Weed, R. (1985). *Chronology of chemical and biological weathering of cold desert sandstones in the Dry Valleys, Antarctica*. MS thesis. Department of Geology, University of Maine.

White, D.C. (1983). Analysis of microorganisms in terms of quantity and activity in natural environments. *Microbes in their Natural Environments*. (Ed. by J.H. Slater, R. Whittenbury & J.W.I. Wimpenny), pp. 37–66. Cambridge University Press, Cambridge.

Whittaker, R.H. (1975). *Communities and Ecosystems*, 2nd Edn. Macmillan, New York.

Wynn-Williams, D.D. (1990). Ecological aspects of Antarctic microbiology. *Advances in Microbial Ecology* (Ed. by K.C. Marshall), Vol. 11, pp. 71–146. Plenum Press, New York.

Microbial processes and initial stabilization of fellfield soil

D. D. WYNN-WILLIAMS
British Antarctic Survey, Natural Environment Research Council, High Cross, Madingley Road, Cambridge CB3 0ET, UK

SUMMARY

1 Fellfield substrata are usually flexible or mobile and normally contain free water except in very cold deserts. Disruption by freezing and thawing with its resultant hydrostatic pressures causes particulate sorting and patterning of the ground.
2 Abiotic weathering and exfoliation of rock initiate fellfield soil formation, while endo- and chasmolithic microbial communities accelerate these processes and provide an inoculum as a precursor to stabilization.
3 Although not visually apparent, microbes may be the first colonists of all virgin substrata and may initiate or escalate plant succession.
4 Phototrophic microbial crusts stabilize the fellfield soil surface by binding mineral particles in a cyanobacterial and algal filament–mucigel matrix.
5 Epifluorescence microscopy and TV image analysis of undisturbed, moist frost-polygon fines at Signy Island, maritime Antarctica, suggested that scarce microbial resources may be conserved by the soil crust which is composed of a separable mosaic of heterogeneous microbial 'rafts'.
6 A cloche enhanced ambient conditions for microbial growth, (including a mean temperature elevation of 3·1°C above ambient), resulting in an increase in surface area of microbial colonization of soil from 5 to 74% in three successive growing seasons.
7 The relatively rapid growth rate of fellfield cyanobacteria and their sensitivity to environmental factors makes them valuable indicators of the short- and long-term effects of climatic change.

INTRODUCTION

It is natural to think of a terrestrial substratum as a solid, immobile foundation for biological colonization. Fellfield habitats (cold climate mineral substrata having an open, vegetation community if plants are present at all) are far from immobile. They are subjected to changes which produce various physical waves of expansion and contraction, in turn creating patterned and rippled ground. These movements may be spatially small and temporally slow but are fundamentally disruptive to microbial attempts at colonization. Nevertheless, the microbiota is frequently the primary colonizing group of virgin fellfield habitats in the Antarctic (Wynn-

Williams 1990a) and other perturbed polar, Alpine and desert habitats (Cameron 1969, 1971, 1974; Friedmann 1980).

Parent rock is frequently unsuitable for colonization until it has been weathered by both physical and chemical attack. However, when microbes do establish, biological weathering can be an important component in rock breakdown, such as in exfoliation by endolithic microbial communities (Friedmann 1982) in the McMurdo Dry Valleys region (hereafter referred to as the Ross Desert) and by chasmolithic communities (Fig. 1a) in less hostile parts of Antarctica (Broady 1981, 1986). These microbial communities also provide an inoculum for raw mineral soils or 'lithosols' at the life-supporting limits of fellfields.

Sandy soils of the Ross Desert (Fig. 1b) in places form barchan sand-dunes (Miotke 1985) which barely support microbial life (Cameron 1972; Horowitz *et al.* 1972). This environment has a unique combination of conditions preclusive to life. These include virtual absence of free water, exceptionally strong and persistent katabatic winds and highly unstable mineral grains with negligible cementation other than by permafrost or superficial ice crusts (Miotke 1985). However, where the particles are less unstable, the ground becomes patterned not by freeze–thaw action but by thermal stress cycles which result in polygons delineated by sand wedges (Fig. 1c; Berg & Black 1966). Although containing marginally more water than the sand-dunes, these patterned soils are still near the limit of life support, which has led to their selection as a testing ground for the past (Viking) and future life-detection systems for Mars probes (Cameron *et al.* 1971, 1976; Klein 1979; McKay 1986).

Soils of coastal continental Antarctica (Boyd *et al.* 1966; Campbell & Claridge 1987) and the maritime Antarctic islands are disrupted not only by hot–cold cycles but also by wet–dry and freeze–thaw cycles. When such ground is level, particle sorting results in stone circles and polygons (Fig. 1d), while solifluction lobes and stone stripes are formed on slopes (Chambers 1967). The stones and soil particles of these habitats are frequently turned and inverted, presenting a perpetual challenge to potential colonizers (Walton 1984).

POTENTIAL MICROBIAL COLONISTS

Before or during the initiation of primary succession in the botanical sense, there is always a primary microbial succession, initially invisible to the eye. Phototrophic, chemolithotrophic and heterotrophic microbes colonize rocks and lithosols and contribute to their weathering as well as stabilizing particle substrata.

The microbial colonists of endolithic (Vestal 1991) and chasmolithic (Broady 1981) communities (lichens, fungi, yeasts, algae, cyanobacteria and bacteria) do

FIG. 1. (*Opposite and over page.*) Fellfield habitats. (a) Chasmolithic niches and exfoliating rock, Red Rock Ridge, Antarctic Peninsula. The scale marker is 7 cm long. (b) Cold desert, Barwick/Balham/Victoria Valley Junction, Ross Desert. (c) Sand wedge polygons (*c.* 20 m in diameter) near Beacon Valley, Ross Desert. The helicopter shadow indicates scale. (d) Frost-sorted polygon, Signy Island. The ice axe is *c.* 70 cm long.

(a)

(b)

(c)

(d)

FIG. 1. *Contd.*

not contribute to stabilization so much as to soil creation and inoculation by acting as a local microbial propagule bank. However, these are the beginnings of the soil stabilization process. Some of these microbes have been found in soil adjacent to endolithic communities on Linnaeus Terrace, Ross Desert (J. R. Vestal, pers. comm.) but their development into a soil community is limited by the availability of free water or, more probably, by the humidity of the microenvironment.

The successful establishment of other extraneous aerial colonists also depends on moisture and suitable growth temperatures at the microenvironment level. This has resulted in the development of distinct, diverse microbial and cryptogamic communities in geothermal areas on the summits of Mount Erebus (3270 m a.s.l.) and Mount Melbourne (2733 m a.s.l.) derived primarily from the global airspora (Broady 1984; Broady *et al.* 1987).

Vishniac & Hempfling (1979) have described an exceptionally xerotolerant yeast flora in the Ross Desert whose cells may contain up to 14% trehalose as a compatible solute. Such molecules are capable of protecting cell membranes and proteins against molecular disruption as the osmotic potential of the cytoplasm increases with progressive desiccation (Brown 1978). Vishniac & Klinger (1986) and Klinger & Vishniac (1988) have drawn attention to the importance of water potential as an essential factor to be considered when assessing the colonization and stabilization potential of xeric habitats. They showed Ross Desert yeasts to be exceptionally tolerant of such conditions. Abyzov *et al.* (1987) have reported the isolation of dormant but viable spore-forming bacteria in 12 000 year old zones of ice cores at −55°C drilled in the Antarctic ice cap at Vostok (78°S, 107°E). Among other strains they found *Pseudomonas fluorescens*-like groups 3250 years old. In permafrost cores up to 380 m deep drilled at Ross Island (77°30′S, 167°40′E), Cameron & Morelli (1974) described similarly ancient active flagellated bacteria and cyanobacteria resembling *Calothrix*, which has been found as a fossil elsewhere. Although it is difficult to eliminate the possibility of penetration of contaminants along deceptively self-annealing hairline cracks formed during the drilling process (D. A. Peel, pers. comm.), it is probable that these findings provide evidence of prolonged microbial survival in a freeze-dried anabiotic state. Similarly, Cameron & Blank (1966) reported desert algae viable after 80–100 years desiccation. Desiccation *per se* is therefore not a limiting factor for survival of potential microbial colonists of Antarctic fellfield soils. However, the existence of mummified seals up to 800 years old (Dort 1981) in the Dry Valleys of the Ross Desert indicates that actual surface microbial activity, and therefore colonization and stabilization potential, is severely limited by the prevailing absence of free water and low humidity.

Where water is available in the Ross Desert, cyanobacteria, especially *Nostoc commune* and *Phormidium* spp., are evident as crusts or felts on the soil (Boyd *et al.* 1966; Broady 1986, 1989). Cameron & Devaney (1970) showed by optical and scanning electron microscopy that these algal–cyanobacterial crust communities were able to incorporate and stabilize soil surface particles during the brief

Antarctic summer. However, their growth is transient and the crusts are frequently blown away as the ground temperature falls below freezing point and the ice evaporates or sublimes out of their cells. This erratic and stochastic dispersal does, however, provide a source of inoculum for downwind sites. Cyanobacterial felts of benthic origin may also fulfil this function after upward migration through the ice cap of meromictic saline lakes of the Ross Desert (Parker *et al.* 1982). Preformed cyanobacterial foci may thus be made available for soil stabilization of Ross Desert fellfield soils under favourable microclimatic conditions.

As long ago as 1907, Fritsch showed that algae were pioneers of primary succession and Booth (1941) emphasized their significance for erosion control. More recently, algae and cyanobacteria have been implicated in the stabilization of a diverse range of xeric (albeit less extreme) habitats. The green sand-dunes by the Maybelle River, north-eastern Alberta, contain Oscillatoriaceae and Nostocaceae which promote the growth of *Hudsonia tomentosa* (Nelson *et al.* 1986). Cyanobacterial soil crusts have also been shown to enhance seedling establishment in arid soils of Utah (St Clair *et al.* 1984).

The importance of desert algal crusts has been reviewed by Cameron & Blank (1966) as a prelude to exobiological programmes (see Wynn-Williams 1990a). Soil crust stabilization activity is not restricted solely to algae and cyanobacteria as fungi and actinomycetes have also been implicated in sand stabilization (Forster 1979) and capsulated bacteria are associated with all these communities.

Soil colonization research by the British Antarctic Survey has focused on the maritime Antarctic (Holdgate 1964). Wetter soils, as found in this region, are capable of abiotic soil crust formation (Evans & Buol 1968), but this is unstable and transient. Frost heave and desiccation both lead to disruption of the soil surface (Chambers 1967). The significance of microbial activity for soil stabilization was investigated at an exposed and recently deglaciated site with patterned ground on Jane Col, Signy Island (60°43′S, 45°35′W), South Orkney Islands (Smith 1987).

COLONIZATION PROCESSES

The investigation was based on a hypothetical colonization process, summarized in Fig. 2, which shows the pathways and variables that the succession may follow. Both the duration of environmental conditions (beneficial or detrimental) and regression are significant components of the process.

Epifluorescence microscopy (EFM) with selective optical filters and a photofading retarding mountant were used for direct examination of intact soil crusts and quantification of the microbiota by TV image analysis (TVIA; Wynn-Williams 1985, 1988). This showed that although bacteria were abundant in the community and fungi were present, the dominant colonizers of mineral fines in the centre of frost polygons (Fig. 1d) were cyanobacteria and algae. Bacteria formed a biofilm over rock particles and the phototrophs formed a crust over the surface of the soil.

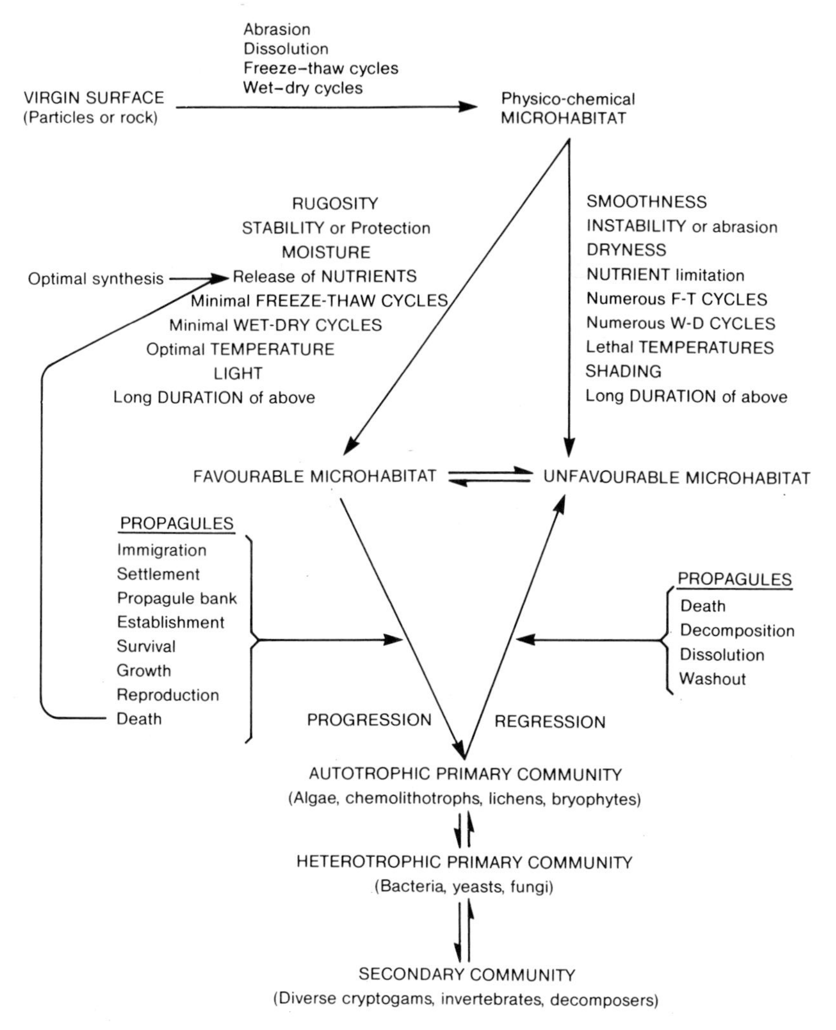

FIG. 2. Hypothetical colonization processes for maritime Antarctic fellfields (after Wynn-Williams 1986).

Two dominant morphological forms were evident: filamentous trichomes of cyanobacteria and aggregates of unicellular eukaryotic algae. Both groups had conspicuous mucilagenous sheaths. It was noticeable, however, that the distribution of these organisms was frequently heterogeneous and that aggregates resembling 'rafts' in a loose mosaic were prevalent (Fig. 3). The significance of crumb structure for temperate soil fertility is a well-known fact. Observations at Jane Col suggested that surface 'raft' mosaic concept may be equally significant for the stabilization of frost-sorted fellfields (Wynn-Williams 1990b), not only for the microbiota but also for lichens (Walton 1993) and bryophytes (Smith 1993).

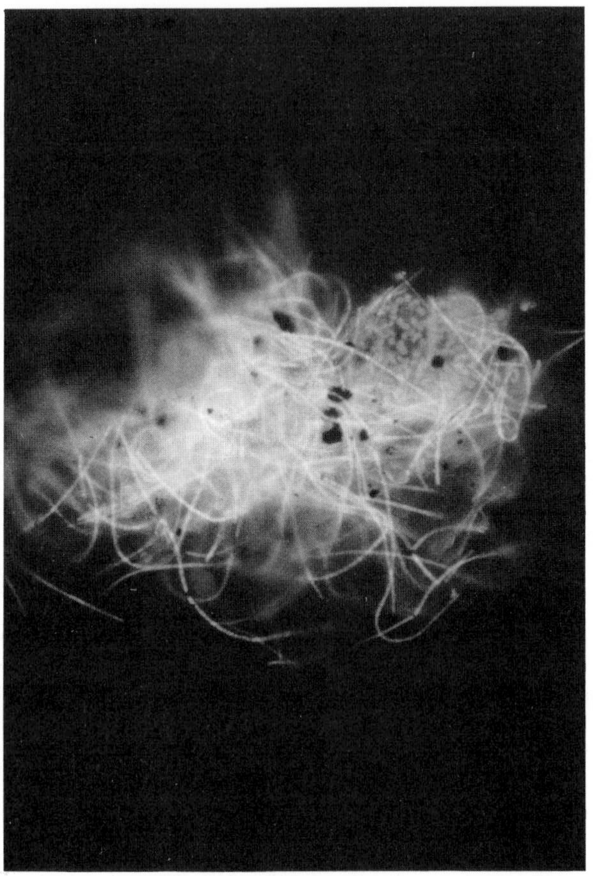

FIG. 3. Algal–cyanobacterial raft seen by autofluorescence on an undisturbed fellfield fines crust at Jane Col, Signy Island. The dominant cyanobacterial filaments are *c*. 7 μm wide.

The raft mosaic concept

The raft analogy reflects the essentially superficial location of the phototroph assemblages, which are a physical and metabolic focus of living activities and drift over the mobile substratum. The significance of the microbial raft as a colonizing and stabilizing unit lies in the tenuous existence of the microbiota. The short growing season, limited light, low temperatures for growth and limiting nutrients all lead to a need to conserve structural and nutritive resources. If all the individual cells of the crust are stretched and damaged during the disruption of the soil surface by the biologically irresistible physical forces of frost heave, much cytoplasm and nutrient will be lost. However, if the resistance is concentrated on the focus of the raft so that only peripheral filaments are disrupted, the resources of the colonizer unit may be conserved (Fig. 4). The raft mosaic could re-coalesce on release of the tension in the convex soil surface during melting of the ice crystals in the profile.

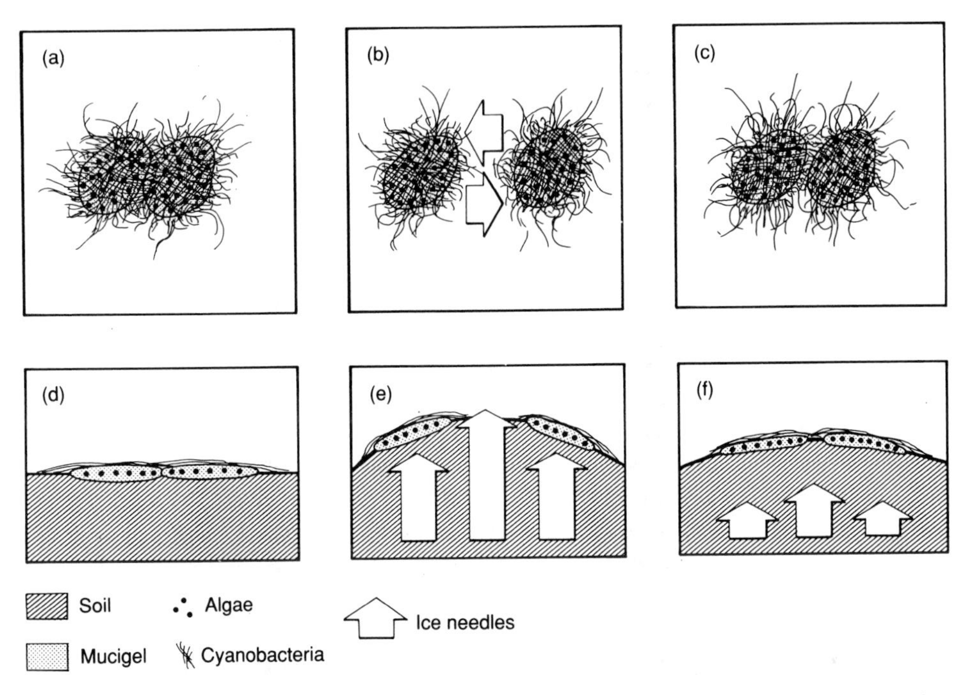

FIG. 4. Schematic diagram of fellfield fines raft structure and resilience to disruption by ice needles. (a)–(c) Surface views; (b)–(d) equivalent profiles. (a) and (d) Initial; (b) and (e) disruption after frost-heave by ice needles; (c) and (f) coalescence during melt.

Integrity of microbial rafts

The size and structure of microbial rafts will depend on the grain size of the substratum, its water content and the size, morphology, metabolism and mucilage production of the colonists. Filamentous mucilage-sheathed trichomes of cyanobacteria are abundant in fellfield soil crusts (Wynn-Williams 1988), but mucilaginous aggregates of eukaryotic algal unicells are frequently associated with them. Occasionally, convoluted trichomes of *Nostoc* spp. in globose masses of mucilage are prevalent. Polysaccharide mucilage is therefore an integral feature of this system. Not only does it act as an adhesive between cells and the substratum (Geesey 1982), but it also protects the cells themselves. Its 98% water content acts as a buffer against desiccation and it provides a matrix for the concentration of nutrients for transfer to the heterotrophic bacteria, which are also an integral component of the system. This aquatic environment can also be an important zone of exoenzyme activity. Although Greenfield (1989) has shown that nitrogen is unlikely to be generally limiting at Jane Col due to atmospheric and meltwater NH_4 input, nitrogen may become locally limiting at specific sites in the predominantly oligotrophic fellfield ecosystem. The role of mucilage in restricting oxygen diffusion to nitrogen-fixing oscillatorian cyanobacteria will therefore promote nitrogen fixation if required. Its concomitant role in maintaining the C:N ratio by excretion of excess carbon as polysaccharide may also be significant in sustaining the metabolic balance of the stabilizing microbiota. Certain polyanionic heteropolysaccharides synthesized by bacteria and cyanobacteria act as bioflocculants (Fattom & Shilo 1984a). These are hydrophobic molecules which interact with clay particles and may be instrumental in nucleating rafts in a way analogous to the precipitation of minerals during the creation of benthic cyanobacterial felts in water bodies (Fattom & Shilo 1984b).

The nucleation of the rafts and their physical integrity is only useful in soil stabilization if the microbial components remain viable. During prolonged dormancy, under dried or freeze-dried conditions, it is probable that compatible solutes with xero- and cryoprotectant properties sustain cell membrane integrity under extreme osmotic stress. Not only trehalose but also polyols have these properties. Arabitol and ribitol are associated with the survival of exposed Antarctic lichens (Tearle 1987) while other polyols are associated with cryotolerance in microarthropods of Antarctic fellfields (Cannon & Block 1988). The production of such molecules may be seasonal (Tearle 1987), under the control of environmentally triggered enzyme switching mechanisms for different metabolic pathways.

The glaciers of Signy Island develop extensive cover of snow algae. These originate in late winter snow when free moisture is available and the cells and their products are precipitated onto the soil surface as the snow melts. Although snow algae, predominantly *Chlamydomonas antarcticus* (Kawecka 1986), and the resulting cryoconite communities derived from them (Smith 1993), are not dominant in the fellfield microbiota, their heterogeneous population may act as a propagule bank for the nucleation of 'rafts'. Their input of nitrogen (Greenfield 1989), carbon (especially mucilage) and other nutrients from senescing cells will certainly be beneficial for the soil microbiota in early spring.

HETEROGENEITY OF SOIL MICROBIAL CRUST

To investigate the distribution of the soil crust 'mosaic', 15 mm diameter soil cores were sampled at 5 cm intervals in a 625 cm^2 quadrat over the central area of fine soil in a frost polygon at Jane Col. The phototrophic microbial colonizers were quantified by TVIA (Wynn-Williams 1988). The resulting data included the area of colonization of the soil surface and the length of the propagules responsible for the colonization. The results (Fig. 5) show heterogeneity of cyanobacterial and algal cover on a centimetre scale and a close correlation between cover and length of colonizer. The crust was dominated by filaments of the oscillatorian *Phormidium autumnale*. However, observations of two plastic cloches established over polygon fines at this site to enhance ambient conditions (see below) showed that the crusts on different polygons may be of very different microbial composition. After three annual growing seasons, the soil surface in one cloche was dominated by *Phormidium* spp. (Broady *et al.* 1984) while the other was dominated by globose, convoluted *Nostoc* assemblages. An intensive study of 65 frost-sorted soil polygons at this site revealed constant micro-algal communities within polygons, but con-

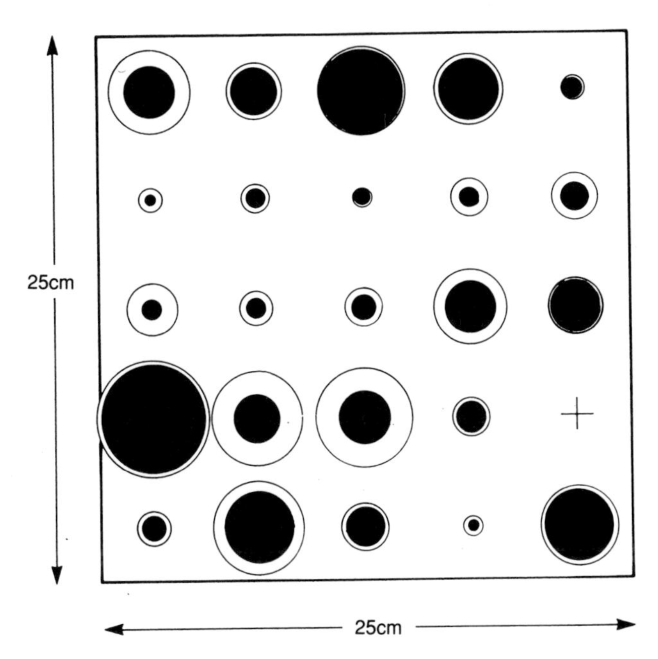

FIG. 5. Heterogeneity of autofluorescent algal–cyanobacterial cover of mineral fines in a $625\,cm^2$ quadrat at the centre of a soil polygon, Jane Col, Signy Island. Quantification was by TVIA of EFM displays at ×100 magnification. The diameter of the black discs is proportional to the total area of cell cover, while the outer circle is proportional to the constituent cell length per unit area.

siderable differences in species composition between polygons; even those in close proximity (Davey & Rothery, in press).

ENHANCED MICROCLIMATE IN CLOCHES

In an experiment to enhance the ambient microclimatic conditions of the soil crust on fellfield fines, transparent polystyrene cloches (57 × 29 × 15 cm high) were emplaced at the Jane Col site. During the summer period January to March 1988, the mean integrated ground surface temperature determined by temperature of integrating spheres (±SD for $n = 3$) within a cloche was 7·6 ± 1·7°C relative to 4·4 ± 1·2°C in an adjacent exposed control plot. The average ground temperature in the cloche was therefore 3·2°C above ambient. After three growing seasons, five replicate cores were taken from each of the cloche and the control plot for quantification by EFM and TVIA. After this period, the microbial crust on fines outside the cloche comprised an open community of predominantly filamentous cyanobacteria amounting to a total cover of 4·8% (Table 1). However, the equivalent cover on fines in the adjacent cloche had escalated to a fairly uniform mesh of 74% coverage. The cover consisted of an 'understorey' of fine (*c.* 2 μm diameter) oscillatorian filaments intimately associated with the soil surface, overlain by a 'canopy' of coarser oscillatorian filaments (*c.* 7 μm diameter). The abundance

TABLE 1. The response after 3 years of algal and cyanobacterial colonizers of fellfield soil polygon fines dominated by filamentous cyanobacteria to improved environmental conditions in a plastic cloche located at Jane Col, Signy Island, relative to an exposed adjacent control site

Variable	Mean ± SD		Mean difference	Percentage change
	Control	Cloche†		
Temperature (°C)*	4·4 ± 1·2	7·6 ± 1.7	3·2	+73
Area colonized (%)	4·8 ± 1·5	73·9 ± 8·0	69·1	+2163
Total length of cells per area ($\mu m^2 \, mm^{-2}$)	3·9 ± 1·7	17·9 ± 7·4	14·0	+358
Cell length (μm)	23·3 ± 1·7	78·0 ± 45·2	52·7	+208

* Integrated surface temperatures (January–March 1988) determined using Ambrose cells by the method described in Walton (1982).
† All values are significantly different from the controls ($P < 0{\cdot}01$).

of Oscillatoriaceae and *Nostoc* spp. imply potential nitrogen fixation under physically favourable circumstances (Sprent 1993). It is possible that the loss of cells and nutrients during the disintegration of such an abundant microbial crust by occasional frost-heave was not of overall significance for the community over the period studied. The two-layered phototrophic structure, resembling a forest ecosystem in miniature, would be a potential grazing zone for flagellate protozoa (Hughes & Smith 1989) and other fellfield microfauna (Block 1984).

The intra-cloche temperature elevation is of the same order as that predicted for the 'global warming' effect by the year 2050. Assuming a middle range scenario of a 0·3°C increase per decade, the predicted elevation in summer temperature is in the range 0·9–1·3°C (while that for winter is 3·6–4·3°C). The predicted mean elevation is *c.* 1·8°C (World Meteorological Organization 1988). Svensson (1980) showed that Arctic psychrotolerant methanogens had very high values for Q_{10} in the temperature range just above freezing point. However, C. Howard-Williams (pers. comm.) reports Q_{10} values of only 2·0 for continental Antarctic micro-organisms. Those of the maritime Antarctic are more intermediate at $Q_{10} = c.$ 3 (Wynn-Williams 1980). However, the primary influence of the elevated ambient temperature in the cloche experiment is more likely to be a reduction in the number of freeze–thaw cycles and a prolonged growing season. Moreover, the enclosure of the predominantly wet soil by the cloche sustains a more consistent (but not necessarily elevated) moisture regime and protects the crust from rain splash and the effects of wind. Photosynthetic carbon fixation will be stimulated by prolonged penetration of photosynthetically active radiation because of a potentially thinner snow cover during the spring and autumn. The polystyrene cloches cut out 95% of ultraviolet light (R. I. Lewis Smith, pers. comm.) which might otherwise bleach the pigments of the vulnerable superficial phototrophs. Negative influences on crust development would include restricted (but not eliminated) aerial propagule and nutrient input, and decreased meltwater input from snow. Despite these variables, it is evident that if the quality of the environment of the soil crust is enhanced, the phototrophic microbiota shows great resilience to the environmental change. The result is a proliferation of growth analogous to that of the bryophyte community in nearby cloches established to investigate cryptogamic propagule banks (Smith 1993). Soil surface cyanobacteria and algae have potential value as readily quantifiable, sensitive indicators of environmental change, especially on a small scale.

CONCLUSIONS

Antarctic soils are eminently suitable for studies of early colonization and succession processes because of their low species diversity, which minimizes interactions and optimizes the chances of detecting biotic responses to environmental change. Moreover, the long time-scale of these processes permits a higher resolution of successional stages. Finally, the reduction of microbial metabolism by freezing, with concurrent preservation of habitat integrity permits sampling of undisturbed communities for perturbation studies *in situ* and in microcosms, to replicate or exaggerate environmental factors monitored in the field and predicted for the future.

Our understanding of the resilience of this microbial ecosystem, especially the raft mosaic concept and proven enhancement of colonization, now requires refining by selective elimination of environmental variables both in the field and in laboratory microcosms. A dependence of cryptogamic succession on primary microbial colonization has yet to be proved for the Antarctic, although this has been shown elsewhere (St Clair *et al.* 1984; Nelson *et al.* 1986). It will also be of value to compare colonization of this 'badlands' environment, including Antarctic volcanoes (Cameron & Benoit 1970; Broady 1984; Smith 1984; Broady *et al.* 1987) with a less hostile and more accessible virgin substratum, such as sub-Arctic volcanoes (Brock 1973) and temperate or tropical volcanic substrata (del Moral 1993; Whittaker 1993), which have a more diverse propagule input and faster growth rates.

ACKNOWLEDGMENTS

I thank my colleagues in British Antarctic Survey at Signy Island and at BAS Headquarters for their help and discussion of the work described in this paper.

REFERENCES

Abyzov, S.S., Bobin, N.Y. & Kudryashov, B.B. (1987). Anabiosis of microorganisms in the Antarctic ice sheet. *Anabiosis of Cells* (Ed. by Y.E. Beker), pp. 43–54, Academy of Sciences of Latvian SSR, August Kirkhensteyn Institute of Microbiology, Riga.

Berg, T.E. & Black, R.F. (1966). Preliminary measurements of growth of nonsorted polygons, Victoria Land, Antarctica. *Antarctic Research Series*, **8**, 61–108.

Block, W. (1984). Terrestrial microbiology, invertebrates and ecosystems. *Antarctic Ecology*, 1 (Ed. by

R.M. Laws), pp. 164–236. Academic Press, London.

Booth, W.E. (1941). Algae as pioneers in plant succession and their importance in erosion control. *Ecology*, **22**, 38–46.

Boyd, W.L., Staley, J.T. & Boyd, J.W. (1966). Ecology of soil microorganisms of Antarctica. *Antarctic Research Series*, **8**, 125–59.

Broady, P.A. (1981). The ecology of chasmolithic algae at coastal locations of Antarctica. *Phycologia*, **20**, 259–72.

Broady, P.A. (1984). Taxonomic and ecological investigations of algae on steam-warmed soil on Mt Erebus, Ross Island, Antarctica. *Phycologia*, **23**, 257–71.

Broady, P.A. (1986). Ecology and taxonomy of the terrestrial algae of the Vestfold Hills. *The Vestfold Hills: An Antarctic Oasis* (Ed. by J. Pickard), pp. 165–202. Academic Press, Sydney.

Broady, P.A. (1989). Broadscale patterns in the distribution of aquatic and terrestrial vegetation at three ice-free regions on Ross Island, Antarctica. *Hydrobiologia*, **172**, 77–95.

Broady, P.A., Garrick, R. & Anderson, G. (1984). Culture studies on the morphology of ten strains of Antarctic Oscillatoriaceae (cyanobacteria). *Polar Biology*, **2**, 233–44.

Broady, P.A., Given, D., Greenfield, L.G. & Thompson, K. (1987). The biota and environment of fumaroles on Mount Melbourne, northern Victoria Land. *Polar Biology*, **7**, 97–113.

Brock, W.T. (1973). Primary colonization of Surtsey, with special reference to blue-green algae. *Oikos*, **24**, 239–43.

Brown, A.D. (1978). Compatible solutes and extreme water stress in eukaryotic micro-organisms. *Advances in Microbial Physiology*, **17**, 181–242.

Cameron, R.E. (1969). Cold desert characteristics and problems relevant to other arid lands. *Arid Lands in Perspective* (Ed. by W.G. McGinnies & B.J. Goldman), pp. 167–205. American Association for the Advancement of Science, Washington DC.

Cameron, R.E. (1971). Antarctic soil microbial investigations. *Research in the Antarctic* (Ed. by L.O. Quam & H.D. Porter), pp. 137–89. American Association for the Advancement of Science, Washington DC.

Cameron, R.E. (1972). Microbial and ecological investigations in Victoria Dry Valley, Southern Victoria Land, Antarctica. *Antarctic Research Series*, **20**, 195–260.

Cameron, R.E. (1974). Application of low latitude microbial ecology to high latitude deserts. *Polar Deserts and Modern Man* (Ed. by T.L. Smiley & J.H. Zumberge), pp. 71–90. University of Arizona Press, Tucson.

Cameron, R.E. & Benoit, R.E. (1970). Microbial and ecological investigations of recent cinder cones, Deception Island – A preliminary report. *Ecology*, **51**, 802–9.

Cameron, R.E. & Blank, G.B. (1966). Desert algae: soil crusts and diaphanous substrata as algal habitats. *Jet Propulsion Laboratory, California Institute of Technology, Technical Report*, **32–971**, 1–41.

Cameron, R.E., Conrow, H.P., Gensel, D.R., Lacy, G.H. & Morelli, F.A. (1971). Surface distribution of microorganisms in Antarctic Dry Valley soils: A Martian analog. *Antarctic Journal United States*, **6**, 211–3.

Cameron, R.E. & Devaney, J.R. (1970). Antarctic soil algal crusts. A scanning electron and optical microscope study. *Transactions of the American Microscopical Society*, **80**, 264–73.

Cameron, R.E., Honour, R.C. & Morelli, F.A. (1976). Antarctic microbiology – preparation for Mars life detection, quarantine, and back contamination. *Extreme Environments; Mechanisms of Microbial Adaptation* (Ed. by M.R. Heinrich), pp. 57–82. Academic Press, New York.

Cameron, R.E. & Morelli, F.A. (1974). Viable microorganisms from ancient Ross Island and Taylor Valley drill cores. *Antarctic Journal United States*, **9**, 113–5.

Campbell, I.B. & Claridge, G.G.C. (1987). *Antarctic Soils, Weathering Processes and Environment.* Elsevier, Amsterdam.

Cannon, R.J.C. & Block, W. (1988). Cold tolerance of microarthropods. *Biological Reviews*, **63**, 23–77.

Chambers, M.J.G. (1967). Investigations of patterned ground at Signy Island, South Orkney Islands: III. Miniature patterns, frost heaving and general conclusions. *British Antarctic Survey Bulletin*, **12**, 1–22.

Davey, M.C. & Rothery, P. (in press). Primary colonization by microalgae in relation to spatial

variation in edaphic factors on Antarctic fellfield soils. *Journal of Ecology*, **81**.
del Moral, R. (1993). Mechanisms of primary succession on volcanoes: A view from Mount St Helens. *Primary Succession on Land* (Ed. by J. Miles & D.W.H. Walton), pp. 79–100. Blackwell Scientific Publications, Oxford.
Dort, W. (1981). The mummified seals of Southern Victoria Land, Antarctica. *Antarctic Research Series*, **30**, 123–54.
Evans, D.D. & Buol, S.W. (1968). Micromorphological study of soil crusts. *Soil Science Society of America*, **32**, 19–22.
Fattom, A. & Shilo, M. (1984a). *Phormidium* J-1 bioflocculant: production and activity. *Archiv fur Mikrobiologie*, **139**, 421–6.
Fattom, A. & Shilo, M. (1984b). Hydrophobicity as an adhesion mechanism of benthic cyanobacteria. *Applied and Environmental Microbiology*, **47**, 135–43.
Forster, S.M. (1979). Microbial aggregation of sand in an embryo dune system. *Soil Biology and Biochemistry*, **11**, 537–43.
Friedmann, E.I. (1980). Endolithic microbial life in hot and cold deserts. *Origins of Life*, **10**, 233–45.
Friedmann, E.I. (1982). Endolithic microorganisms in the Antarctic cold desert. *Science*, **215**, 1045–53.
Fritsch, F.E. (1907). Role of algal growth on the colonization of new ground and in the determination of scenery. *Geographical Journal*, **30**, 531–48.
Geesey, G.G. (1982). Microbial exopolymers: Ecological and economic considerations. *American Society of Microbiology News*, **48**, 9–14.
Greenfield, L.G. (1989). Water soluble substances in terrestrial Antarctic plants and microbes. *New Zealand Natural Science*, **16**, 21–30.
Hughes, J. & Smith, H.G. (1989). Temperature relations of *Heteromita globosa* Stein in Signy Island fellfields. *University Research in Antarctica* (Ed. by R.B. Heywood), pp. 117–22. British Antarctic Survey, Cambridge.
Holdgate, M.W. (1964). Terrestrial ecology in the maritime Antarctic. *Biologie Antarctique* (Ed. by R. Carrick, M. Holdgate & J. Prevost), pp. 181–94. Hermann, Paris.
Horowitz, N.H., Cameron, R.E. & Hubbard, J.S. (1972). Microbiology of the Dry Valleys of Antarctica. *Science*, **176**, 242–5.
Kawecka, B. (1986). Ecology of snow algae. *Polish Polar Research*, **7**, 407–15.
Klein, H.P. (1979). The Viking Mission and the search for life on Mars. *Reviews of Geophysics and Space Physics*, **17**, 1655–62.
Klinger, J.M. & Vishniac, H.S. (1988). Water potential of Antarctic soils. *Polarforschung*, **58**, 231–8.
McKay, C.P. (1986). Exobiology and future Mars missions: The search for Mars' earliest biosphere. *Advances in Space Research*, **6**, 12, 269–85.
Miotke, D.D. (1985). Die Dunen im Victoria Valley, Victoria Land, Antarktis. Ein Beitrag zur aolischen Formung in extrem kalten Klima. *Polarfurschung*, **55**, 79–125.
Nelson, S.D., Bliss, L.C. & Mayo, J.M. (1986). Nitrogen fixation in relation to *Hudsonia tomentosa*: a pioneer species in sand dunes, northeastern Alberta. *Canadian Journal of Botany*, **64**, 2495–501.
Parker, B.C., Simmons, G.M. Jr, Wharton, R.A. Jr, Seaburg, K.G. & Love, F.G. (1982). Removal of organic and inorganic material from Antarctic lakes by aerial escape of blue-green algal mats. *Journal of Phycology*, **18**, 72–8.
Smith, R.I.L. (1987). The bryophyte propagule bank of Antarctic fellfield soils. *Symposia Biologica Hungarica*, **35**, 233–45.
Smith, R.I.L. (1984). Colonization and recovery by cryptogams following recent volcanic activity in Deception Island, South Shetland Islands. *British Antarctic Survey Bulletin*, **62**, 25–51.
Smith, R.I.L. (1993). The role of bryophyte propagule banks in primary succession: Case-study of an Antarctic fellfield soil. *Primary Succession on Land* (Ed. by J. Miles & D.W.H. Walton), pp. 55–77. Blackwell Scientific Publications, Oxford.
Sprent, J.I. (1993). The role of nitrogen fixation in primary succession on land. *Primary Succession on Land* (Ed. by J. Miles & D.W.H. Walton), pp. 209–19. Blackwell Scientific Publications, Oxford.
St Clair, L.L., Webb, B.L., Johansen, J.R. & Nebeker, G.T. (1984). Cryptogamic soil crusts, Enhancement of seedling establishment in disturbed and undisturbed areas. *Reclamation and*

Revegetation Research, **3**, 129–36.

Svensson, B.H. (1980). Carbon dioxide and methane fluxes from the ombrotrophic parts of a Subarctic mire. *Ecological Bulletin*, **30**, 235–50.

Tearle, P.V. (1987). Cryptogamic carbohydrate release and microbial response during spring freeze–thaw cycles in Antarctic fellfield fines. *Soil Biology and Biochemistry*, **19**, 381–90.

Vestal, J.R. (1993). Cryptoendolithic communities from hot and cold deserts: Speculation on microbial colonization and succession. *Primary Succession on Land* (Ed. by J. Miles & D.W.H. Walton), pp. 5–16. Blackwell Scientific Publications, Oxford.

Vishniac, H.S. & Klingler, J. (1986). Yeasts in the Antarctic deserts. *Perspectives, in Microbial Ecology, Proceedings of Fourth International Symposium on Microbiology Ecology, Ljubljana* (Ed. by F. Megusar & M. Cantar), pp. 46–51. Slovene Society for Microbiology, Ljubljana.

Vishniac, H.S. & Hempfling, W.P. (1979). Evidence of an indigenous microbiota (yeast) in the Dry Valleys of Antarctica. *Journal of General Microbiology*, **112**, 301–14.

Walton, D.W.H. (1982). Instruments for measuring biological microclimates for terrestrial habitats in polar and high alpine regions: a review. *Arctic and Alpine Research*, **14**, 275–86.

Walton, D.W.H. (1984). The terrestrial environment. *Antarctic Ecology*, 1 (Ed. by R.M. Laws), pp. 1–60. Academic Press, London.

Walton, D.W.H. (1993). The effects of cryptogams on mineral substrates. *Primary Succession on Land* (Ed. by J. Miles & D.W.H. Walton), pp. 33–53. Blackwell Scientific Publications, Oxford.

Whittaker, R.J. & Bush, M.B. (1993). Dispersal and establishment of tropical forest assemblages, Krakatoa, Indonesia. *Primary Succession on Land* (Ed. by J. Miles & D.W.H. Walton), pp. 147–60. Blackwell Scientific Publications, Oxford.

World Meteorological Organization (1988). World Climate Programme. Developing policies for responding to climatic change. *Summary of Workshops and Discussions held in Villach 28 September–2 October 1987 and Bellagio 9–13 November 1987 under the auspices of the Beijer Institute, Stockholm.* WMO/TD No. 225. (WMO/WCIP1), World Meteorological Organization Geneva.

Wynn-Williams, D.D. (1980). Seasonal fluctuations in microbial activity in Antarctic moss peat. *Biological Journal of the Linnean Society*, **14**, 11–28.

Wynn-Williams, D.D. (1985). Photofading retardant for epifluorescence microscopy in soil micro-ecological studies. *Soil Biology and Biochemistry*, **17**, 739–46.

Wynn-Williams, D.D. (1986). Microbial colonisation of Antarctic fellfield soils. *Perspectives in Microbial Ecology, Proceedings of the Fourth International Symposium on Microbial Ecology, Ljubljana* (Ed. by F. Megusar & M. Cantar), pp. 191–200. Slovene Society for Microbiology, Ljubljana.

Wynn-Williams, D.D. (1988). Television image analysis of microbial communities in Antarctic fellfields. *Polarforschung*, **58**, 239–50.

Wynn-Williams, D.D. (1990a). Ecological aspects of Antarctic microbiology. *Advances in Microbial Ecology*, **11**, 71–146.

Wynn-Williams, D.D. (1990b). Microbial colonization processes in Antarctic fellfield soils — an experimental overview. *Proceedings of NIPR Symposium on Polar Biology*, **3**, 164–78.

The effects of cryptogams on mineral substrates

D. W. H. WALTON
British Antarctic Survey, Natural Environment Research Council, High Cross, Madingley Road, Cambridge CB3 0ET, UK

SUMMARY

1 Lichens show chemical and biological interactions with both rock and soil. Crustose epilithic species can cause surface weathering by physical disruption and secondary mineralization by organic acid attack. Endolithic species can dissolve the silicate matrix and induce exfoliative weathering. Terricolous species may bind small stones and soil together, forming a stable island in periglacially active areas.
2 Mosses appear to be able to penetrate rocks along cracks and assist in physical weathering. They also interact by scavenging and accumulating mineral particles. These provide a core around which a stable island can grow and the particles may be chemically attacked by exudates from the moss.
3 Algae, fungi and bacteria all have significant roles in chemical weathering of minerals and constitute the primary colonizers of all bare rock and soil.
4 The significance of lichen and moss interactions is yet to be quantified at the ecosystem level. It seems likely that for primary succession cryptogam weathering is mainly facilitative, whilst microbial weathering is facultative.

INTRODUCTION

The lower plants are found in all regions of the world but are usually most prominent in high latitude and high altitude communities. Areas of unstable ground, either sand or mineral soil subject to periglacial disturbance, often appear to be colonized by mosses and lichens while both of these groups are prominent in bare rock habitats and often on bare ground all over the world. Algae are also found in these sites, frequently growing in cracks or actually inside the rock fabric, or forming a mat on the surface of bare damp mineral soil.

The early assumption by Clements (1916) that site modification was essential to the development of succession was exemplified by his belief that mosses and lichens were not only the primary colonizers on rocky substrates, but necessary precursors for the establishment of the later communities. With the exceptions of some bare rock and the polar and Alpine communities where succession stops at an entirely cryptogamic assemblage, this is not generally true, but the idea took quite some time to die. In Arctic and Alpine habitats, as well as in temperate ones, some flowering plants appear to be able to establish themselves directly into bare ground and even into cracks in rocks, providing a focus around which cryptogams later establish. Even in these situations bacteria, micro-algae and fungi almost certainly play the important primary role, a feature which has been largely ignored by ecologists.

The contribution of all these groups of lower organisms to pedogenesis has been the subject of some disagreement. The situation is not clear-cut. Mosses and lichens do play a role in primary succession but their importance varies in different habitats and in most, if not all cases these cryptogams follow on from the real primary colonizers: bacteria, fungi and algae. Is there then an interaction between these early colonizers and their substrate which provides the necessary conditions for succession to begin? If so, biological weathering may be a key determinant of the rate of change in the earliest stages of primary succession.

Biological weathering by cryptogams is now an established fact, involving both biogeochemical and biogeophysical processes. These necessarily interact with the more general physical and chemical weathering processes which are well documented elsewhere (e.g. Yatsu 1988; Lerman & Meybeck 1989). Although the organisms themselves have often been the subjects of intensive ecological study over long periods there is generally little data about their interaction with substrates. The most detailed studies so far undertaken have focused on lichens and it is these that this chapter will examine in particular. Recent reviews by Topham (1977), Jones & Wilson (1985) and Jones (1988) have all dealt with some aspects of this field. There is also clear evidence that the algae (Marathe & Chaudhari 1975), fungi (Cromack *et al.* 1979) and bacteria (Wagner & Schwartz 1967) play significant roles in certain types of biological weathering. In particular, the biodegradation of silicate minerals, the major component of the matrix of most rocks, may be much more extensive than was previously suggested (Dacey *et al.* 1981).

This chapter will examine three questions:

1 What is the evidence for interactions between cryptogams and substrates?
2 How extensive and important are any of these interactions?
3 How do they relate to primary succession?

CRYPTOGAMS IN PRIMARY SUCCESSION

Studies on bare soils, bare rocks, spoil heaps and man-made substrates have all shown that mosses and lichens can often be among the initial colonizers, but normally only after microbial establishment has occurred. It seems likely that bacteria are the first to arrive at any surface, with or closely followed by fungi and algae. In many instances either mosses or phanerogams are the next to establish, although in some particular instances lichens may appear along with mosses.

Topham (1977), in a general review of the importance of lichens in succession, notes that many successions in mesic or humic environments do not involve lichens to any significant extent. Even in xeric habitats mosses may compete for the same niches; in northern temperate sites *Andraea* and *Rhacomitrium* are often the first cryptogamic colonizers on rock. There is also little need to accept the old generalization that crustose species precede foliose and fruticose forms. In Antarctica *Usnea* and *Umbilicaria* can both establish before crustose species, while in Greenland *Umbilicaria* and nitrophilous species were found to precede crustose species (Beschel & Weideck 1973). Longton (1988) provides the most recent review of bryophytes and lichens in the polar regions.

The classic study of colonization of newly exposed glacial moraine was made by Cooper (1939) at Glacier Bay, Alaska. His data show that although both moss (*Rhacomitrium* spp.) and lichen (*Stereocaulon tomentosum*) appeared at the earliest stage they were accompanied by at least seven phanerogams. More recently a study in northern Norway on glacial moraine (Worsley & Ward 1974) showed colonization by mosses and grasses within 1 year but epilithic lichens (*Umbilicaria, Lecidea, Alectoria, Rhizocarpon*) only after 12 years. Even longer was required (up to 30 years) for epilithic lichens on Alpine moraines (Spence 1981) where the distribution was controlled not by aspect but by lithology and boulder size. The greatest cover developed on gneiss and schist with the least on quartzite and sedimentary rocks. Mosses appeared rapidly on the ground but only slowly on the rocks and their distribution was apparently affected by aspect. On the new volcanic island of Surtsey, off Iceland, colonization of lava flows was principally by mosses (Fridriksson 1975), while on the lavas of Hawaii the lichen *Stereocaulon vulcani* is the most abundant pioneer species (Jackson & Keller 1970).

Thus, bryophytes and lichens are closely implicated in the establishment of new plant communities on bare ground and rock, but their importance varies from site to site. In some instances successional development is limited and the climax communities are composed wholly or principally of cryptogams. For example, Kubiena (1948) described a primary succession on dolomitic limestone in Austria. An endolithic species, *Verrucaria calsiseda*, and an epilithic species, *Squammaria crassa*, colonized first and were in turn overgrown by the moss *Grimmia orbicularis*. Under the moss cushions a protorendzina-like soil accumulated which in turn was colonized by micro-arthropods. It is in these instances that the role of mosses and lichens is most obvious but, together with algae, bacteria and fungi, they also play a largely unappreciated role in almost all primary successions.

Microbial colonization has been the subject of a great deal of detailed research but since little of it has been published in the ecological literature it has largely been ignored in discussions of colonization and succession. Indeed, since much of the published work is experimental, using artificial substrates, it is often difficult to transfer the conclusions directly to the heterogeneity of a natural habitat.

PHYSICAL AND CHEMICAL INTERACTIONS

Lichens grow on a wide variety of substrates, both natural and man-made. Of particular interest in the present context are rock, bare soil and man-made mineral substrates. It is clear from ecological data that some lichens are highly substrate specific, while others are much less restricted (Brodo 1974). While some species appear to be chemically limited (e.g. calcicolous, corticolous) others may be restricted by surface texture (Garty & Galva 1974).

An investigation of the effects of surface texture in colonization by saxicolous species used slate (Armstrong 1978, 1981). After 6 years it appeared that for *Parmelia glabratula*, *P. conspersa* and *Buellia aethalea* the key to establishment was the presence of cracks on the surface of the slate. Colonization began at the top of the slate and spread downwards, a pattern consistent with dissemination of propagules through a water film. Armstrong (1981) suggested that the matching of propagule size to microsite was an important element in colonization and could be related to the range of propagule types used by many lichens.

Biogeophysical attack

Syers & Iskandar (1973) provide a good historical review of the early literature and note that the ability of lichens to disintegrate rocks had been reported as early as 1856. Although Fry (1924, 1927) was not the first to describe the physical breakdown of the rock surface by lichens she was the first to investigate the mechanism experimentally.

She suggested that for the crustose species which she investigated (*Lecidea confluens, L. plana, Lecanora sordida* [= *L. rupicola*], *L. atra, L. sulphurea, L. parella, Rhizocarpon geographicum, Aspicilia alpina, A. calcarea* [= *Lecanora calcarea*]) disintegration of the substrate normally occurred at the margins of the thallus, and especially below the apothecia. The physical properties of the substrate — especially hardness, crystallinity, foliation, etc. — were important determinants of the extent of disintegration. Her experiments showed that shale and schist were easy to fragment but obsidian was difficult. She was able to simulate this mechanical damage using drying gelatin films as an analogue for thalli (Fry 1924). From this she concluded that mechanical disruption preceded chemical weathering. Although she was unable to offer any specific data on chemical alteration she commented on changes in the appearance of rock fragments beneath some species. These seminal studies generated little interest for many years.

Physical disruption can be divided into two processes: rhizine or hyphal penetration into the substrate and thallus expansion/contraction cycles. The depth of rhizine/hyphal penetration appears to depend on both rock type and lichen species; Topham (1977) noted 19 mm for *Verrucaria marmorea*, 16 mm for *Caloplaca heppiana* in limestone and 10 mm for *Baeomyces rufus* in schist. It appears that data on this are sparse.

Penetration, not surprisingly, is along planes of least resistance utilizing foliations in mica schist (Walton 1985) and cleavage planes and crystal boundaries in granite (Bachman 1904) and limestone (Syers 1964). Hyphal penetration below crustose species can probably utilize all the microfractures since the minimum width is that of a single hypha. Rhizines, being organized bundles of hyphae, are much larger and presumably less able to exploit all openings. The physical significance of rhizine/hyphal penetration is assumed to be twofold. First, it provides an anchor point for the expanding and contracting thallus as it gains and loses water. Second, the rhizines/hyphae themselves must expand and contract with water availability within the microcracks, assisting in the extension of microfractures within the substrate, although this has yet to be proved.

Expansion and contraction of foliose thalli is directly related to humidity changes. Fig. 1 illustrates both the extent of areal change for an Antarctic foliose lichen (*Umbilicaria decussata*) and the rapidity with which contraction takes place. The area of rock periodically covered and uncovered by the thalli is clearly subjected to a different weathering regime to rock with no lichen cover. However, for many foliose species, such as this one, and for fruticose species direct physical attack on the rock surface must presumably be confined to the hapteron, as reported by Syers (1964) for some species on limestone. On the other hand, in foliose species attached by rhizines and in crustose species where the thallus is attached much more intimately to the rock surface by penetrating hyphae a considerable direct pulling force must be generated during contraction and it may be this which loosens the surface fragments for incorporation within the thallus.

As the lichen grows it appears to pluck rock fragments from the surface and incorporate them within the thallus as Ascaso *et al.* (1976) have shown for *Rhizocarpon geographicum* and *Parmelia conspersa* and Walton (1985) for *Lecidella bullata* (Fig. 2). The ability to disrupt the surface of the rock is presumably dependent on depth and type of penetration, thickness of the thallus and the frequency of wet/dry cycles. There appears to have been no study made of this.

Epilithic, chasmolithic and endolithic lichens (see Golubic *et al.* (1981) for terminology) can be considered separately, but there is little data to indicate which type has the greater physical weathering effect. Caution must be exercised in drawing conclusions. It is significant that the endolithic species in the Antarctic which produce exfoliative weathering of sandstone do so by chemical dissolution rather than physical disruption (Freidmann 1982). On limestone, Syers (1964) found that the endolithic and some epilithic species with thin thalli which colonized first generally had poorly developed rhizines, while the epilithic species with thicker thalli which appeared later had more extensive networks of rhizines and a deeper penetration. This also seems true for the chasmolithic species (Fig. 3). Endolithic species in the British flora are mainly calcicolous (e.g. *Opegrapha calcarea, Petractis clausa, Catillaria lenticularis, Thelidium decipiens*), but there are some calcifuges such as *Lecidea auriculata*, *L. diducens* and *Sarcogyne simplex* growing on acidic granites and schists. Other species have epilithic thalli, but the perithicia are immersed in pits dissolved in the substrate, e.g. *Verrucaria sphinctrina*.

Lichens also act in some instances as binding agents for mineral materials. Their low growth rates and poor competitive abilities relative to higher plants normally put them at a disadvantage in colonizing soil surfaces. However, some species, e.g. *Placopsis contortuplicata*, are surprisingly successful on dry sites with

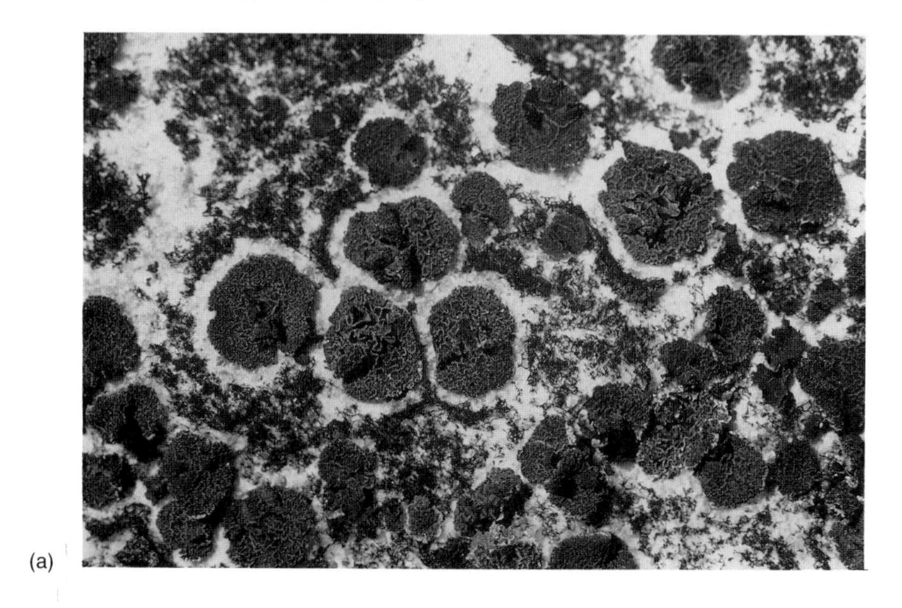
(a)

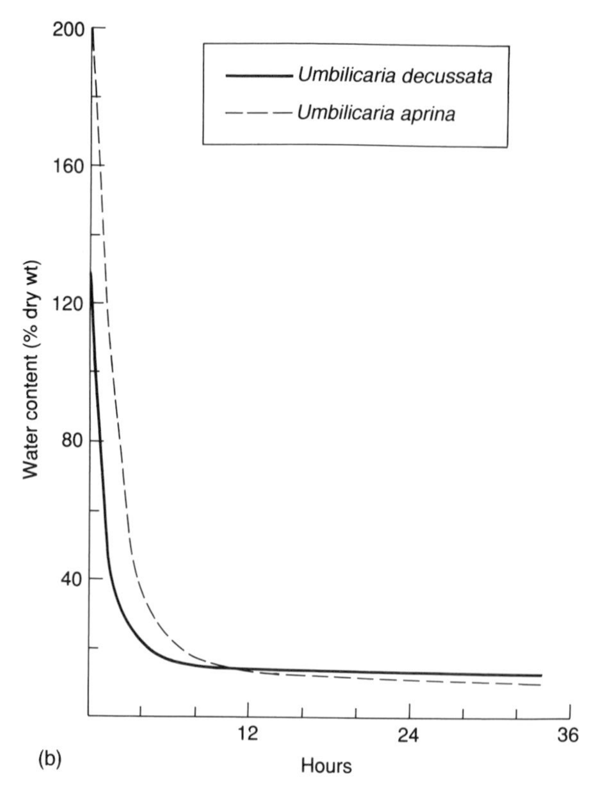

FIG. 1. (a) Dry *Umbilicaria decussata* thalli at Casey Station, Antarctica. When fully hydrated the thalli expand to cover the light-coloured areas. (Photo R.I. Lewis Smith.) (b) Rate of water loss from *U. decussata* and *U. aprina* (from Smith 1988.)

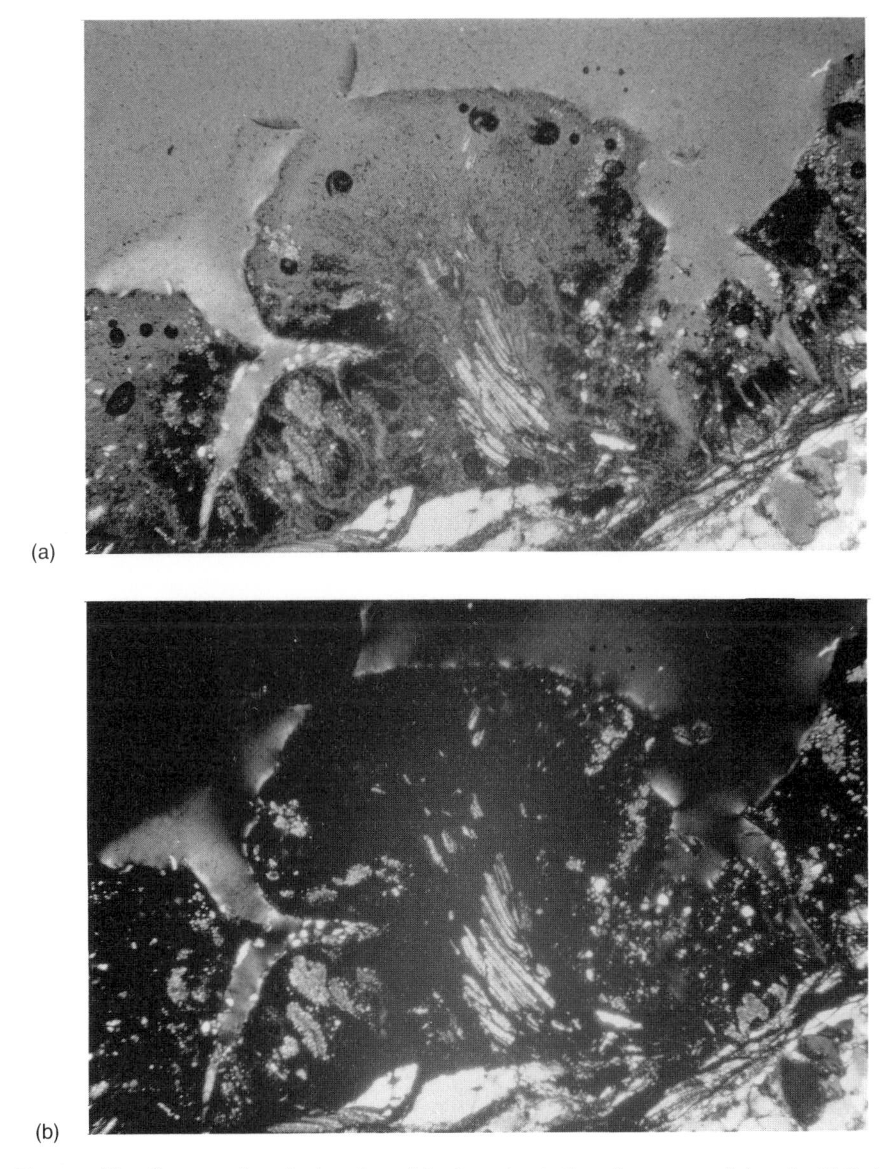

FIG. 2. The direct uptake of mineral particles into the thallus of a crustose lichen *Lecidella bullata*. Biotite fragments are visible (centre) as a stack of lamella in (a) normal and (b) polarized light.

low nutrients and surface disturbance. An important feature of many of the pioneer soil and rock species is the presence of a blue-green alga as phycobiont, allowing atmospheric nitrogen to be fixed. *Stereocaulon, Placopsis, Usnea, Umbilicaria, Peltigera, Collema*, etc. all have *Trebouxia* as phycobiont.

Some mosses are also involved in physical attack on softer rocks. The rhizoids

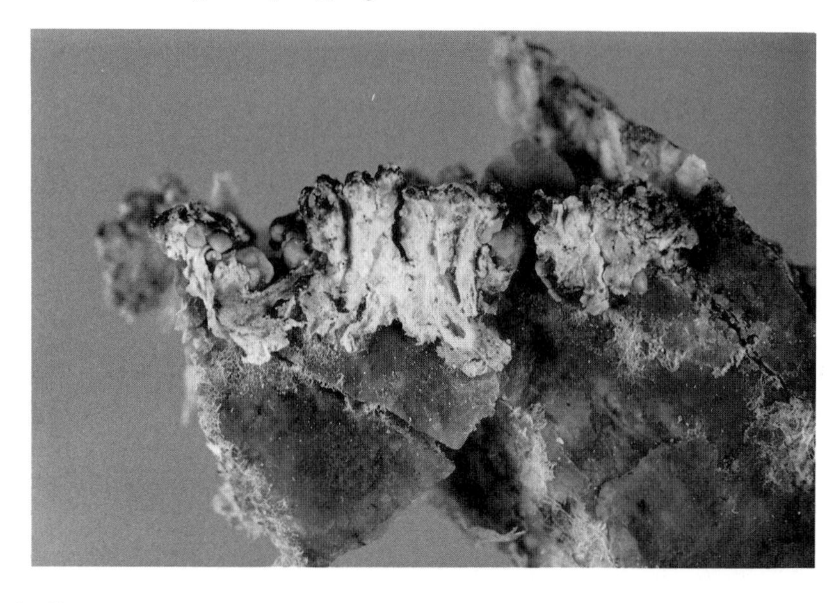

Fig. 3. Chasmolithic *Rhizoplaca melanophthalina* growing in a protected crevice on an isolated nunatak at 85°S in the Theil Mountains, Antarctica. Note the extensive network of rhizines (photo British Antarctic Survey).

of *Grimmia apocarpa* penetrate limestone up to a depth of 10 mm (Syers 1964), while both *Tortella tortuosa* and *Camptothecium sericeum* show more limited rhizoidal penetration.

Biogeochemical attack

Lichens

Chemical attack on the substrate by lichens was suggested a long time ago but early theories all proposed that carbonic acid, derived both from rain and from the lichen's own respiration, was the responsible agent. This proved not to be the case and attention then focused on organic acids of which oxalic acid, secreted by the mycobiont, appeared to be the one most frequently present. Calcicolous species in particular have high levels of calcium oxalate (Syers *et al.* 1967), but proving the causal connection took some time. Examination of the effect of *Pertusaria corallina* on basalt (Jones *et al.* 1980) showed chemical attack on plagioclase feldspar and ferromagnesian minerals, and the complete breakdown of ferruginous minerals giving a ferrihydrite as a product. Culture of the mycobiont showed significant excretion of calcium oxalate and incubation of mineral grains with the oxalate-producing fungus *Aspergillus niger* reproduced many of the dissolution and etching effects seen at the basalt/lichen interface.

To test the generality of this reaction Wilson *et al.* (1981) studied the weathering of serpentinite by *Lecanora atra*; with virtually no calcium in the rock, crystalline magnesium oxalate dihydrate was formed and the electron microprobe study of thin section showed magnesium depletion in the rock underlying the thallus. A second study using *Pertusaria corallina* on manganese ore (Wilson & Jones 1984) showed the formation of crystalline manganese oxalate dihydrate.

These studies were not the only ones to recognize the importance of oxalic acid. Ascaso *et al.* (1982) showed that *Caloplaca callopisma, Diploschistes ocellatus, Squamarina oleosa* and *Protoblasemia testacea* all produced calcium oxalate in the thallus from the attack of oxalic acid on calcite. In one iron-rich rock ferric oxalate was also found.

In a British study of the weathering effects of epilithic species Viles (1987) looked at *Caloplaca* and *Verrucaria* on limestone. She found no evidence of physical weathering beneath the thalli and no difference between the genera. She suggested that the phycobiont and mycobiont produced different forms of chemical weathering: circular pits up to 2 mm deep caused by algae, and holes through individual crystals attributed to hyphal penetration. Etching of feldspar crystals she considered was possibly due to lichen acids. Her conclusion that weathering was better developed in drier situations agrees with both Klappa (1979) and Danin & Gerson (1983).

The most detailed investigation so far of surface changes on individual minerals by chemical attack has characterized a number of specific interactions (Jones *et al.* 1981). Preferential etching of discontinuities in chemical content, cleavage planes or specific crystalline faces produce characteristic etch pits, trenches, etc. when viewed by scanning electron microscope (SEM). Examination of biotite by microprobe analysis confirmed the previous finding that weathering extends inwards from the edge of the flake, but SEM pictures also show apparent etch pits and blistering on the basal cleavage faces. A second study (Wilson & Jones 1983) provided further illustrations of weathering features and a more detailed mineralogical discussion of them. Muxart & Blanc (1979) have also reported fungal boreholes and chemical etching of calcite and dolomite crystals associated with endolithic lichens.

Lichens also produce a group of substances called 'lichen acids', polyphenolic compounds of limited solubility in water. Schatz *et al.* (1956) had suggested that these might be important chelating compounds, forming soluble complexes with iron and aluminium. This opened up a possible route for the weathering of silicate minerals and Schatz (1963) was able to demonstrate the reaction experimentally. Ground-up *Parmelia conspersa*, *P. stenophylla* and *Umbilicaria arctica* all reacted with suspensions of mica and granite, as did two pure lichen acids: physodic and lobaric. Syers (1969) confirmed this for *P. conspersa* and showed that the chelating ability was similar to that of fumaprotocetraric acid.

The more extensive studies on chelating ability by Iskandar & Syers (1972) tested six lichen compounds — atranorin and salazinic, stictic, evernic, roccelic and lecanoric acids — with basalt, granite and biotite. The solubility of the compounds was confirmed as was their ability to complex cations. The theme was pursued by Williams & Rudolph (1974) in a study using iron chelation as a measure of biological weathering by lichen mycobionts and free-living fungi. Mycobionts of *Caloplaca holocarpa*, *Lecanora dispersa*, *Cladonia cristellata* and *C. squamosa* showed no activity in culture but squamitic acid extracted from *C. squamosa* did show low activity. In contrast, many of the fungi, especially *Penicillium* and *Cephalosporium*, showed considerable chelation ability attributable to citric acid secretion.

All this evidence of the role of lichen acids persuaded Ascaso & Galvan (1976) to look at the reactions between four lichen compounds — atranorin, norstictic, stictic and usnic acids — and crushed granite and gneiss and their primary minerals (albite, ortose, biotite, muscovite, quartz). Oxalic acid was used as a control. All lichenic substances showed chelating ability but this was much less than that of oxalic acid and was not correlated with solubility. More importantly they noted structural changes, especially in biotite, and the formation of new minerals (montmorillonite and halloysite). They were then able to confirm chemical and morphological changes in the substrate below thalli of *Parmelia conspersa, Rhizocarpon geographicum* and *Umbilicaria pustulata* (Ascaso *et al.* 1976). Clay minerals, amorphous gels and goethite (iron oxide) were identified in the interface.

Independent confirmation of the structural changes induced by chemical weathering was given by Hallbauer & Jahns (1977) who examined quartz grains apparently deeply etched by the mycobiont of *Dimelaena oreina*. This was significant since quartz was believed to be largely resistant to chemical weathering.

Later studies on metamorphic rocks (Galvan *et al.* 1981) used *Parmelia conspersa, Parmelina tiliacea*, *Lasallia putulata* and *Ramalina protecta*. These showed that the same species of lichens induced greater alterations in metamorphic than in igneous rocks. Chlorites and feldspars were always altered, while micas were preferentially colonized and accumulated under the thalli. Evidence for the formation of amorphous gels and goethites was again found in the lichen/rock interface.

In the Antarctic cryptoendolithic lichen communities calcium oxalate has also been identified (Weed 1985). It had already been suggested that it might play a significant role in the movement of iron within the rock matrix and be implicated in the dissolution of silica leading eventually to exfoliative weathering (Friedmann 1982).

A few crustose lichens (*Acarospora sinopica, Rhizocarpon oederi*, *Lecidea dicksonii*, *L. silacea*) appear to be limited to iron-rich rocks where they are apparently able to mobilize the iron-forming ferric oxide crystals in the thalli. A comparison of the interaction between rock type and a single lichen species was undertaken in the tropical regions. *Acarospora sinopica* is abundant on several rock types in South Africa and has been shown to cause silica depletion in conglomerate, arkose and dolomite (Kerr & Zavada 1989), which the investigators attributed to localized pH mediation allowing the use of molecular diffusion pathways into the rock matrices. All three rock types were affected but not to the same extent.

Algae

Algae have been seen as important constituents of the soil ecosystem for a long time, but were not seen to have a role in weathering. However, since they are important colonizers, (especially in deserts and the polar regions) of rock fissures (Broady 1981a), the underside of rocks (Broady 1981b), the rock matrix (Friedman & Galun 1974) and the soil surface (Metting 1981) it is essential to assess the available data again.

In most temperate, Alpine and polar sites rock/algae interactions are found in sublithic, chasmolithic and endolithic habitats. Epilithic algae are generally restricted to wetter sites and are thus not as common. In addition there are the terricolous algae growing on and in the soil which may play a significant role in the aggregation of soil particles (Bailey *et al.* 1973).

Vincent (1989) has reviewed the lithic algal communities for the Antarctic. Endolithics are the most restricted and, as well as *Trebouxia* as the phycobiont of endolithic lichens, the endolithic algae *Hemichloris antarctica, Chroococcidiopsis* and *Gleocapsa* are also found. Chasmolithic algae are much more widespread and are probably found in any calcareous rocks, as well as in many sandstones, granites, gneisses and granodiorite. They occupy both vertical fissures and horizontal cracks and there is often a zonation of species presumably related to light availability. The sublithic community is the most common, especially under translucent pebbles but is also found around the edges of dark rocks. The dominant species are cyanobacteria and the species diversity is almost always greater than in the other lithic habitats. There is no direct evidence of *in situ* biological weathering by these algae but, since they produce organic acids, there seems no reason not to suppose that slow chemical weathering does occur.

Chasmolithic algal communities are found in temperate latitudes but it is terricolous species that are the more important. Algal mats, often of *Prasiola crispa* or *Phormidium* spp., formed on the soil surface provide a similar cementation action to terricolous lichens, but for smaller particles of soil. Since the algal mats are only annual any biological weathering at this particulate level must be of limited duration.

Fungi

The importance of fungi as active agents in soil formation has been known for a long time. They produce a variety of organic acids of which the most important in weathering terms are citric and oxalic. The study by Webley *et al.* (1963) established that there was a direct correlation between an increasing degree of colonization of rocks by cryptogams and an increasing number of micro-organisms. These were isolated even from the centre of weathered rocks. Many of them could attack silica. Some of these isolates were tested on a range of mineral types: geothite, leucite, nepheline, olivine, wollastonite, biotite, muscovite, phlogopite, apophyllite, harmotome, heulandite, natrolite and stilbite (Henderson & Duff 1963). While some minerals were resistant to attack others were significantly weathered. Particularly active isolates were *Aspergillus niger* with oxalic, citric and fumaric acids, *Spicaria* sp. with acetic, oxalic and formic, and *Penicillium* spp. with citric. A key feature of the most active reactions was a pH of less than 3.

Even yeasts isolated from rock surfaces have been shown to solubilize iron and other elements from amphibolite (Rades-Rohkohl *et al.* 1979). Different responses to the same mineral by the same yeast grown on different media suggest that mineral attack may be by a variety of metabolites.

It has been shown in several laboratory studies that organic acids will extract iron from a wide range of minerals (Schalscha *et al.* 1967), and solubilize phosphate from sparingly soluble forms (Silverman & Munoz 1970). The occurrence of natural concentrations of acid salts in the soil has now been linked to intense localized weathering of mineral grains. Mats of the hypogeous fungus *Hysterangium crassum* exude large amounts of oxalic acid, which precipitates in the soil (Cromack *et al.* 1979). Etching of andesite grains and the formation of intergraded clays together with accelerated mobility of iron and aluminium are associated with these mats. Jones *et al.* (1976) have shown that many species in the Mucorales bear external spines of weddellite (calcium oxalate dihydrate) and it has been identified in other fungal genera (Graustein *et al.* 1977).

Penicillium and *Cephalosporium* species also have considerable chelation ability (Williams & Rudolph 1974). More recently, Zykina (1982) has shown that iron and aluminium complexes are formed when biotite and nepheline are decomposed by *Aspergillus niger*, and iron reduction by fungi has been reported from Swedish shales (Napier & Wakerley 1969).

An extreme example of primary colonization similar to that described by Vestal (1993) from the Antarctic is the formation of desert varnish on rocks. The varnish is a dark, clay-rich ferromanganese oxide coating on rocks in arid and semi-arid regions (Cooke & Warren 1973). The coatings have abrupt contact with the underlying rock and no apparent direct mineralogical or chemical affinity. The varnish appears to have a microbiological origin, formed by dermatiaceous hyphomycetes and bacteria (Taylor-George *et al.* 1983), sometimes in direct association with cyanobacteria (Krumbein & Jens 1981). Cultivation of the hyphomycetes shows some to be capable of manganese oxidation. The model proposed by Taylor-George *et al.* (1983) for varnish formation suggests that the microcolonial fungi establish themselves on and in bare rock (Staley *et al.* 1982) and accumulate wind-deposited clay particles and bacteria on their surfaces. As the fungal patches grow remobilized manganese and iron are incorporated within the crust together with (in the more protected habitats) algae and cyanobacteria. The ability of these microcolonial fungi to colonize habitats with high temperatures, very limited water and low nutrients is now established and their adaptations are currently being investigated (Palmer *et al.* 1987).

Bacteria

As long ago as 1890 bacteria were implicated in the weathering of rocks and minerals. The composition of the surface microflora depends on lithology and surface relief as well as on microclimatic conditions. An early microbiological investigation of sandstone (Paine & Linggood 1933) established the presence of viable bacteria inside the rock. Webley *et al.* (1963) showed that bacteria could be found in the centre of weathered rocks, with the highest numbers in serpentine and amphibolite-chlorite. German sandstones show bacterial penetration down to 20 cm, but with no clear relationship between numbers and depth suggesting a very heterogenous distribution (Weirich & Schweisfurth 1985).

Pseudomonas spp. produce 2-ketogluconic acid which is an active chelating agent (Duff *et al.* 1963), while *Bacillus* spp. produce a wide range of other organic acids (see Dacey *et al.* 1981). There are also anaerobic bacteria which can reduce iron, and autotrophic bacteria which can mobilize uranium, manganese, gold, silver, etc. (Berthelin 1988). The literature in this field is too extensive for inclusion in this review.

Mosses

Many of the mosses found growing on rock are obligate or facultative epiliths. Very little has been written on the mineral nutrition of these saxicolous bryophytes. Of relevance is the demonstration of direct uptake of strontium from rock by *Grimmia orbicularis* (Hebrard *et al.* 1974). Rhizoids of *Tortula muralis* have been shown to penetrate at least 5 mm into limestone (Hughes 1982).

Moss cushions can act as scavengers of mineral particles. Dust deposited by wind is one type of material but possibly more important is material derived from precipitation or overwash. Snowmelt and rainstorms can move considerable amounts of fine soil and where this washes over mosses they become coated with fine silts and clay, some of which will be retained within the cushion. Fig. 4 shows a vertical section through a saxicolous moss (*Andreaea*) cushion where mineral material has been accumulated from rare occasions of heavy downwash associated with rapid snowmelt.

For mosses growing on the soil surface, especially in drainage lines, much greater accumulation of mineral soil can occur. In large *Drepanocladus uncinatus* hummocks on Signy Island, Antarctica, growing across a natural gulley, there is normally a very pronounced mineral core. Individual moss stems within the hummock have mineral fragments (mainly biotite) 'cemented' along them. This solid mineral core provides a stable base which keeps the moss hummock from being swept away.

It is worth noting that these isolated moss cushions are of major significance for animals. Nematodes, protozoa, rotifers, tardigrades and many mites and springtails, etc. are found colonizing these organic islands. Not only are they a source of food and a protected microclimate but for many species they provide the ideal habitat for reproduction.

FIG. 4. Vertical section through a cushion of an Antarctic saxicolous moss (*Andreaea gainii*) showing accumulated mineral material.

Artificial substrates

Lichens have been recorded growing on a very wide variety of unnatural substrates (Brightman & Seaward 1977). Those derived from mineral sources are useful in investigating the specificity of lichen/substrate interactions. Clearly, colonization of quarries and mineral spoil heaps is simply an extension of naturally occurring habitats. Dressed stone and sculpture is slightly different since in many cases the surface has been made unnaturally smooth. Yet lichen colonization and direct surface weathering is an important feature on much architectural and other stonework (Seaward *et al.* 1989).

Bricks are another mineral substrate colonized by both foliose and crustose lichens. Brightman (1965) found that surface texture, pH and water-holding capacity all interacted to influence the lichen flora of vertical brick walls.

SIGNIFICANCE OF CRYPTOGAMIC WEATHERING

So far it has been possible to demonstrate that both chemical and physical changes to rock minerals are caused by cryptogams. In temperate and tropical regions, lichens and mosses hardly constitute the predominant group in the flora and there are normally very limited areas of bare rock or ground awaiting colonization. In the high Alpine and polar regions this is not true and there cryptogam weathering may constitute an important component of total weathering. It must, however, be recognized that direct physical, especially in maritime and desert areas, and chemical weathering are probably the principal routes for rock breakdown rather than biological weathering *per se*.

None of these processes operate in isolation. The physical breakdown of rock constantly exposes new surfaces; salt weathering provides new mineral substrates and biological activities aid and abet both of these. Of particular significance is the release of calcium oxalate into the soil environment. This has a major effect on biological and geochemical processes by increasing the effective solubility of iron and aluminium by several orders of magnitude, by keeping phosphorus available (due to the chelation of iron and aluminium) and by providing readily available calcium.

In the case of lichens the crustose species appear to be the group most significantly involved in direct breakdown, but chelation reactions resulting from leachates are likely to be attributable to foliose and fruticose species as well. Sedimentary rocks appear to be more easily attacked than metamorphic or igneous rocks. The relationships between mineral composition, rock texture, permeability, cleavage and schistosity are of crucial importance in determining the type of both biogeophysical and biogeochemical weathering but the available data are still very limited.

The importance of cryptogam weathering to primary succession is at least fourfold:

1 Biophysical attack at and just below the rock surface, and biochemical attack of particles incorporated within a lichen or algal thallus comminutes the parent rock to provide an increased surface area for further disintegration by environmental processes.

2 Penetration by rhizines, hyphae and rhizoids almost certainly assists in micro-fracturing of the rock, expanding access channels for other organisms (bacteria, fungi, algae) to colonize and provide increased protective niches.

3 Biogeochemical weathering both releases unavailable elements from the rock, which when leached, are available in other parts of the ecosystem, and provides new secondary mineral substrates for attack.

4 Lichens, mosses and algae establishing on rock surfaces or on unstable soil act as foci for the development of more complex miniature ecosystems.

What is missing at present is an assessment of the rate of natural biological weathering and the amounts of material made available. The micaceous minerals have been better studied experimentally than most and illustrate the difficulties of transferring the results of laboratory experiments to natural ecosystems. Using organic acids and a shaken slurry of biotite Boyle *et al.* (1974) were able to show considerable rates of cation release; for example, up to 20% of total potassium after 48 hours in oxalic acid. This was almost certainly due to the continual mechanical disintegration of the biotite flakes during shaking, exposing new weathering surfaces. In non-agitated systems an amorphous, ion-depleted edge forms around particles and the rate of cation release quickly falls off.

Huang & Keller (1970) used five mineral types and four organic acids to show solubility was higher in 0·01 M acid than in distilled water or weak carbonic acid.

Their calculations suggested that weathering of this type could result in a reactive layer up to 3000 Å deep in the mineral particles, and that the differential dissolution effects on silicon and aluminium were consistent with particular geological processes. While all of this helps to substantiate the chemical reactions involved in weathering it is difficult to see any immediate usefulness in determining rates of biological weathering.

RELEVANCE TO PRIMARY SUCCESSION

Substrate specificity is an important feature of the ecology of any species. For the primary species of any succession colonizing bare rock or mineral particles the substrate framework is usually silica or calcium. Much of the decomposition of these is mediated by micro-organisms and thus, in this respect at least, biological activity controls the cycling rate of many major elements.

For some groups of organisms, primary colonizers have been characterized as 'weedy' species with little substrate specificity, and this apparently applies to some of the lithophytic lichens. Yet studies on saxicolous lichens colonizing deglaciated moraine in the Arctic (Fahselt *et al.* 1988) have shown these communities to be of comparable complexity to those on older substrata containing lichens, mosses and phanerogams.

The ability of organisms to interact with a substrate is a fundamental part of the process of colonization and establishment. Those habitats preferentially colonized must have some advantage over surrounding habitats and temperature (Fahselt *et al.* 1988), water (Kappen 1985) and roughness (Smith 1988) have all been implicated. The ability of the propagules to adhere firmly in an acceptable niche is as important for a lichen establishing on bare rock as for a moss on bare soil. Yet the challenge of quantifying the microtopography of habitat structure for mosses and lichens has hardly been attempted (Alpert 1991)

The changes wrought in the substrate during the subsequent growth of the plant have several effects. Some directly affect the microclimate of the substrate: increased shading lessens temperature amplitude, a change in reflectance will have consequent effects on radiative heating, an increased duration of moisture retention will affect wet/dry cycles. Others affect the stability of the substrate: increased chance of microcracking in rocks, solubilization of rock fabric, retention of soil particles from overwash and precipitation. Yet others affect not only the chemistry of the supporting substrate but ion availability elsewhere in the ecosystem: mineralization by organic acid attack, leaching of chelating agents into the soil.

Finally, the establishment of an organic base in an inorganic milieu provides a focus for invertebrate colonization (Gerson & Seaward 1977; Gerson 1982) and the potential for secondary colonization by epiphytic plants.

FUTURE RESEARCH

Biogeochemistry is now an acceptable subject and it is in this field that much of the important progress will be made. There is a growing literature, driven in part by commercial interests and conservation requirements, on microbial interactions with specific minerals which is establishing important baseline data on reaction rates, penetration into stone, etc. for certain species (e.g. Gauri & Gwinn 1984). There are still few data on the microbial populations of weathered rocks, the relationship with mineral type and time since weathering began. Equally, changes in populations (both numbers and diversity) with the early stages of succession are also missing.

The application of microbial models may be useful in understanding natural colonization patterns for other organisms. Caldwell *et al.* (1981) examined the initial phases of microbial growth on a surface in an attempt to improve on previous models using exponential growth equations but with no allowances for attachment delays, re-inoculation, etc. Their model integrated attachment and growth so that growth is either exponential (if rapid compared with attachment) or nearly linear (when the reverse is true). They successfully tested the model using *Thermothrix thiopara* as a model organism (Brannan & Caldwell 1982) and subsequently developed the model further (Caldwell *et al.* 1983; Kiefte & Caldwell 1983). It still remains to be tested in situations of nutrient competition, predation and emigration.

The few data on the depth of penetration of colonizing organisms into rocks need extending. The mechanism of penetration must be better documented. Are bacteria and fungi carried into microcracks by freezing/thawing or wetting/drying fronts? How extensive is microcracking anyway in various rock types? Does the expansion of rhizines and rhizoids actually generate enough pressure to induce rock fracture?

Even for interactions on the surface of rocks the available data are still preliminary. The principal propagule type for successful colonization has yet to be established for most lichens and for many mosses. Even the most recent field studies on lichen reproduction, e.g. Fahselt *et al.* (1989), have only characterized the range of reproductive modes shown by established species and not which of these modes is the most successful for a given environment. As far as the process of colonization is concerned, aside from the reasonable suggestion that surfaces to be colonized must not be completely smooth, there is no general description of what features constitute an acceptable primary habitat. Cracks, of various sizes and alignments, have been identified as important but only in qualitative terms. Several ways forward are possible. One approach is to look for ecophysiological characterization of successful pioneer species with particular reference to carbon balance. A second is to attempt to characterize the microtopography of habitats using variables such as slope, aspect, height, etc. especially if there is a geometric repeatability in habitat structure (see Yarranton 1967; Yarranton & Beasleigh 1968). One other way forward in this field might be through the experimental use of artificial substrates which have standardized surface textures, chemical composition and water-holding capacity. Bricks and some types of dressed stone are obvious possibilities here, and would build on the existing data on masonry deterioration.

The increasing interest in both chasmolithic and endolithic communities is welcome since in these extreme habitats many of the interactions are simplified. The detailed studies on the Antarctic (Vestal 1993) and desert communities (Krumbein 1969) are providing models for future testing in more complex systems.

There is still much to learn in this field and many of the advances will come from interdisciplinary investigations.

REFERENCES

Alpert, P. (1991). Microtopography as habitat structure for mosses on rocks. *Habitat Structure – The Physical Arrangement of Objects in Space* (Ed. by S.S. Bell, E.D. McCoy & H.R. Mushinsky), pp. 120–40. Chapman & Hall, London.

Armstrong, R.A. (1978). The colonisation of a slate rock surface by a lichen. *New Phytologist*, **81**, 85–8.

Armstrong, R.A. (1981). Field experiments on the dispersal, establishment and colonisation of lichens on slate rock surfaces. *Environmental and Experimental Botany*, **21**, 115–20.

Ascaso, C. & Galvan, J. (1976). Studies on the pedogenetic action of lichen acids. *Pedobiologia*, **16**, 321–31.

Ascaso, C., Galvan, J. & Ortega, C. (1976). The pedogenic action of *Parmelia conspersa, Rhizocarpon geographicum* and *Umbilicaria pustulata*. *Lichenologist*, **8**, 151–71.

Ascaso, C., Galvan, J. & Rodriguez-Pascual, C. (1982). The weathering of calcareous rocks by lichens. *Pedobiologia*, **24**, 219–29.

Bachmann, E. (1904). Die Beziehungen der Kieselflechten zu ihrem Substrat. *Bereichte Deutsche Botanic Gesellschaft*, **22**, 101–4.

Bailey, D., Mazurak, A.P. & Rosowski, J.R. (1973). Aggregation of soil particles by algae. *Journal of Phycology*, **9**, 99–101.

Berthelin, J. (1988). Microbial weathering processes in natural environments. *Physical and Chemical Weathering in Geochemical Cycles* (Ed. by A. Lerman & M. Meybeck), pp. 33–59. Kluwer, Dordrecht.

Beschel, R.E. & Weideck, A. (1973). Geobotanical and geomorphological reconnaisance in west Greenland, 1961. *Arctic and Alpine Research*, **5**, 311–9.

Boyle, J.R., Voigt, G.K. & Sawhney, B.L. (1974). Chemical weathering of biotite by organic acids. *Soil Science*, **117**, 42–5.

Brannan, D.K. & Caldwell, D.E. (1982). Evaluation of a proposed surface colonisation equation using *Thermothrix thiopara* as a model organism. *Microbial Ecology*, **8**, 15–21.

Brightman, F.H. (1965). The lichens of Cambridge walls. *Nature Cambridgeshire*, **8**, 45–50.

Brightman, F.H. & Seaward, M.R.D. (1977). Lichens of man-made substrates. *Lichen Ecology* (Ed. by M.R.D. Seaward), pp. 253–93. Academic Press, London.

Broady, P.A. (1981a). The ecology of chasmolithic algae at coastal locations of the Antarctic. *Phycologia*, **20**, 259–72

Broady, P.A. (1981b). The ecology of sublithic terrestrial algae at the Vestfold Hills, Antarctica. *British Phycological Journal*, **16**, 231–40.

Brodo, I.M. (1974). Substrate ecology. *The Lichens* (Ed. by V.E. Ahmadjian & M.E. Hale), pp. 401–39. Academic Press, London.

Caldwell, D.E., Brannan, D.K., Morris, M.E. & Betlach, M.R. (1981). Quantisation of microbial growth on surfaces. *Microbial Ecology*, **7**, 1–11.

Caldwell, D.E., Malone, J.A. & Kieft, T.L. (1983). Derivation of a growth rate equation describing microbial surface colonisation. *Microbial Ecology*, **9**, 1–6.

Clements, F.E. (1916). *Plant Succession: Analysis of the Development of Vegetation*. Carnegie Institute, Washington DC.

Cooke, R.U. & Warren, A. (1973). *Geomorphology in Deserts*. Batsford, London.

Cooper, W.S. (1939). A fourth expedition to Glacier Bay, Alaska. *Ecology*, **20**, 130–59.

Cromack, K., Sollins, P., Graustein, W.C., Speidel, K., Todd, A.W., Spycher, G., Li, C.Y. & Todd, R.L. (1979). Calcium oxalate accumulation and soil weathering in mats of the hypogenous fungus

Hysterangium crassum. Soil Biology and Biochemistry, **11**, 463–8.

Dacey, P.W., Wakerley, D.S. & Le Roux, N.W. (1981). *The biodegradation of rocks and minerals with particular reference to silicate minerals*. Report LR 380(ME). Warren Spring Laboratory, Stevenage.

Danin, A. & Gerson, R. (1983). Weathering patterns in hard limestone and dolomite by endolithic lichens and cyanobacteria: supporting evidence for eolian contribution to Terra Rossa soil. *Soil Science*, **136**, 213–7.

Duff, R.B., Webley, D.M. & Scott, R.O. (1963). Solubilization of minerals and related materials by 2-ketogluconic acid producing bacteria. *Soil Science*, **95**, 105–14.

Fahselt, D., Maycock, P. & Svoboda, J. (1988). Initial establishment of saxicolous lichens following recent glacial recession in Sverdnip Pass, Ellesmere Island, Canada. *Lichenologist*, **20**, 253–68.

Fahselt, D., Maycock, P. & Wong, P.Y. (1989). Reproductive modes of lichens in stressful environments in central Ellesmere Island, Canadian High Arctic. *Lichenologist*, **21**, 343–53.

Fridriksson, S. (1975). *Surtsey: Evolution of Life on a Volcanic Island*. Butterworths, London.

Friedmann, E.I. (1982). Endolithic microorganisms in the Antarctic cold desert. *Science*, **215**, 1045–53.

Friedmann, E.I. & Galun, M. (1974). Desert algae, lichens and fungi. *Desert Biology*, vol. 2 (Ed. by G.W. Brown), pp. 165–212. Academic Press, New York.

Fry, E.J. (1924). A suggested explanation of the mechanical action of lithophytic lichens on rocks (shale). *Annals of Botany*, **38**, 175–96.

Fry, E.J. (1927). The mechanical action of crustaceous lichens on substrata of shale, schist, gneiss, limestone and obsidian. *Annals of Botany*, **41**, 437–60.

Galvan, J., Rodriguez, C. & Ascaso, C. (1981). The pedogenic action of lichens in metamorphic rocks. *Pedobiologia*, **21**, 60–73.

Garty, J. & Galva, M. (1974). Selectivity in lichen–substrate relationships. *Flora*, **163**, 530–36.

Gauri, K.L. & Gwinn, J.A. (Eds) (1984). *Proceedings of Fourth International Congress on the Deterioration and Preservation of Stone Objects*, July 7–9, 1982. University of Louisville, Kentucky.

Gerson, U. (1982). Bryophytes and invertebrates. *Bryophyte Ecology* (Ed. by A.J.E. Smith), pp. 291–332. Chapman & Hall, London.

Gerson, U. & Seaward, M.R.D. (1977). Lichen–invertebrate associations. *Lichen Ecology* (Ed. by M.R.D. Seaward), pp. 69–120. Academic Press, London.

Golubic, S. Freidmann, E.I. & Schneider, J. (1981). The lithobiontic ecological niche, with special reference to microorganisms. *Journal of Sedimentary Petrology*, **51**, 457–78.

Graustein, W.C., Cromack, K. & Sollins, P. (1977). Calcium oxalate: occurrence in soils and effect on nutrient and geochemical cycles. *Science*, **198**, 1252–4.

Hallbauer, D.K. & Jahns, H.M. (1977). Attack of lichens on quartzitic surfaces. *Lichenologist*, **9**, 119–22.

Hebrard, J.P., Foulguier, L. & Grauby, A. (1974). Approche expérimentale sur les possibilités de transfert du ^{90}Sr d'un substrat solide à une mousee terrestre: *Grimmia orbicularis*. *Bulletin de Société Botanique de France*, **121**, 235–50.

Henderson, M.E.K. & Duff, R.B. (1963). The release of metallic and silicate ions from minerals, rocks and soils by fungal activity. *Journal of Soil Science*, **14**, 236–46.

Huang, W.H. & Keller, W.D. (1970). Dissolution of rock forming silicate minerals in organic acids: simulation of the first stage weathering of fresh mineral surfaces. *American Mineralogist*, **55**, 2076–94.

Hughes, J.G. (1982). Penetration of rhizoids of the moss *Tortula muralis* Hedw. into well cemented oolitic limestone. *International Biodeterioration Bulletin*, **18**, 43–6.

Iskandar, I.K. & Syers, J.K. (1972). Metal complex formation by lichen compounds. *Journal of Soil Science*, **23**, 255–65.

Jackson, T.A. & Keller, W.D. (1970). A comparative study of the role of lichens and 'inorganic' processes in the chemical weathering of recent Hawaiian lava flows. *American Journal of Science*, **269**, 446–66.

Jones, D. (1988). Lichens and pedogenesis. *CRC Handbook of Lichenology*, vol. 3. (Ed. by M. Galun), pp. 109–24. CRC Press Inc., Baton Rouge, FL.

Jones, D., McHardy, W.J. & Wilson, M.J. (1976). Ultrastructure and chemical composition of spines in Mucorales. *Transactions of the British Mycological Society*, **66**, 153–7.

Jones, D. & Wilson, M.J. (1985). Chemical activity of lichens on mineral surfaces — a review. *International Biodeterioration Bulletin*, **21**, 99–104.

Jones, D., Wilson, M.J. & Tait, J.M. (1980). Weathering of a basalt by *Pertusaria corallina*. *Lichenologist*, **12**, 277–89.

Jones, D., Wilson, M.J. & McHardy, W.J. (1981). Lichen weathering of rock-forming minerals: application of scanning electron microscopy and microprobe analysis. *Journal of Microscopy*, **124**, 95–104.

Kappen, L. (1985). Vegetation and ecology of ice-free areas of northern Victoria Land, Antarctica. 2. Ecological conditions in typical microhabitats of lichens on Birthday Ridge. *Polar Biology*, **4**, 227–36.

Kerr, S. & Zavada, M.S. (1989). The effect of the lichen *Acarospora sinopica* on the elemental composition of three sedimentary rock substrates in South Africa. *Bryologist*, **92**, 407–10.

Kiefte, T.L. & Caldwell, D.E. (1983). A computer simulation of surface microcolony formation during microbial colonisation. *Microbial Ecology*, **9**, 7–13.

Klappa, C.F. (1979). Lichen stromatolites: criterion for subaerial exposure and a mechanism for the formation of laminar calcretes (caliche). *Journal of Sedimentary Petrology*, **49**, 387–400.

Krumbein, W.E. (1969). Uber den Einfluβ der Mikroflora auf die exogene Dynamik. *Geologische Rundschau*, **58**, 333–65.

Krumbein, W.E. & Jens, K. (1981). Biogenic rock varnishes of the Negev Desert (Israel): an ecological study of iron and manganese transformation by cyanobacteria and fungi. *Oecologia*, **50**, 25–38.

Kubiena, W.L. (1948). *Entwicklungslehre des Bodens*. Wein.

Lerman, A. & Meybeck, M. (Eds) (1989). *Physical and Chemical Weathering in Geochemical Cycles*. Kluwer, Dordrecht.

Longton, R.E. (1988). *Biology of Polar Bryophytes and Lichens*. Cambridge University Press, Cambridge.

Marathe, K.V. & Chaudhari, P.R. (1975). An example of algae as pioneers in the lithosphere and their role in rock corrosion. *Journal of Ecology*, **63**, 65–9.

Metting, B. (1981). The systematics and ecology of soil algae. *Botanical Review*, **47**, 195–312.

Muxart, T. & Blanc, P. (1979). Contribution a l'etude de l'alteration differentielle de la calcite et de la dolomite dans les dolomies sous l'action des lichens. Premiere observations au microscope optic et au MEB. *Actes Symposium Int. U I S*, 165–74.

Napier, E. & Wakerley, D.S. (1969). Fungal growth and iron reduction in Swedish alum shales. *Nature*, **223**, 289–90.

Paine, S. & Linggood, F. (1933). The relationship of microorganisms to the decay of stone. *Philosophical Transactions of the Royal Society, Series B*, **222**, 97–127.

Palmer, F.E., Emery, D.R., Stemmler, J. & Staley, J.T. (1987). Survival and growth of microcolonial rock fungi as affected by temperature and humidity. *New Phytologist*, **107**, 155–62.

Rades-Rohkohl, E., Hirsch, P. & Franzle, O. (1979). Neutron activation analysis for the demonstration of amphibolite rock-weathering activity of a yeast. *Applied and Environmental Microbiology*, **38**, 1061–8.

Schalscha, E.B., Appelt, H. & Shatz, A. (1967). Chelation as a weathering mechanism. I. Effect of complexing agents on the solubilisation of iron from minerals and granodiorite. *Geochemica Cosmochimica Acta*, **31**, 587–96.

Schatz, A. (1963). Soil micro-organisms and soil chelation, the pedogenic action of lichens and lichen acids. *Agricultural and Food Chemistry*, **11**, 112–8.

Schatz, V., Schatz, A., Trelawny, G.S. & Barth, K. (1956). Significance of lichens as pedogenic (soil forming) agents. *Proceedings of the Pennsylvanian Academy of Science*, **30**, 62–9.

Seaward, M.R.D., Giacobini, C., Giuliani, M.R. & Roccardi, A. (1989). The role of lichens in the biodeterioration of ancient monuments with particular reference to Central Italy. *International Biodeterioration Bulletin*, **25**, 49–55.

Silverman, M.P. & Munoz, E.F. (1970). Fungal attack on rock: solubilization and altered infra-red spectra. *Science*, **169**, 985–7.

Smith, R.I.L. (1988). Aspects of cryptogam water relations at a continental Antarctic site. *Polarforschung*, **58**, 139–53.

Spence, J.R. (1981). Comments on the cryptogam vegetation in front of glaciers in the Teton Range. *Bryologist*, **84**, 564–8.

Staley, J.T., Palmer, F. & Adams, J.B. (1982). Microcolonial fungi: common inhabitants on desert rocks? *Science*, **215**, 1093–5.

Syers, J.K. (1964). *A study of soil formation on Carboniferous Limestone with particular reference to lichens as pedogenic agents*. PhD thesis. Durham University.

Syers, J.K. (1969). Chelating ability of fumaroprotocetraric acid and *Parmelia conspersa*. *Plant and Soil*, **31**, 205–8.

Syers, J.K. & Iskandar, I.K. (1973). Pedogenetic significance of lichens. *The Lichens* (Ed. by V. Ahmadjian & M.E. Hale), pp. 225–48. Academic Press, London.

Syers, J.K., Birnie, A.C. & Mitchell, B.D. (1967). The calcium oxalate content of some lichens growing on limestone. *Lichenologist*, **3**, 409–14.

Taylor-George, S., Palmer, F.E., Staley, J.T., Borns, D.J., Curtiss, B. & Adams, J.B. (1983). Fungi and bacteria involved in desert varnish formation. *Microbial Ecology*, **9**, 227–45.

Topham, P.B. (1977). Colonization, growth, succession and competition. *Lichen Ecology* (Ed. by M.R.D. Seaward), pp. 31–68. Academic Press, London.

Vestal J.R. (1993). Cryptoendolithic communities from hot and cold deserts: Speculation on microbial colonization and succession. *Primary Succession on Land* (Ed. by J. Miles & D.W.H. Walton), pp. 5–16. Blackwell Scientific Publications, Oxford.

Viles, H. (1987). A quantitative scanning electron microscope study of evidence for lichen weathering of limestone, Mendip Hills, Somerset. *Earth Surface Processes and Landforms*, **12**, 467–73.

Vincent, W.F. (1989). *Microbial Ecosystems of Antarctica*. Cambridge University Press, Cambridge.

Walton, D.W.H. (1985). A preliminary study of the action of crustose lichens on rock surfaces in Antarctica. *Antarctic Nutrient Cycles and Food Webs* (Ed. by W.R. Siegfried, P.R. Condy & R.M. Laws), pp. 180–5. Springer-Verlag, Berlin.

Wagner, M. & Schwartz, W. (1967). Geomicrobiological studies. VIII. Effect of bacteria on the surface of rocks and minerals and their role in erosion. *Zeitschrift für Allgemeine Mikrobiologie*, **7**, 33–52.

Webley, D.M., Henderson, M.E.K. & Taylor, I.F. (1963). The microbiology of rocks and weathered stones. *Journal of Soil Science*, **14**, 102–12.

Weed, R. (1985). *Chronology of chemical and biological weathering of cold desert sandstones in the Dry Valleys, Antarctica*. MSc thesis. University of Maine.

Weirich, G. & Schweisfurth, R. (1985). Extraction and culture of microorganisms from rock. *Geomicrobiology Journal*, **4**, 1–20.

Williams, M.E. & Rudolph, E.D. (1974). The role of lichens and associated fungi in the chemical weathering of rock. *Mycologia*, **66**, 649–60.

Wilson, M.J. & Jones, D. (1983). Lichen weathering of minerals: implications for pedogenesis. *Residual Deposits: Surface Related Weathering Processes and Materials* (Ed. by R.C.L. Wilson), pp. 5–12. Geological Society of London, London.

Wilson, M.J. & Jones, D. (1984). The occurrence and significance of manganese oxalate in *Pertusaria corallina*. *Pedobiologia*, **26**, 373–9.

Wilson, M.J., Jones, D. & McHardy, W.J. (1981). The weathering of serpentine by *Lecanora atra*. *Lichenologist*, **13**, 167–76.

Worsley, P. & Ward, M.R. (1974). Plant colonisation of recent annual moraine ridges at Austre Okstindbreen, North Norway. *Arctic and Alpine Research*, **6**, 217–30.

Yarranton, G.A. (1967). A quantitative study of the bryophyte and macrolichen vegetation of the Dartmoor granite. *Lichenologist*, **3**, 392–408.

Yarranton, G.A. & Beasleigh, W.J. (1968). Towards a mathematical model of limestone pavement vegetation. I. Vegetation and microtopography. *Canadian Journal of Botany*, **46**, 1591–9.

Yatsu, E. (1988). *The Nature of Weathering: An Introduction*. Sozosha, Tokyo.

Zykina, L.V. (1982). Decomposition of minerals under the effect of *Aspergillus niger*. *Soviet Soil Science*, **14**, 56–62.

The role of bryophyte propagule banks in primary succession: Case-study of an Antarctic fellfield soil

R. I. LEWIS SMITH
British Antarctic Survey, Natural Environment Research Council, High Cross, Madingley Road, Cambridge CB3 0ET, UK

SUMMARY

1 The role of bryophytes in primary succession is highlighted and their general omission from most studies of the early stage of colonization is emphasized.
2 Bryophyte colonists develop from sexual and asexual propagules deposited over a long period from both local and distant provenances. Some may rapidly establish new plants, while others remain dormant indefinitely on or beneath the surface of the substratum. The viable component of these diaspores, the soil propagule bank, constitutes a reservoir of potential colonists equivalent to the seed bank of higher plants.
3 An environmental stimulus or suite of stimuli may activate the dormant viable propagules into developing as new plants. Before this, microbial modification of the soil surface is usually required to bind and stabilize soil particles and provide a nutrient base.
4 Laboratory and field experiments on maritime Antarctic soils are used to illustrate aspects of the bryophyte propagule bank.
5 The importance of ice fields as a sink for spores and vegetative propagules is stressed. Their release in meltwater onto terrestrial habitats near the ice margins is of particular importance in the colonization of newly exposed substrata.
6 The possible effects of global warming, especially in polar regions, on these propagule banks, on the rate of colonization and on the species composition of the developing communities is considered.

INTRODUCTION

The initial stages in the process of primary succession have been largely overlooked in most studies of colonization and community development. So often, ecologists have been interested specifically in the sequence of changes as they relate to higher plants and animals (e.g. Gray *et al.* 1987); this is primarily because they are easier to observe and experiment with and, in practical terms, they are considered to be of far greater value to applied science. Remarkably few investigations have been directed towards the true nature of primary colonization, probably because it is essentially a cryptic process involving micro-organisms. Interaction between microbiologists and ecologists is rare and integrated research programmes involving both have been very few. Nevertheless, it is fundamental to the understanding and analysis of the dynamics of any biological process to investigate and assess its initial stages.

THE ROLE OF BRYOPHYTES IN PRIMARY SUCCESSION

Fig. 1 illustrates a simplified flow diagram of the dynamics of primary succession. The process comprises three major stages: immigration, colonization and establishment, and development (which may include several phases). Primary colonization is considered here to be essentially a microbial phase in which a short succession of bacterial, fungal, cyanophyte and chlorophyte populations sequentially develop on a virgin substratum. In favourable situations (e.g. moist, sheltered fine soil) this succession may be rapid, while on less favourable surfaces (e.g. dry, exposed smooth rock) it may be protracted. These pioneer colonists are believed to be crucial in stabilizing otherwise mobile surfaces (Wynn-Williams 1986, 1993) and in laying the foundation of an organic and nutrient-enriched medium in which more complex organisms may become established. These secondary colonists are generally bryophytes (in particular, short acrocarpous mosses), microlichens (especially crustose and squamulose forms) and herbs (especially short grasses and caespitose forbs). Exceptions to this sequence do occasionally occur; for instance, the establishment of crustose lichens directly on bare rock, herbs on foreshore sand and tree saplings on industrial spoil tips. However, in none of these examples has an investigation been made of possible microbial progenitors, which may have permitted such pioneer colonization. Once these secondary colonists have established an open patchwork of transient populations, conditions are generally suitably advanced, in terms of substrate stability, nutrients and foci for the trapping and establishment of the tertiary colonists. When this phase is reached a population

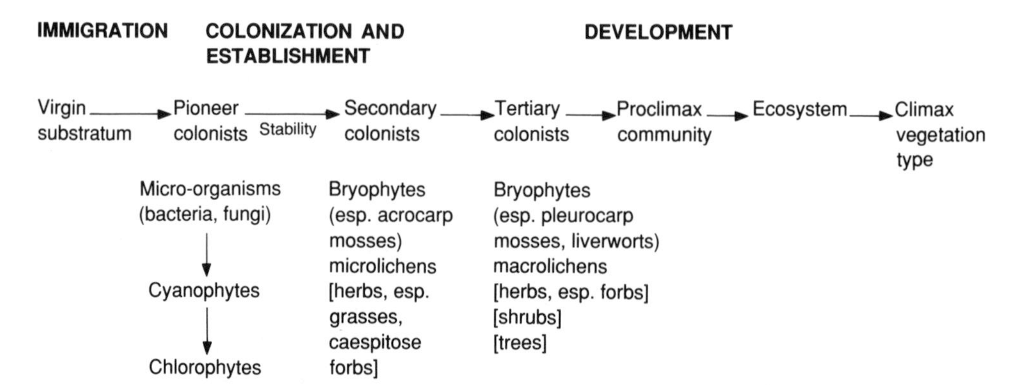

FIG. 1. Chronosequence of microbial and plant components in a generalized primary succession. (Life forms in parenthesis are absent from Antarctic fellfield ecosystems.)

explosion of both taxa and life-forms may ensue. These include further bryophytes (particularly pleurocarpous mosses, and liverworts), microlichens (especially as epiphytes on acrocarpous mosses), macrolichens (especially foliose and fruticose species), herbs (especially perennial forbs), shrubs and trees (both of which may support a succession of cryptogamic epiphytes) (e.g. Polunin 1936; Stork 1963; references in Matthews 1992; Crouch 1993). Once this stage has been reached the proclimax community has been initiated. Its rate of development towards a fully integrated and functioning ecosystem, and ultimately a climax vegetation type, is dependent on interactions within the biota and between it and various environmental criteria.

Because most studies of succession tend to commence at the stage where higher plants become established (i.e. as tertiary colonists), it is not surprising that there have been many studies of buried seed banks and of their importance in determining the composition of colonizing vegetation and subsequent community structure (see reviews by Harper 1977; Roberts 1981; Thompson 1987; Leck *et al.* 1989). However, most of these studies concern secondary successions. Early stages in a primary succession must rely heavily on the equivalent of such seed banks, namely, the propagule banks of the lower organisms which comprise the primary and secondary colonists. This account focuses on only one component of this diverse and complex reservoir of life — that of viable bryophyte soil propagule banks and of their importance as pools of potential colonists in the early development of plant communities.

The existence and role of cryptogamic diaspores in the soil has been largely ignored in most studies of succession (e.g. van der Valk 1992). Consideration is rarely given to the role of the microbial and lower plant groups, although a few studies have made particular reference to the role of bryophytes (e.g. for liverworts, Griggs 1933; for mosses, Eggler 1959; Magnusson & Fredriksson 1974), cyanobacteria (e.g. Schwabe 1974; Nelson *et al.* 1986) and green algae (e.g. Booth 1941; Broady 1982) in the earliest visible stages of colonization of new surfaces (especially volcanic ash and scoria, glacial detritus, fluvial deposits and sand-dunes). Aspects of this process have been reviewed for the polar regions by Longton (1988). Nevertheless, very few have examined or even considered the cause of such invasions, and references to buried propagules or spore rain are minimal. Investigations of the bryophyte propagule content of soils are limited to those of Leck (1980; tundra soil, Alaska), Furness & Hall (1981; reservoir mud, England), During & ter Horst (1983; chalk grassland soil, The Netherlands), Clymo & Duckett (1986; *Sphagnum* peat, England), During *et al.* (1987; various soils, Spain), Smith (1987) and Smith & Coupar (1987; both fellfield soils, Antarctica). However, only Furness & Hall, Smith and Smith & Coupar relate to primary successions, the remainder to secondary successions.

This account is illustrated by examples taken from studies at an Antarctic site (Smith 1985). Since 1980 the British Antarctic Survey Fellfield Ecology Research Programme has been investigating aspects of primary colonization and community development at Signy Island (*c.* 20 km^2), South Orkney Islands (60°43′S, 45°36′W) by an integrated approach involving the interactions of soil and microclimate, micro-organisms, cryptogams and invertebrates. The locality is ideal for such a study, with a simple ecosystem lacking in complex organisms such as vascular plants and macrofauna; tropic levels and interactions are consequently very much reduced.

CASE-STUDY: JANE COL FELLFIELD SITE, SIGNY ISLAND

The concept of bryophyte propagule bank dynamics is illustrated in Fig. 2. Although it relates specifically to an Antarctic scenario, this model may be readily modified to accommodate the more complex interactions and pathways which occur in other biomes, including those of high Arctic and Alpine regions. The stages are largely self-explanatory. The term 'diaspore' includes both sexual and asexual propagules capable of giving rise to new plants. The driving force in the process is the activation of dormant propagules by a stimulus which, as yet, is largely unknown and unquantified. Germination of spores or development of vegetative structures may be a response to a single stimulus, or to a combination of stimuli, of which temperature and moisture are probably the most important;

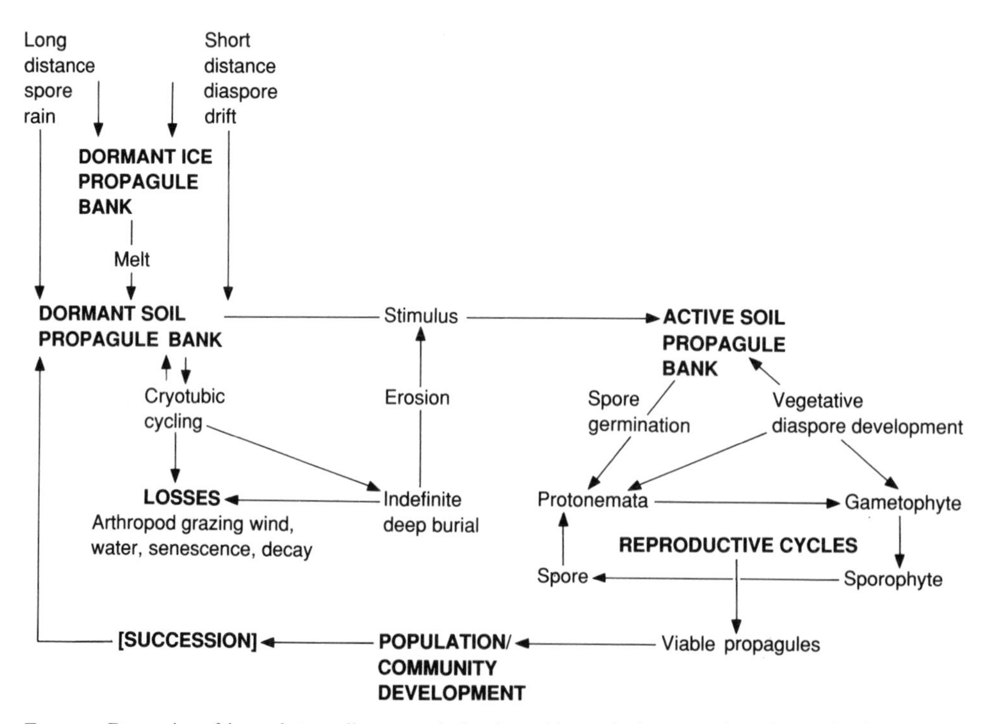

FIG. 2. Dynamics of bryophyte soil propagule banks, with particular regard to Antarctic sites.

although light quality, organic or inorganic nutrients, substrate stability and texture, may be equally important. Cryoturbic cycling can be both a diel and seasonal perturbation which continually buries and exposes propagules. The frequent formation and thawing of needle ice are responsible for much of this transfer of material through a vertical profile of up to 10 cm (see Davey *et al.* 1992), but it is only while the propagules are exposed at the surface that they may become activated. Depending on the habitat, once a population or community has been initiated, the assemblage of species often remains static, having already reached what may be considered as a climax ecosystem. However, fellfields comprise the most complex terrestrial ecosystem in Antarctica and probably experience a gradual shift in dominance by mosses and invasion by increasing numbers of lichens; such temporal changes may be construed as a simple succession.

This investigation of bryophyte propagule banks was conducted at Jane Col on Signy Island. Here, as in many regions of the maritime Antarctic, mean summer air temperatures have been rising since *c.* 1950 and the margins of the island's ice cap and glaciers have been receding and diminishing in thickness. This process has been accelerating during the past decade and recently exposed terrain reveals many examples of former vegetation and relictual periglacial microtopography (Fenton 1982; Smith 1990).

Jane Col is a recently deglaciated inland site of *c.* 8 ha, at *c.* 150 m altitude, which emerged from beneath a receding ice field between *c.* 1950 and 1980 (Smith 1985). The site comprises fairly level terrain ranging from areas of shattered schist boulders, through coarse and fine gravels to fine silts and clays. Much of the site is subjected to cryoturbic disturbance and sorted circles and solifluction stripes abound. Most of the site is visibly unvegetated, but observations over the past 25 years have indicated relatively rapid colonization by cyanobacteria, mosses and lichens on the more stable surfaces.

Several culture experiments have been carried out to examine the floristic composition, sequence of pioneer colonists, and rate of establishment resulting from diaspores on and beneath the soil surface, in order to assess the colonization potential of the soil propagule bank (Smith 1987; Smith & Coupar 1987).

Laboratory culture experiments

Methods

Experiment A. At each of five subsites at Jane Col fine mineral skeletal soil from the upper 0.5 cm of individual unvegetated periglacial features (one per subsite) was placed in 4 cm diameter by 2 cm deep sterilized polystyrene Petri dishes and their lids sealed with adhesive tape in the field to avoid possible contamination. Within a few hours of collection the plates were stored in darkness at −20°C until required for culture in England several months later. Details of the sub-sites are as follows:

1 Subsite 1: an accumulation of dark mineral detritus washed down from the adjacent ice slope at the south side of the site *c.* 25 m from the nearest sparse visible vegetation.
2 Subsite 2: fine soil from the centre of a sorted circle towards the south side of the site and surrounded by very sparse bryophytes *c.* 3 m away.
3 Subsite 3: as subsite 2 towards the centre of the site.
4 Subsite 4: fine soil from the centre of a sorted gravel stripe *c.* 25 m from the nearest visible very sparse bryophyte or lichen colonies and *c.* 35 m uphill of the receding ice edge at the north-east of the site.
5 Subsite 5: fine soil from the centre of a sorted circle at the west side of the site, and with relatively dense (*c.* 10–20% cover) bryophytes and lichens around the stable margin of and beyond the circle.

Prior to culturing the soils on a thermogradient incubator similar to that described by Grime & Thompson (1976; see also Smith & Coupar 1987), the plates were thawed slowly at 2°C for several days. Four replicates of each soil in their original Petri dishes, were then arranged along five parallel aluminium bars 70 cm long and 8 cm wide, each of which provided a temperature gradient from 2 to 25°C. The mean temperature of the soil in the plates was checked periodically and the actual culture conditions for each pair of replicate dishes varied by *c.* ±0·25°C, although the range over the width of the dishes on the bars was almost 1°C. For each of six temperature bands one pair of plates was kept moist by the addition of deionized water as required, and another pair was treated weekly with a nutrient solution ('Phostrogen') containing N (10% of dry weight), P (10%), K (27%), Mg (1·3%), Fe (0·4%), Mn (trace) diluted 0·5 $g\,l^{-1}$. A 16 hours day^{-1} photoperiod of *c.* 175 $\mu mol\,s^{-1}\,m^{-2}$ photosynthetically active radiation was provided by two overhead high intensity quartz iodide lamps. Long-wave radiation from these lamps did not affect the incubation temperatures which were controlled from below. The plates were cultured for 23 weeks and the percentage cover afforded by different bryophytes visually estimated at intervals (usually every 2 weeks). The identities of all taxa were determined when the experiments were terminated.

Experiment B. In a separate experiment (Smith & Coupar 1987) soil was sampled at depths of 0·1–1·0 cm and at 4–5 cm from the barren centre of a sorted circle near the centre of the Jane Col site. Soils were also sampled from four other fellfield sites on the island. The soils were stored in the dark at −10°C for 1 week before culturing in 4 cm diameter by 2·5 cm deep Petri dishes to a depth of 1 cm on the thermogradient incubator at Signy Island. There were four replicates for each of four soil surface temperatures (5, 10, 15, 20°C) which were maintained continuously, two being treated with distilled water and two with dilute Phostrogen as described above. Incubation was for 12 weeks, with a 16 hour light/8 hour dark regime; irradiance was again *c.* 175 $\mu mol\,s^{-1}\,m^{-2}$. Observations were made at 2-week intervals.

Results

Experiment A. Development of bryophyte colonies from spores or vegetative propagules on or buried just below the surface on the various soil plates is illustrated in Fig. 3 (see also Smith 1987). In each series an initial film of cyanobacteria and/or coccoid and filamentous green algae appeared, followed by bryophytes on many of the plates throughout the range of temperatures tested, although the amount of growth and diversity of species was much less at 2, 5 and 25°C. The optimum temperature for most species was between 10 and 20°C, a range commonly attained for several hours per day on many summer days at the soil surface in the field. Development also usually commenced earlier at these intermediate temperatures, although the first appearance of protonemata on some plates occurred after 2 weeks and the first gametophytes by the third week (notably *Ceratodon* cf. *purpureus* and *Bartramia patens*). The abundance of shoots increased steadily with time on most soils, but in a few instances at the highest and lowest temperatures the shoots died after a few weeks. Maximum growth (as percentage cover) and biomass was achieved at 10–20°C by several species after *c*. 10–15 weeks but slightly later at the more extreme temperatures.

The cultures revealed that even soils remote from any established vegetation (subsites 1 and 4) contain a pool of viable propagules. Although the greatest production of shoots occurred on soils associated with neighbouring vegetation, notably subsite 5, the greatest diversity of taxa (eight of 12 recorded on all the plates) was on soil from subsite 1, indicating the importance of the ice cap as a reservoir of viable propagules, which can eventually reach terrestrial substrata during the course of spring and summer melt. The principal effect of enriching replicate samples with NPK was to accelerate development, i.e. protonemata and gametophytes appeared earlier than in the deionized water treatment. However, increasing the nutrient status did not increase biomass nor species diversity.

Experiment B. The time taken for each species or taxon to appear on the soil dishes and, in the case of the bryophytes, the approximate maximum number of shoots produced after 12 weeks for each of the four temperatures are given in Table 1. Although cyanobacteria were not prominent on the Jane Col samples, they were the earliest visible organisms to become established on soils from several other sites (not described or discussed here). These often appeared by week 2, even at the lowest temperature, while green algae also became quickly established, particularly at the higher temperatures; a similar trend was followed by moss protonemata. Although 17 bryophyte species were recorded *in situ* at the Jane Col site, only five developed in the soil cultures. Once again, 10–15°C yielded the most species and nutrient treatment did not increase species diversity, although it enhanced development slightly for some species. However, no plants developed from the subsurface samples, indicating an essentially sterile (in terms of bryophytes) zone only a short distance below the active surface. Cyanobacteria, algae and protonemata were also remarkably scarce at this site.

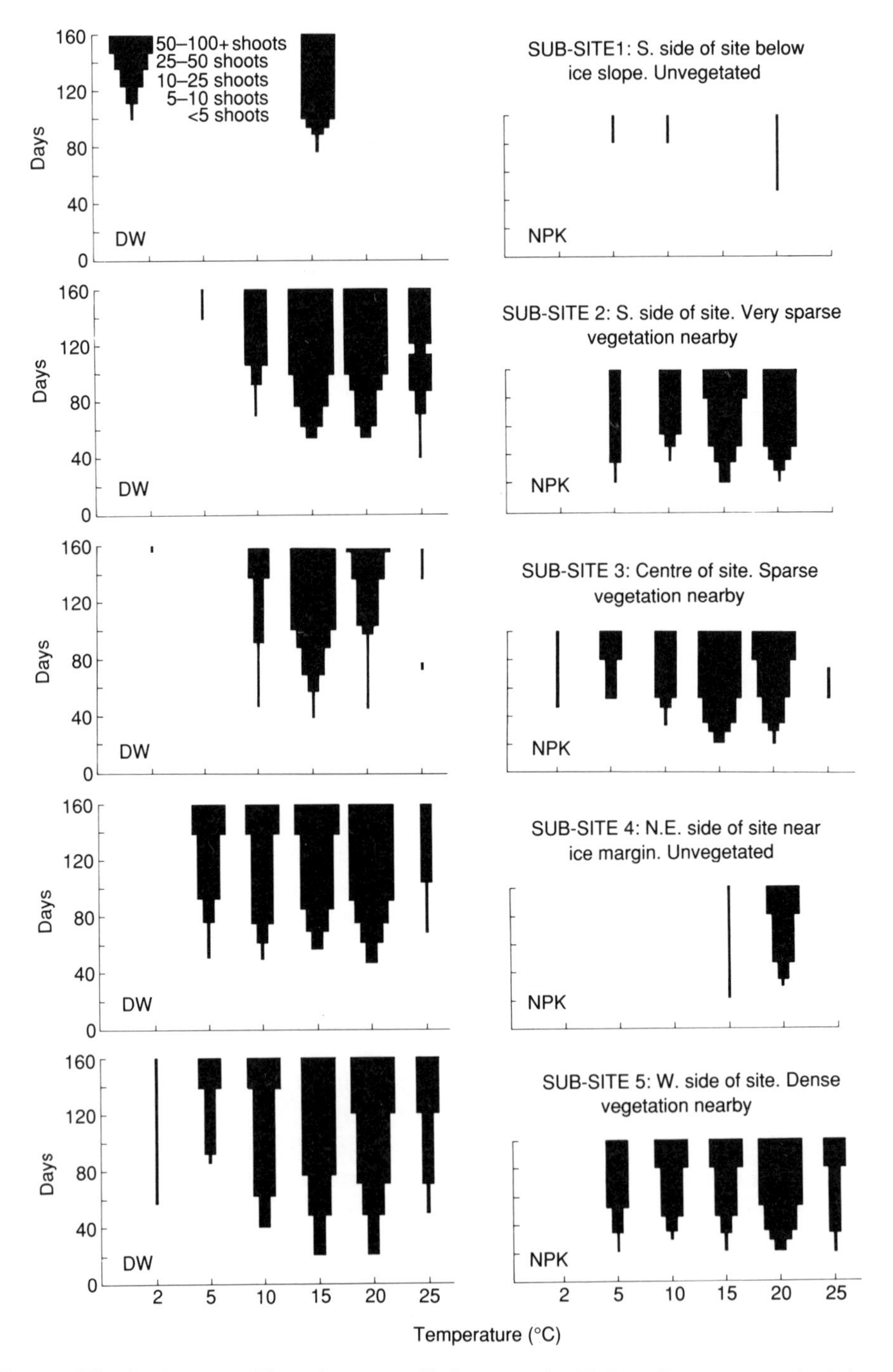

FIG. 3. The development of bryophytes on soils (0–5 mm depth) from five unvegetated subsites at Jane Col, fellfield site, Signy Island, incubated at a range of temperatures. DW: soils treated with deionized water; NPK: soils treated with a nitrogen, phosphorus and potassium-rich nutrient solution. The key at top left indicates shoot density. Modified from Smith (1987).

Table 1. Earliest appearance and shoot density after 12 weeks of bryophytes recorded on surface and subsurface unvegetated soil cultured at four temperatures (°C) with two treatments

		Surface soil								Subsurface soil							
		DW				NPK				DW				NPK			
	Species	5°	10°	15°	20°	5°	10°	15°	20°	5°	10°	15°	20°	5°	10°	15°	20°
Jane Col	Cyanobacteria*	—	—	—	—	—	—	—	—	—	—	—	—	—	—	—	—
	Green algae*	—	—	—	—	—	—	—	2	—	—	—	—	—	—	—	2
	Protonemata	—	2	—	—	—	—	—	2	—	—	—	—	—	—	—	6
	Cephaloziella exiliflora	—	—	—	—	2^{1}	—	—	—	—	—	—	—	—	—	—	—
	Ceratodon cf. *purpureus*	—	4^{3}	2^{4}	2^{5}	12^{1}	4^{4}	2^{4}	2^{4}	—	—	—	—	—	—	—	—
	Pohlia nutans	4^{1}	—	8^{1}	12^{1}	—	10^{1}	10^{1}	—	—	—	—	—	—	—	—	—
	Tortula princeps	—	10^{1}	—	—	—	—	—	—	—	—	—	—	—	—	—	—
	Tortula saxicola	—	—	10^{1}	—	—	2^{1}	—	—	—	—	—	—	—	—	—	—
	No. bryophyte spp.	1	2	3	2	2	3	2	1	0	0	0	0	0	0	0	0
	Total bryophyte spp.	4				4				0				0			
Marble Knolls	Cyanobacteria*	2	2	2	2	2	2	2	2	2	2	2	2	4	4	2	2
	Green algae*	—	8	8	8	12	6	6	6	4	4	4	4	6	4	4	4
	Protonemata	6	4	2	2	4	2	2	2	8	2	2	4	4	2	2	2
	Amblystegium sp.	—	12^{1}	4^{2}	2^{3}	—	10^{1}	10^{1}	4^{3}	—	—	—	2^{2}	—	12^{1}	—	4^{2}
	Bryum algens	—	—	6^{2}	2^{3}	—	—	2^{3}	4^{3}	—	—	8^{1}	8^{1}	—	—	8^{1}	8^{1}
	Bryum argenteum	—	4^{1}	—	—	—	—	—	—	—	—	—	—	—	—	—	—
	Cephaloziella exiliflora	—	—	2^{1}	2^{1}	—	6^{1}	10^{1}	2^{1}	—	—	—	—	—	—	—	—
	Ceratodon cf. *purpureus*	—	—	—	—	—	—	—	10^{1}	—	—	10^{1}	—	—	—	—	—
	Distichium capillaceum	—	10^{1}	4^{2}	—	—	12^{1}	6^{1}	6^{1}	—	—	—	—	—	10^{1}	—	4^{3}
	Drepanocladus uncinatus	—	—	—	2^{2}	—	2^{1}	—	—	—	—	—	—	—	—	—	—
	Encalypta patagonica	—	12^{1}	10^{1}	10^{1}	—	10^{1}	10^{1}	12^{1}	—	12^{1}	10^{1}	—	—	10^{1}	10^{1}	—
	Marchantia berteroana	—	12^{1}	—	10^{1}	—	—	—	—	—	—	—	—	—	—	—	—
	Pottia austro-georgica	—	—	10^{2}	2^{4}	12^{1}	8^{2}	2^{5}	4^{3}	—	10^{1}	6^{4}	8^{3}	—	12^{1}	8^{4}	6^{5}
	Pterygoneurum ovatum	—	—	—	10^{1}	—	—	10^{1}	10^{1}	—	—	—	—	—	—	10^{1}	—
	Tortula filaris	—	10^{1}	6^{1}	—	—	8^{1}	6^{1}	—	—	—	—	—	—	—	—	—
	Tortula princeps	—	—	—	—	—	—	—	—	—	8^{2}	6^{1}	—	—	—	—	—
	Tortula saxicola	10^{1}	2^{4}	2^{5}	4^{2}	10^{1}	2^{5}	2^{5}	8^{1}	8^{1}	8^{3}	6^{4}	6^{1}	—	4^{3}	6^{3}	2^{1}
	No. bryophyte spp.	1	7	8	9	2	8	9	9	1	4	6	4	0	5	5	5
	Total bryophyte spp.	12				11				7				7			

Values 2–12: weeks from commencement of experiment.
Superscripts 1–5: bryophyte shoot density. 1 = <5 shoots, 2 = 5–10 shoots, 3 = 11–25 shoots, 4 = 26–50 shoots, 5 = 51–100 shoots.
DW: deionized water treatment; NPK: nutrient treatment. * Visible to the naked eye.

In contrast, an identical series of soils taken from Marble Knolls fellfield site, a lowland calcareous site with comparatively extensive cryptogamic vegetation (including 22 species of bryophytes), yielded 14 bryophyte taxa (12 on the surface deionized water treatment, 11 on the surface NPK treatment and seven on each of the subsurface sample treatments). Here, there is a much greater deposition of propagules onto areas of bare soil, and clearly a relatively high proportion of these are transported down the soil profile by precipitation and freeze–thaw activity. Also, the comparatively diverse flora proliferating on the culture plates reflects the preponderance of the island's calcicolous species exhibiting 'colonist' and various 'shuttle' growth strategies, as proposed by During (1979).

In situ *culture experiments*

Growth from the soil propagule bank

Method. Culturing soil in the laboratory demonstrated the existence of a pool of propagules in or on the soil. An experiment was then designed to enhance conditions *in situ* to stimulate the development of the potential community lying dormant in the soil. In 1985 six unvegetated subsites within the Jane Col site were selected (the four described here being subsites 1–4 from where the soils in Experiment A were sampled) and two clear, ventilated polystyrene cloches (55 × 28 cm) placed over the soil at each subsite. Adjacent unprotected control plots of the same dimensions were marked with pegs. One cloche and one control plot per subsite was treated regularly with NPK nutrient solution, while the adjacent pair received no treatment. The development of vegetation has been recorded annually (see Fig. 4).

Subsite 1 is close to the receding ice slope margin and subsites 2 and 3 are progressively farther across the col and consequently their surfaces have been exposed longer. While subsite 1 is *c.* 10 years old, the postulated ages of 2 and 3 are *c.* 20 and 30 years, respectively. Subsite 4 is probably also only *c.* 10 years old.

Results. After only 1 year the previously totally barren soil at subsite 1 had *c.* 20% cover of moss beneath both cloches; by year 2 this had increased to *c.* 40% by January and to 75% by March, and by year 3 to *c.* 90%, with numerous populations of several species of bryophytes forming a fully integrated community. When examined in detail in year 4 (see Table 2), there was also a well-established invertebrate fauna of protozoans, tardigrades, nematodes and the springtail *Cryptopygus antarcticus* living among the thick layer of moss turf. In the nutrient-treated cloche several populations of *Bryum algens* had produced sporophytes with exceedingly long setae (Fig. 5). These had also been produced in year 3. On Signy Island this species is very rarely fertile. Besides the favourable microclimate, perhaps the nutrient enrichment stimulated sexual reproduction. No visible development of cyanobacteria, algae, bryophytes or invertebrates occurred in the

TABLE 2. Percentage cover in 0·15 m^2 plots after 4 years of bryophytes (and other cryptogams) developed from the soil propagule bank *in situ* at Jane Col fellfield site

	Subsite 1				Subsite 2				Subsite 3				Subsite 4			
Species	Cl^-	C^-	Cl^+	C^+	Cl^-	C^-	Cl^+	C^+	Cl^-	C^-	Cl^+	C^+	Cl^-	C^-	Cl^+	C^+
Andreaea gainii	—	—	—	—	—	—	—	—	<1	—	—	—	—	—	—	—
Bartramia patens	1	—	1	—	—	—	—	—	—	—	—	—	—	—	—	—
Bryum algens	5	—	20*	—	3	—	5	—	1	—	1	—	—	—	—	—
Calliergon sarmentosum	—	—	—	—	—	—	—	—	—	—	<1	—	—	—	—	—
Campylium polygamum	—	—	—	—	<1	—	1	—	—	—	—	—	—	—	—	—
Cephaloziella exiliflora	—	—	—	—	—	—	—	<1	1	—	1	—	—	<1	—	—
Ceratodon purpureus	80	—	65	—	50	3	30	5	35	5	25	20	15	3	8	1
C. purpureus protonemal felt	5	—	10	—	—	—	—	—	—	—	—	—	—	—	—	—
Dicranaceae	—	—	—	—	<1	—	<1	—	—	—	—	—	—	—	—	—
Drepanocladus uncinatus	<1	—	<1	—	—	<1	<1	—	18	2	10	<1	<1	—	1	—
Pohlia cruda	—	—	—	—	—	—	—	—	—	—	<1	—	—	—	—	—
Pohlia nutans	—	—	—	—	<1	1	—	<1	10	1	5	1	1	—	2	<1
Polytrichum alpinum	<1	—	—	—	—	—	—	—	—	—	—	—	—	<1	—	—
Polytrichum juniperinum	—	—	—	—	<1	—	<1	—	—	—	1	—	—	—	—	—
Tortula princeps	—	—	—	—	—	—	—	—	<1	—	<1	—	<1	—	<1	—
Massalongia carnosa	—	—	—	—	—	—	—	<1	—	<1	—	1	—	—	—	—
Placopsis contortuplicata	—	—	—	—	<1	—	—	<1	—	—	—	—	—	—	—	—
Parmeliella sp.	—	—	—	—	—	<1	—	<1	—	—	—	<1	—	—	—	—
Grey crustose lichen	—	—	—	—	—	—	—	—	—	<1	—	<1	—	—	—	—
Pink crustose lichen	—	—	—	—	—	<1	—	<1	—	—	—	—	—	—	—	—
Prasiola crispa	—	—	—	—	—	—	—	<1	—	—	—	—	—	—	—	—
Bare ground	10	100	5	100	50	97	65	95	35	92	60	80	85	97	90	99

Cl: Cloche; C: control; $^-$: no treatment, $^+$: NPK nutrient added weekly in summer; * : with sporophytes.

(a)

(b)

FIG. 4. Growth of bryophytes from soil propagule bank *in situ* at Signy Island, maritime Antarctic. (a) Jane Col fellfield site showing subsites 1–5; subsite 1 is close to a receding ice slope (February 1985). (b) Nutrient treated and untreated cloches, with adjacent control plots on unvegetated glacial detritus at the margin of the island's receding ice cap, Subsite 1. (c) Soil surface beneath untreated cloche at commencement of experiment (January 1985), Subsite 1. (d) As (c) with *c*. 90% moss cover (February 1989).

(c)

(d)

unprotected control plots, nor anywhere else throughout this subsite (*c.* 30 × 10 m). By 1992 numerous small colonies of *Ceratodon purpureus* had become established within this area, i.e. about 12 years after it first became exposed by the receding ice margin. However, there is clearly a large viable propagule bank in this soil which is readily activated by increasing the soil and near-surface air temperature

FIG. 5. Fruiting population of *Bryum algens* in a closed turf dominated by *Ceratodon* cf. *purpureus*, cultured *in situ* in cloche at Subsite 1 (see Fig. 4d).

and aiding the retention of moisture in the soil (and later in the moss turf). Fig. 6 illustrates the diel fluctuation of temperature and relative humidity over 3 days in late summer in one of the cloches and in a control plot. At the end of this period the mean moisture content of unvegetated soil (0–1 cm depth) within the cloches was 64% (of dry weight) compared with only 11% in the control plots.

The older but very sparsely vegetated subsites produced higher numbers of bryophyte species within the cloches, although the total cover afforded after 4 years was fairly low, suggesting a lower incidence of soil propagules than at subsite 1. On the other hand, the older substrata (subsites 2 and 3), both in the cloches and controls, had tiny thalli of several crustose lichens, many of which were already visible at the commencement of the experiment. Subsite 4, being young, yielded six bryophyte species but no lichens, and cover was very low.

These results (particularly for subsite 1) clearly indicate a large input of viable propagules from meltwater running off the ice cap. Incubating samples of soil from cryoconite holes on the ice cap yielded several species of alga and moss gametophytes as well as protozoans tardigrades, rotifers and nematodes. The biota of such cryomicrocosms has been discussed by Wharton *et al.* (1985). Above Jane Col the ice field supports a dense snow alga flora in late summer and live and dead algal cells become incorporated in mineral debris blown onto the ice. This material is washed onto ice-free terrain during the spring and summer melt, giving rise to a relatively nutrient- and propagule-rich accumulation of debris below the ice field (cf. Bonde 1969; Teeri & Barrett 1975; Miller & Ambrose 1976).

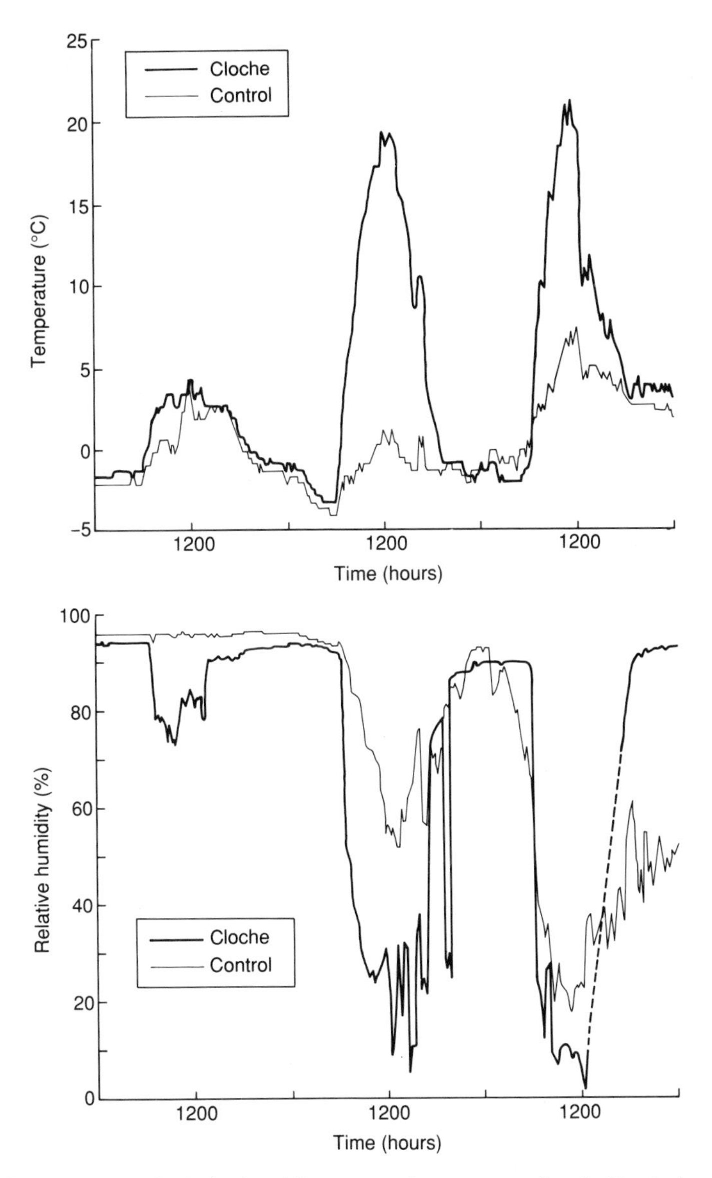

FIG. 6. Temperature and relative humidity at 2 cm above moss surface inside cloche and above bare soil in control plot over 3 days (3–5 March 1989) at Subsite 1, Jane Col fellfield site, Signy Island. Data recorded at 5 min intervals using a Grant Instruments Squirrel Datalogger.

Nature of propagules

These experiments have clearly demonstrated that unvegetated soil contains a diverse and often abundant reservoir of bryophyte propagules. So far, no detailed examination of the soil has been made to identify the nature of these propagules and any hypothesis regarding their origins and dispersal must remain speculative. However, there is little doubt that the products of both sexual and asexual reproduction are involved and that, while the great majority originate locally, there is evidence of a tiny element of the propagule bank flora having a trans-oceanic provenance (see below).

Spores

Less than 20% of the Signy Island bryoflora produce sporophytes (Webb 1973; Convey & Smith, in press) and, of these, only seven are both common and widespread, and produce capsules in abundance. Of these, only three (*Bartramia patens*, *Encalypta patagonica* and *Pottia austro-georgica*, and all of very sporadic occurrence) appeared in the soil cultures. Tests have shown that spores of these species readily germinate under favourable moisture and temperature conditions, while germination and sporeling development are virtually unknown in the three dominant fellfield genera (*Andreaea*, *Dicranoweisia* and *Schistidium*). Despite their dominance in the fellfield ecosystem and their high spore production, some as yet unknown suite of conditions is required for their germination and development.

Vegetative propagules

By far the most important type of propagules are vegetative structures, of which various forms exist in the maritime Antarctic bryoflora. Of these, the most common are detached leaves and stem apices which, in many species, readily develop rhizoids when they reach a moist substratum (see also Bayfield 1976). There is even subfossil evidence of this mode of dispersal and establishment in bryophytes (Miller 1985). In the Antarctic flora deciduous apices are typical of *Bryum algens* and *B. argenteum*, while more specialized structures include leaf tip and axillary rhizoids (*Pseudoleskea* sp.?), leaf and thallus gemmae (*Barbilophozia hatcheri, Cephaloziella exiliflora, Lophozia propagulifera, Marchantia berteroana*), leaf axil bulbils (*Pohlia* cf. *bulbifera*), deciduous leaf apices (*Bartramia patens, Sarconeurum glaciale*), rhizoid buds (*Bryum argenteum*), rhizoidal gemmae (*Ceratodon purpureus*) and rhizoidal tubers (*Bryum* cf. *algens*).

Growth from shoot fragments

Method

To test the ability of such vegetative propagules to colonize bare ground and establish populations, the apical 0·5 cm of shoot aggregations of 12 fellfield moss species were macerated using a microhomogenizer. The fragments were sown on soil *in situ* at two subsites (near 2 and 3) on Jane Col. Each species was cultivated within 30 cm^2 open-ended cylindrical perspex tubes pressed into the soil. At each subsite one series was covered by a clear perspex cloche and a duplicate series left unprotected. No nutrients were applied. Approximately 200 fragments per species were sown in each of the four treatments. Their progress was assessed annually.

Results

After 4 years (Table 3), establishment in the control plots was generally sparse or non-existent, except for *Dicranoweisia grimmiacea* and *Drepanocladus uncinatus*, the former being an abundant cushion-forming lithophyte, the latter a species principally occupying wet habitats, but with a wide ecological amplitude. Considerably greater success was recorded in the cloche experiments. Results from both subsites were almost identical, implying that the higher temperatures and soil moisture were beneficial to the survival and development of the fragments and confirming the capability of many species to establish new individuals from apical shoot fragments in favourable habitats.

DISCUSSION

Studies done at a recently deglaciated site at Signy Island, maritime Antarctic, have revealed that both visually unvegetated soils and barren areas near older vegetated substrata possess a diverse flora of viable propagules. In the pioneer stage of colonization microbita play a major role in providing favourable conditions for the secondary colonists, of which bryophytes are often the principal or earliest floristic components. In an extensive study of soil polygons, Davey & Rothery (in press) have shown that microalgal species diversity is low. Species composition within individual polygons is constant but varies markedly between polygons. The

TABLE 3. Establishment after 4 years of mosses from *c.* 200 leaf and stem fragments sown on soil *in situ* at Jane Col fellfield site

Species	Mean no. live shoots per 30 cm²	
	Cloche	Control
Andreaea depressinervis	4	1
Andreaea gainii	2	0
Andreaea regularis	4	0
Bartramia patens	6	1
Dicranoweisia grimmiacea	6*	4
Drepanocladus uncinatus	6	5
Polytrichum alpinum	3	2
Pottia austro-georgica	3	0
Racomitrium austro-georgicum	4	1
Schistidium antarctici	6	1
Tortula princeps	5	2
Tortula saxicola	4	2

1 = 1–5 shoots; 2 = 6–10; 3 = 11–25; 4 = 26–50; 5 = 51–100; 6 = >100; * with sporophytes.

filamentous cyanobacterium *Phormidium autumnale* is ubiquitous and provides the largest component of the microbiota biovolume. Davey & Clarke (1991) found that most of the microalgal biomass is concentrated near the surface of the soil, but that few algae occur on the actual surface, suggesting this may be a desiccation-avoidance strategy. *P. autumnale* is the dominant species and Wynn-Williams (1988) and Davey (1988) have shown that mucilage secreted by the filamentous mats of this cyanophyte is important in binding unicellular algae and inorganic particles to produce stabilized 'rafts' on the generally mobile fines (Wynn-Williams 1993). It is hypothesized that these relatively stable organically- and nutrient-enriched rafts serve as foci where the reservoir of bryophyte (and other) propagules germinate and develop to initiate the process of secondary colonization.

In polar and Alpine regions a wide range of live material is deposited on ice caps and glaciers, stored and released at times of melt to provide a relatively rich pool of potential colonists close to the margins of receding ice fields. An example of this on Signy Island is the recent establishment of *Polytrichum piliferum*, first recorded in 1989, on soil which first became exposed at the foot of a receding ice field in the early 1980s. Intensive searches specifically for this prominent moss during the previous 25 years had indicated that this species was absent from the South Orkney Islands. The sudden appearance of numerous colonies suggests that spores had either been deposited on the ice and washed onto the soil as the ice melted, or had fallen directly onto the soil; a third possibility is that spores had remained viable in the soil propagule bank since before the ice field developed (probably pre-Little Ice Age *c.* 100–250 years ago) when the species may have been a component of the island's bryoflora but subsequently became extinct. In 1992 a small colony of *Polytrichum longisetum* was discovered at the same site. This species is a new record for the Antarctic.

Much attention has been given to the reproductive biology of bryophytes (see Mogensen 1981, Longton & Miles 1982, and reviews by Longton & Schuster 1983, Longton 1988), and to morphological development following spore germination (see review by Nehira 1983). Numerous studies have also been made of the effects of individual physical and chemical factors on germination and development in controlled laboratory experiments. However, remarkably little is known of the requirements of bryophyte spores or vegetative propagules *in situ* to promote germination and establishment. In a recent study Miles & Longton (1990) compared the germination success of spores of four mosses in the field. They showed that while germination failed completely in a long-lived perennial, it was highly successful in an annual fugitive species; two short-lived perennials species gave intermediate results. These experiments therefore confirm the reproductive criteria proposed by During (1979) in defining different categories of life-history strategies in mosses.

Similarly, it is not possible to state with any conviction what factors may be responsible for the enhanced germination, growth and biomass production recorded under cloches in field experiments. They may be a response to a single stimulus or to a combination of stimuli, of which increased soil and air temperature and soil moisture (but not relative humidity) are probably the most important. UVB is significantly reduced by the cloches (by *c.* 95%), so that light quality may be critical. However, a current field experiment is testing the effects of enhanced UVB, using UVB-transparent and UVB-opaque cloches, on propagule development. Since the cloches are ventilated it is unlikely that there is a significant increase in CO_2 concentration, which might otherwise enhance growth. Certainly, mean integrated air temperatures near the soil surface within the cloches are *c.* 3–5°C higher than over the control plots, with surface temperatures during the day rising well above those in the open (see Fig. 6). A key factor may also be that cloches can maintain above-freezing temperatures for several consecutive days, unlike the unprotected soil surface where a diurnal freeze–thaw cycle invariably exists. Brief continuous periods above a certain temperature threshold may be sufficient to promote germination and the initial stages of development.

Another factor which may be important in the development of gametophytes is chemical exudate produced by the juvenile plants themselves. Knoop (1984) has shown that in some moss species gametophytes in early stages of development release a complex of organic compounds into the substratum, which promote shoot and branch formation. Furthermore, it has been demonstrated that chemical interactions between juvenile mosses may be a determining factor in the eventual composition of the developing bryophyte community (Watson 1981). Development may be enhanced by pioneer microbial colonists which increase the concentration of organic nutrients leached into the soil (Tearle 1987; Greenfield 1989). Roser *et al.* (1992) have shown that Antarctic lichens, bryophytes and algae accumulate polyols and sugars, although most snow algae had low concentrations. Melick & Seppelt (1992) demonstrated significant loss of soluble carbohydrates from the mosses. However, it seems probable that the accumulations of the dead cells of the dense communities of red snow algae (dominated by *Chlamydomonas nivalis*) on Signy Island must contribute an important organic input to the almost totally inorganic soil when it is first exposed by receding icefields. Almost certainly, different types of propagules will have different developmental requirements.

The experiments discussed here have demonstrated that unvegetated skeletal soils contain a pool of viable propagules of a variety of microbial and plant taxa, almost all of which occur as components of the corresponding site communities. This suggests that the propagules are of local provenance and that their dissemination, at least for most taxa, is probably over only a few metres; although at the Jane Col site the occurrence of some calcicolous mosses must have involved immigration from at least several hundred metres away.

Evidence of long-distance spore dispersal comes from the soil incubation experiments in which, very rarely, a species develops which is not known to occur in the Antarctic. At least two exotic mosses (*Campylopus* cf. *canescens* and *Leptobryum* cf. *pyriforme*) and one basidiomycete (*Coprinus* sp.) have been cultured from the Jane Col soil from below the ice slope (subsite 1; Smith 1987). Each is common in southern South America. Other exotic sporomorpha in Antarctic moss peat have been reported by Churchill (1973) and Kappen & Straka (1988), while pollen of numerous southern South American phanerogams,

FIG. 7. Unique moss community associated with fumarole vents near the summit (450 m) of Mount Pond, Deception Island, South Shetland Islands. It is dominated by *Philonotis acicularis* and *Dicranella hookeri*, neither moss known from elsewhere in the Antarctic biome but both occurring in southern South America. These species exemplify transoceanic dispersal to and the spore rain over the Antarctic from lower latitude landmasses.

and other diaspores, have been identified in samples of firn snow from Signy Island (Smith 1992). The problems of long-distance dispersal of bryophyte spores, particularly in the Southern Hemisphere, have been discussed in detail by van Zanten (1978) and van Zanten & Pocs (1981); Mogensen (1981) and Convey & Smith (in press) have considered spore dispersal potential, in relation to spore size, in Arctic and Antarctic bryophyte populations respectively.

Even more convincing evidence of a viable bryophyte spore rain over the Antarctic is seen in areas where there is volcanic activity. At each of the four major active sites around the Antarctic, geothermal soils support a diversity of bryophyte species, several of which occur nowhere else in the biome. At Deception Island, South Shetland Islands (Smith 1984, 1988), and in the South Sandwich Islands (Longton & Holdgate 1979), heated ground associated with fumaroles is colonized by well-developed communities of such species, again all common in southern South America (Fig. 7). Comparable sites at the summits of Mount Erebus and Mount Melbourne in the Ross Sea region also support unique floras (Broady 1984; Broady *et al.* 1987). Clearly, there is a rain of spores over the Antarctic, a large proportion of which is trapped in ice. If they are successful in reaching a soil surface, either directly or in meltwater from ice caps or glaciers, they may lie dormant indefinitely until conditions become favourable for germination. Such conditions are afforded by geothermal activity — or, experimentally, by relatively high culture temperatures. Little is known about bryophyte spore or vegetative propagule longevity (Sussman 1965) but reports of the existence of viable seeds in Arctic permafrost (Porsild *et al.* 1967) and bacteria in an Antarctica ice core (Abyzov *et al.* 1987), both many thousands of years old, and a preliminary study of regeneration from moribund moss shoots which may be several centuries old and only now are being re-exposed by receding ice (author, unpublished data), point towards the possibility of ice being an exceedingly important refugium for viable propagules. The theory of survival of organisms through Ice Ages or lesser glacial episodes as successive populations of mature individuals on isolated nunataks or other ice-free refugia, has many weaknesses. There seems to be a much greater likelihood of organisms remaining in perpetuity as viable propagules buried in ice or in frozen soil which, when conditions become favourable, may permit the recolonization of the newly exposed substrata.

Postscript

The experiments using cloches raise a potentially interesting and timely question: what influence might the 'greenhouse effect' have on propagule banks in polar regions? Global warming is gaining momentum and it is predicted that this trend will be considerably greater in the polar regions than elsewhere (World Meteorological Organization 1988; Houghton *et al.* 1990). Indeed, Signy Island has been experiencing increasing temperatures since *c.* 1950 (Smith 1990). This could have significant ecological implications, because a rise in mean summer air temperature of 2–3°C (cf. cloche temperature) may be sufficient to trigger germination and establishment of some of the exotic components of the soil propagule bank in the climatically sensitive peripheral Antarctic regions. Furthermore, the rate of colonization, population and community development and dispersal may be expected to accelerate. This could lead to the development of new communities, changes in existing communities and to the possibility of additional cryptogams and even vascular plants becoming established if their diaspores can overcome the dispersal barriers. Furthermore, the effects on germination and growth of the increasing level of UVB radiation reaching the soil and vegetation, caused by the 'ozone hole' phenomenon, are as yet unknown; but are being tested at Signy Island. The exceptional development of vegetation illustrated by the cloche experiments, in which UVB is virtually excluded, may be no more than a temperature and moisture effect, but reduced UVB could enhance the development of soil propagule banks. What then might be the effect of the increased ultraviolet radiation now being experienced? Ironically, in ecological terms, it could cancel the 'greenhouse effect', at least in high latitudes.

ACKNOWLEDGMENTS

I am indebted to D. J. Wright and H. MacAlister for their assistance with and maintenance of the field experiments at Jane Col, Signy Island. I also appreciate the helpful comments of Drs H. J. During and R. E. Longton.

REFERENCES

Abyzov, S.S., Bobin, N.Y. & Kudryashov, B.B. (1987). Anabiosis of microorganisms in the Antarctic Ice Sheet. *Anabiosis of Cells* (Ed. by Y.E. Baker), pp. 43–54. Academy of Sciences of Latvian SSR, August Kirkhensteyn Institute of Microbiology, Riga.

Bayfield, N.G. (1976). Effects of substrate type and microtopography on establishment of a mixture of bryophytes from vegetative fragments. *The Bryologist*, **79**, 199–207.

Bonde, E.K. (1969). Plant disseminules in wind-blown debris from a glacier in Colorado. *Arctic and Alpine Research*, **1**, 135–9.

Booth, W.E. (1941). Algae as pioneers in plant succession and their importance in erosion control. *Ecology*, **22**, 38–46.

Broady, P.A. (1982). Green and yellow-green terrestrial algae from Surtsey (Iceland) in 1978. *Surtsey Research Progress Report*, **9**, 13–32.

Broady, P.A. (1984). Taxonomic and ecological investigations of algae on steam-warmed soil on Mt. Erebus, Ross Island, Antarctica. *Phycologia*, **23**, 257–71.

Broady, P.A., Given, D., Greenfield, L. & Thomson, K. (1987). The biota and environment of fumaroles on Mt. Melbourne, Northern Victoria Land. *Polar Biology*, **7**, 97–113.

Churchill, D.M. (1973). The ecological significance of tropical mangroves in the early Tertiary floras of southern Australia. *Special Publications of the Geological Society of Australia*, **4**, 79–86.

Clymo, R.S. & Duckett, J.G. (1986). Regeneration of *Sphagnum*. *New Phytologist*, **102**, 589–617.

Convey, P. & Smith, R.I.L. (in press). Investment in sexual reproduction by Antarctic mosses. *Oikos*.

Crouch, H.J. (1993). Plant distribution patterns and primary succession on a glacier foreland: A comparative study of cryptogams and higher plants. *Primary Succession on Land* (Ed. by J. Miles & D.W.H. Walton), pp. 133–45. Blackwell Scientific Publications, Oxford.

Davey, M.C. (1988). Ecology of terrestrial algae of the fellfield ecosystems of Signy Island, South Orkney Islands. *British Antarctic Survey Bulletin*, **81**, 69–74.

Davey, M.C. & Clarke, K.J. (1991). The spatial distribution of microalgae on Antarctic fellfield soils. *Antarctic Science*, **3**, 257–63.

Davey, M.C., Pickup, J. & Block, W. (1992). Temperature variation and its biological significance in fellfield habitats on a maritime Antarctic island. *Antarctic Science*, **4**, 383–88.

Davey, M.C. & Rothery, P. (in press). Primary colonization by microalgae in relation to spatial variation in edaphic factors on Antarctic fellfield soils. *Journal of Ecology*.

During, H.J. (1979). Life strategies of bryophytes: a preliminary review. *Lindbergia*, **5**, 2–18.

During, H.J. & ter Horst, B. (1983). The diaspore bank of bryophytes and ferns in chalk grassland. *Lindbergia*, **9**, 57–64.

During, H.J., Brugues, M., Cros, R.M. & Lloret, F. (1987). The diaspore bank of bryophytes and ferns in the soil in some contrasting habitats around Barcelona, Spain. *Lindbergia*, **13**, 137–49.

Eggler, W.A. (1959). Manner of invasion of volcanic deposits by plants, with further evidence from Paricutin and Jorullo. *Ecological Monographs*, **29**, 267–84.

Fenton, J.H.C. (1982). Vegetation re-exposed after burial by ice and its relationship to changing climate in the South Orkney Islands. *British Antarctic Survey Bulletin*, **51**, 247–55.

Furness, S.G. & Hall, R.H. (1981). An explanation of the intermittent occurrence of *Physcomitrium sphaericum* (Hedw.) Brid. *Journal of Bryology*, **11**, 733–42.

Gray, A.J., Crawley, M.J. & Edwards, P.J. (Eds) (1987). *Colonization, Succession and Stability*. Blackwell Scientific Publications, Oxford.

Greenfield, L.G. (1989). Water soluble substances in terrestrial Antarctic plants and microbes. *New Zealand Journal of Natural Sciences*, **16**, 21–30.

Griggs, R.F. (1933). The colonization of the Kitmai ash, a new and inorganic 'soil'. *American Journal of Botany*, **20**, 92–113.

Grime, J.P. & Thompson, K. (1976). An apparatus for measurement of the effect of amplitude of temperature fluctuation on the germination of seeds. *Annals of Botany*, **40**, 795–9.

Harper, J.L. (1977). *Population Biology of Plants*. Academic Press, London.

Houghton, J.T., Jenkins, G.J. & Ephraums, J.J. (Eds) (1990). *Climate Change. The IPCC Scientific Assessment*. Cambridge University Press, Cambridge.

Kappen, L. & Straka, H. (1988). Pollen and spores transport into Antarctica. *Polar Biology*, **8**, 173–80.

Knoop, B. (1984). Development in bryophytes. *The Experimental Biology of Bryophytes* (Ed. by A.F. Dyer & J.G. Duckett), pp. 143–76. Academic Press, London.

Leck, M.A. (1980). Germination in Barrow, Alaska, tundra soil cores. *Arctic and Alpine Research*, **12**, 343–9.

Leck, M.A., Parker, V.T. & Simpson, R.L. (1989). *Ecology of Seed Banks*. Academic Press, London.

Longton, R.E. (1988). *Biology of Polar Bryophytes and Lichens*. Cambridge University Press, Cambridge.

Longton, R.E. & Holdgate, M.W. (1979). The South Sandwich Islands: IV. Botany. *British Antarctic Survey Scientific Reports*, **94**, 53 pp.

Longton, R.E. & Miles, C.J. (1982). Studies on the reproductive biology of mosses. *Journal of the Hattori Botanical Laboratory*, **52**, 219–40.

Longton, R.E. & Schuster, R.M. (1983). Reproductive biology. *New Manual of Bryology*. Vol. 1. (Ed. by R.M. Schuster), pp. 386–462. The Hattori Botanical Laboratory, Nichinan, Japan.

Magnusson, S. & Fridriksson, S. (1974). Moss vegetation on Surtsey in 1971 and 1972. *Surtsey Research Progress Report*, **7**, 45–57.

Matthews, J.A. (1992). *The Ecology of Recently-Deglaciated Terrain. A Geoecological Approach to Glacier Forelands and Primary Succession*. Cambridge University Press, Cambridge.

Melick, D.R. & Seppelt, R.D. (1992). Loss of soluble carbohydrates and changes in freezing point of Antarctic bryophytes after leaching and repeated freeze-thaw cycles. *Antarctic Science*, **4**, 399–404.

Miles, C.J. & Longton, R.E. (1990). The role of spores in reproduction in mosses. *Botanical Journal of the Linnean Society*, **104**, 149–73.

Miller, N.G. (1985). Fossil evidence of the dispersal and establishment of mosses as gametophyte fragments. *Monographs in Systematic Botany from the Missouri Botanical Garden*, **11**, 71–8.

Miller, N.G. & Ambrose, L.J.H. (1976). Growth in culture of wind-blown bryophyte gametophyte fragments from arctic Canada. *The Bryologist*, **79**, 55–63.

Mogensen, G.S. (1981). The biological significance of morphological characters in bryophytes: The spore. *The Bryologist*, **84**, 187–207.

Nehira, K. (1983). Spore germination, protonema development and sporeling development. *New Manual of Bryology*, Vol. 1. (Ed. by R.M. Schuster), pp. 343–85. The Hattori Botanical Laboratory, Nichinan.

Nelson, S.D., Bliss, L.C. & Mayo, J.M. (1986). Nitrogen fixation in relation to *Hudsonia tormentosa*: a pioneer species in sand dunes, northeastern Alberta. *Canadian Journal of Botany*, **64**, 2495–501.

Polunin, N. (1936). Plant succession in Norwegian Lapland. *Journal of Ecology*, **24**, 372–91.

Porsild, A.E., Harrington, C.R. & Mulligan, G.A. (1967). *Lupinus arcticus* Wats. grown from seeds of Pleistocene Age. *Science*, **158**, 113–4.

Roberts, H.A. (1981). Seed banks in soils. *Advances in Applied Biology*, **6**, 1–55.

Roser, D.J., Melick, D.R., Ling, H.U. & Seppelt, R.D. (1992). Polyol and sugar content of terrestrial plants from continental Antarctica. *Antarctic Science*, **4**, 413–20.

Schwabe, G.H. (1974). Nitrogen fixing blue-green algae as pioneer plants on Surtsey 1968–1973. *Surtsey Research Progress Report*, **7**, 22–5.

Smith, R.I.L. (1984). Colonization and recovery by cryptogams following recent volcanic activity on Deception Island, South Shetland Islands. *British Antarctic Survey Bulletin*, **62**, 25–51.

Smith, R.I.L. (1985). Studies on plant colonization and community development in Antarctic fellfields. *British Antarctic Survey Bulletin*, **68**, 109–13.

Smith, R.I.L. (1987). The bryophyte propagule bank of Antarctic fellfield soils. *Symposia Biologica Hungarica*, **35**, 233–45.

Smith, R.I.L. (1988). Botanical survey of Deception Island. *British Antarctic Survey Bulletin*, **80**, 129–36.

Smith, R.I.L. (1990). Signy Island as a paradigm of biological and environmental change in Antarctic terrestrial ecosystems. *Antarctic Ecosystems – Ecological Change and Conservation* (Ed. by K.R. Kerry & G. Hempel), pp. 30–48. Springer-Verlag, Heidelberg.

Smith, R.I.L. (1992). Exotic sporomorpha as indicators of potential immigrant colonists in Antarctica. *Grana*, **30**, 313–24.

Smith, R.I.L. & Coupar, A.M. (1987). The colonization potential of bryophyte propagules in Antarctic fellfield sites. *Comité National Français des Recherches Antarctiques*, **58**, 189–204.

Stork, A. (1963). Plant immigration in front of retreating glaciers, with examples from the Kebnekajse area, northern Sweden. *Geografiska Annaler*, **45**, 1–22.

Sussman, A.S. (1965). Physiology of dormancy and germination in the propagules of cryptogamic plants. *Handbuch der Pflanzenphysiologie*, Vol. 15, Pt. 2, (Ed. by W. Ruhland), pp. 933–1025. Springer-Verlag, Berlin; also: Longevity and resistance of the propagules of bryophytes and pteridophytes. Ibid., pp. 1086–93.

Tearle, P.V. (1987). Cryptogamic carbohydrate release and microbial response during spring freeze–thaw cycles in Antarctic fellfield fines. *Soil Biology and Biochemistry*, **19**, 381–90.

Teeri, J.A. & Barrett, P.E. (1975). Detritus transport by wind in a high arctic terrestrial ecosystem. *Arctic and Alpine Research*, **7**, 387–91.

Thompson, K. (1987). Seeds and seed banks. *New Phytologist*, **106** (Suppl.), 23–34.

van der Valk, A.G. (1992). Establishment, colonization and persistence. *Plant Succession. Theory and Prediction* (Ed. by D.C. Glenn-Lewin, R.K. Peet & T.T. Veblen), pp. 60–102. Chapman & Hall, London.

van Zanten, B.O. (1978). Experimental studies on transoceanic long-range dispersal of moss species in the Southern Hemisphere. *Journal of the Hattori Botanical Laboratory*, **44**, 455–82.

van Zanten, B.O. & Pocs, T. (1981). Distribution and dispersal of bryophytes. *Advances in Bryology*, (Ed. by W. Shultze-Motel) **1**, pp. 479–562. J. Cramer, Vaduz.

Watson, M.A. (1981). Chemically mediated interactions among juvenile mosses as possible determinants of their community structure. *Journal of Chemical Ecology*, **7**, 367–76.

Webb, R. (1973). Reproductive behaviour of mosses on Signy Island, South Orkney Islands. *British Antarctic Survey Bulletin*, **36**, 61–77.

Wharton, R.A., McKay, C.P., Simmons, G.M. & Parker, B.C. (1985). Cryoconite holes on glaciers. *Bioscience*, **35**, 499–503.

World Meteorological Organization (1988). WMO/TD No. 225. *World climate programme. Developing policies for responding to climatic change.* A summary of the discussions and recommendations of the workshops held in Villach 28 September–2 October and Bellagio 9–13 November 1987 under the auspices of the Beijer Institut, Stockholm (WMO/WCiP1). Geneva, World Meteorological Organization.

Wynn-Williams, D.D. (1986). Microbial colonization of Antarctic fellfield soils. *Perspectives in Microbial Ecology* (Ed. by F. Megusar & M. Cantar), pp. 191–200. Slovene Society for Microbiology, Ljubljana.

Wynn-Williams, D.D. (1988). Television image analysis of microbial communities in Antarctic fellfields. *Polarforschung*, **58**, 239–50.

Wynn-Williams, D.D. (1993). Microbial processes and initial stabilization of fellfield soil. *Primary Succession on Land* (Ed. by J. Miles & D.W.H. Walton), pp. 17–32. Blackwell Scientific Publications, Oxford.

Mechanisms of primary succession on volcanoes: A view from Mount St Helens

R. DEL MORAL
Department of Botany (KB-15), University of Washington, Seattle, Washington 98195, USA

SUMMARY

1 Early primary succession on the upper slopes of Mount St Helens is described and related to processes found on other volcanoes.
2 The subalpine flora of Mount St Helens has relatively few species with disproportionate representations of taxa relative to surrounding volcanoes. This disharmony results from isolation, the volcano's youth and frequent disturbances.
3 Local extinctions are concentrated on uncommon species with small seeds. Only when a species establishes a viable reproductive population is it relatively safe from immediate local extinction.
4 Species-richness has increased more quickly than cover, indicating that volcanic substrates have sufficient 'safe-sites' for establishment or diverse species, but that general growing conditions are severe. Isolation greatly retards early recovery.
5 Pioneer communities on lahars contain as many species as nearby undisturbed communities, but community structure is drastically different. Large-seeded and well-dispersed species are prominent, while many common dominants of undisturbed communities are uncommon.
6 Physical processes such as erosion and organic nutrient fallout immediately begin to ameliorate site conditions on newly created substrates. Only then can invasion by higher plants begin.
7 Pioneers may facilitate further invasion, but in subalpine habitats they are stress-tolerant species that are inept long-distance dispersers. Facilitation is both species-specific and density-dependent.
8 Long-term studies of volcanic habitats with permanently marked plots will improve understanding of dynamics of colonization and development on harsh substrates. Studies on extensive grids shed light on the dynamics of invasion. Reclamation strategies for other desolate landscapes will thereby be improved.

INTRODUCTION

Primary succession following volcanic eruptions is perhaps the least well-documented form of succession. The continuing study by Fridriksson (1987) of Surtsey is a fine one, but the isolation of this island has produced very slow succession. The excellent analyses of recovery on the Krakatau group (Tagawa *et al.* 1985; Whittaker *et al.* 1989; Whittaker & Bush 1993) do not focus on establishment processes.

This chapter identifies major mechanisms that determine the rate and direction of primary succession on volcanoes and summarizes studies on Mount St Helens, which build on the studies of many other workers on numerous volcanoes. It will also identify those aspects of volcanic primary succession in need of clarification.

Definitions

Primary succession on new or sterilized substrates requires amelioration by processes such as erosion that reveals old substrates or creates favourable germination sites, weathering to release nutrients or remove acid conditions and aerial deposition of organic matter (Edwards 1986b); subsequently, facilitation occurs. Familiar mechanisms include trapping wind-blown soil and seeds, symbiotic nitrogen fixation and drought reduction. While physical amelioration is rarely important in secondary succession (Clements 1936), it is paramount in primary succession (Griggs 1918a; Eggler 1963; del Moral 1983; del Moral & Wood 1988a).

Substrate types

Volcanic substrates vary greatly and strongly affect the rate of succession. Blong (1984) and Bullard (1984) describe six major substrates.

Lava

Hot magma extruded from craters or vents, undergoes succession very slowly. Mosses and lichens dominate succession on exposed lava surfaces (Eggler 1963). Tagawa (1965) found that invasion rates were affected by proximity to surviving vegetation and wind direction. A lava field often has concurrent successions developing at greatly different rates.

Pumice

A vesicular, frothy high silica glass, is ejected in hot masses. It is difficult to invade due to low nutrient status and instability. Colonization requires stable substrates followed by amelioration.

Scoria

Basaltic vesicular ejecta, is denser than pumice and weathers readily. It occurs on volcanoes such as Wizard Island in Crater Lake (Jackson & Faller 1973) and Mount Fuji (Masuzawa 1985).

Pyroclastic flows

Incandescent flows of gases, ash and cinders, are ejected explosively from vents and craters. Invasion is usually slow (Beard 1976), although after a few years, dispersal barriers may inhibit colonization more than substrate conditions.

Lahars

These are flows of water-saturated debris, spawned by rapidly melting glaciers. Hot flows have been postulated (Griggs 1918b). Lahars may entrain seeds or rhizomes that rapidly initiate succession. Mount Shasta (Beardsley & Cannon 1930) and Mount Rainier (Frenzen *et al.* 1988) produced well-documented lahars not associated with an eruption.

Tephra

Includes all airborne materials ejected from a volcano. Only thick deposits kill vegetation. Eruptions at Mount Katmai (Griggs 1918a, 1919, 1933), Parícutin (Eggler 1948) and Mount Usu (Riviere 1982a, b) provided earlier opportunities to study the effects of deep tephra deposits, while Antos & Zobel (1985, 1986) describe less intense tephra impacts.

Factors that influence vegetation development

Primary succession on volcanoes is governed by the degree and type of isolation or impact size, by habitat stress, by the degree and types of facilitative processes, by mitigating factors such as mycorrhizae (Allen 1987; Allen & MacMahon 1988) and by chance factors (Walker & Chapin 1987). These mechanisms and processes have been the focus of my studies on Mount St Helens and will be explored subsequently.

BACKGROUND

Mount St Helens erupted on 18 May 1980, after 130 quiescent years. This volcano, located in south-western Washington State, USA, at 46°12′N, 122° 11′W, produced varied impacts. The once majestic cone lost 400 m and now reaches only to 2550 m. A massive landslide triggered the eruption and loosed a directed blast devastating a wide arc extending up to 18 km north of the crater (Franklin *et al.* 1985, 1988). A series of pyroclastic flows and lahars subsequently covered much of this terrain. Fig. 1 shows the locations of the main study sites. Major substrates with primary succession include: (a) the Plains of Abraham (Fig. 2), a pumice desert (1350 m), that resulted from multiple pumice flows; (b) the Pumice Plains (Fig. 3) above Spirit Lake (1100 m), a result of the avalanche, pyroclastic flows and lahars; (c) lahars (1350 m) that resulted from rapid glacier melt during the

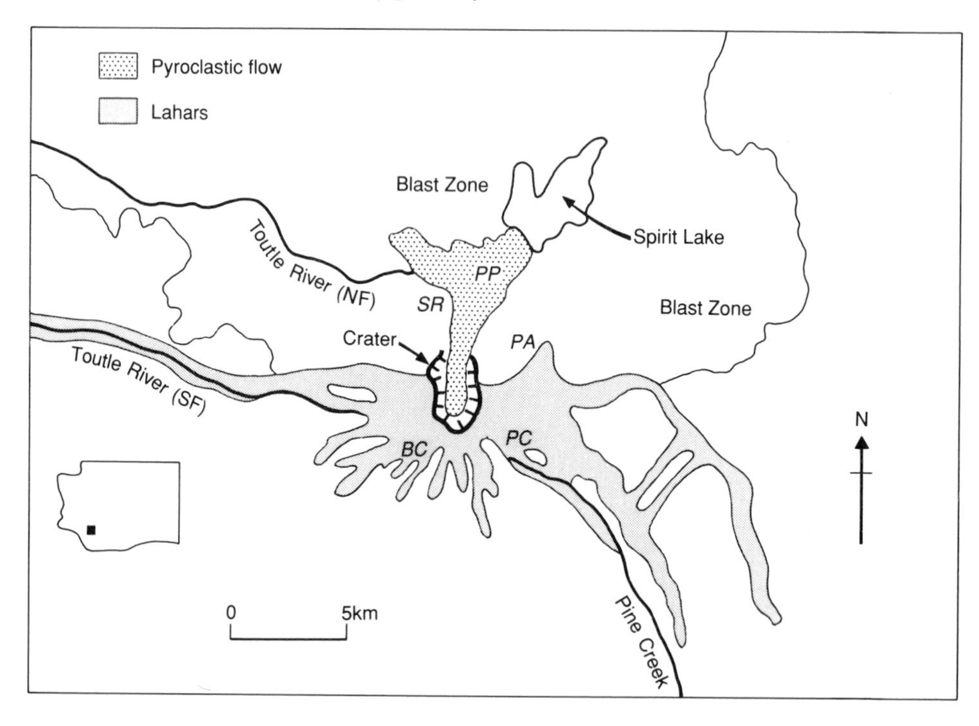

FIG. 1. Location of main study sites. PP = Pumice Pond pyroclastic zone; SR = Studebaker Ridge blast zone; PA = Plains of Abraham pumice zone; PC = Pine Creek scoured zone; BC = Butte Camp lahars.

FIG. 2. Plains of Abraham in 1989. Pumice is coarse, with few nutrients and little water-holding capacity. Vegetation is very sparse.

FIG. 3. The Pumice Plains in 1986. Vegetation is very sparse, but water collects in small depressions. Here, establishment is more likely. A seedling of a *Carex* species has established.

FIG. 4. Conifer seedlings are beginning to establish on lahars at timberline. *Pinus contorta* seedlings are obvious here, but several plants of *Lupinus lepidus* also occur.

FIG. 5. Studebaker Ridge received the directed blast at close range. A single plant of *Lupinus lepidus* occurred in this plot in 1988.

eruption (Fig. 4); and (d) blasted original surfaces on ridges (1250 m) near the crater (Fig. 5). A scoured ridge (1520 m), with a biological legacy, also will be discussed.

The story of Mount St Helens, a young, post-Pleistocene volcano, is still unfolding. The studies are concentrated in the herb-dominated communities of the cone. Nomenclature is that of Hitchcock & Cronquist (1973).

RESULTS

Turnover and extinction

General flora

Even before 1980, the flora of Mount St Helens was limited. The subalpine and Alpine zone of Mount St Helens was about 20% the area of Mount Adams. Families such as Asteraceae were more common than expected while others, such as Brassicaceae, were poorly represented. Isolation from subalpine habitats throughout its existence and frequent eruptions have produced a flora dominated by grasses and composites, and lacking many species that would be expected in comparable habitats on surrounding volcanoes (del Moral & Wood 1988b). Those missing include *Draba aureola, Polemonium elegans, Deschampsia atropurpurea, Anemone* spp., *Dodecatheon jeffreyi*, *Silene* spp., *Cassiope mertensiana* and

Vaccinium deliciosum. New species were noted in each study prior to 1980. The 1980 eruption further reduced the scant subalpine flora by 15% (del Moral & Wood 1988a, b).

Turnover in permanent plots

General descriptive data come from 250-m^2 circular permanent plots located in a wide variety of conditions on Mount St Helens. Within each plot, 24 0·25-m^2 quadrats were used to estimate per cent cover. The large plots were grouped to characterize blast zone vegetation, scours, lahars and tephra. Species-richness is the mean number of species per 250 m^2 plot in these composite samples. Cover is the mean of these same plots.

From 1982 to 1989, local colonization and extinction events were recorded in each quarter of seven permanent lahar plots. Table 1 records several colonization and extinction parameters. The percentage of local extinction is based on the total occurrences in the previous year; the percentage of colonization events is based on total occurrences in a year.

During this time, subpopulations became established 4.6 times more often than they were lost. Thirty of 62 losses were replaced by 1989. *Lupinus lepidus* accounted for 19% of the losses, but most other losses were in the rarer species. Seventeen of 26 species suffered at least one local extinction, only two of which have failed to recover.

The colonization rate declined as species colonized all suitable habitats, but local extinction rates also declined as population sizes increased and conditions became more favourable. Lahar vegetation will not reach equilibrium for at least many decades.

Patterns of early succession

The upper Pumice Plains and Plains of Abraham combine isolation and harsh substrates that sharply reduce successful colonization. These sites remain barren,

TABLE 1. Colonization and extinction on lahars established in 1980

	Year							
Category*	1982	1983	1984	1985	1986	1987	1988	1989
Occurrences	37	45	74	117	179	179	224	277
Colonizations (C)	–	14	35	51	68	19	52	**57**
Extinctions (E)	–	6	6	8	6	19	7	14
C/E	–	2·3	5·8	6·4	11·3	1·0	7·4	4·1
Extinctions year^{-1} (%)	–	16	13	11	5	11	4	6
Colonizations year^{-1} (%)	–	31	47	44	38	11	23	21

* Local extinctions per year are based on the number of occurrences in the previous year. Colonizations per year are based on the number of occurrences in the present year.

though the former has a much higher colonization rate at lower elevations. Colonization is being monitored in grids of 100-m^2 quadrats. Cover is estimated annually using this index: 1, 1–5 plants; 2, 6–20 plants; 3, over 20 plants or 0·25–0·5 m^2; 4, 0·5–1·0 m^2; 5, 1–2 m^2; 6, 3–4 m^2; 7, 5–8 m^2; 8, 9–16 m^2; 9, $>17\,m^2$. Percentage cover was estimated from the index scores using a species-specific conversion.

A 400-quadrat grid on the Plains of Abraham contained 23 species in 1988 and 28 in 1989. Only 26% of the quadrats were inhabited in 1988, but this rose to over 66% a year later. More than 89% of occurrences were by wind-dispersed species in 1988, a value that increased to 95% in 1989. Bird-dispersed species represented less than 1% of the occurrences.

Less severe sites have been monitored closely since 1980. Composite samples describe general patterns of species-richness and cover development.

Fig. 6 shows species-richness summarized annually for lahars, the blast site, the intensively scoured site and, for comparison, a tephra site. Lahars near-colonist sources began to accumulate species in 1982. Blast plots were barren through 1983, having suffered intense heat and severe soil removal. The ridge is isolated from colonizing sources, yet species-richness increases have begun. The scoured ridge lost top soil and several species, but species-richness equilibrated by 1983 as erosion freed species buried by mud and a few species returned. The tephra site equilibrated by 1982 at the latest.

Even after 10 growing seasons, cover on lahars is only about 3% (Fig. 7). Measurable cover on the blasted ridge is sporadic. Values in Fig. 7 overstate

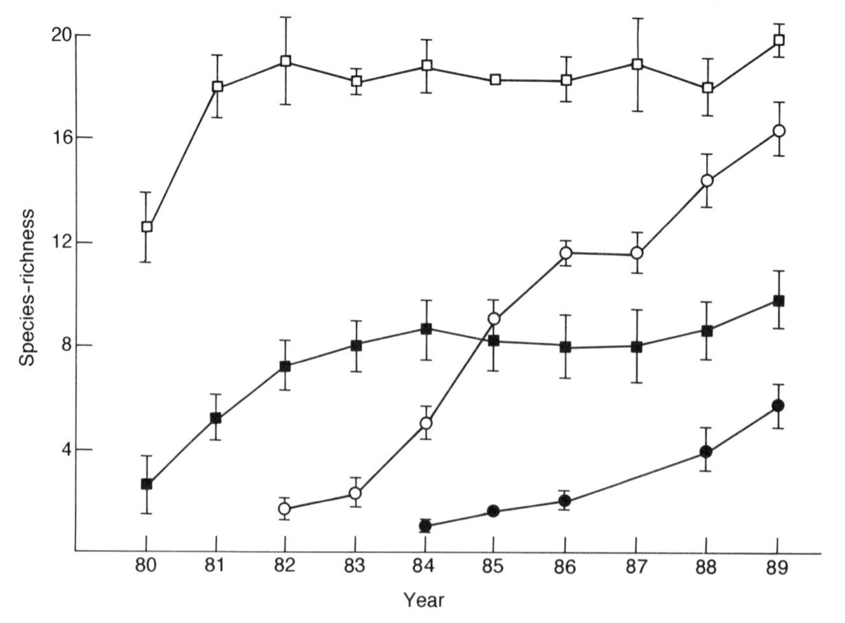

Fig. 6. Species-richness (number per 250 m^2) determined annually in permanent plots. ■——■, scour; ○——○, lahar; ●——●, blasted ridge; □——□, tephra. Vertical lines are SEM.

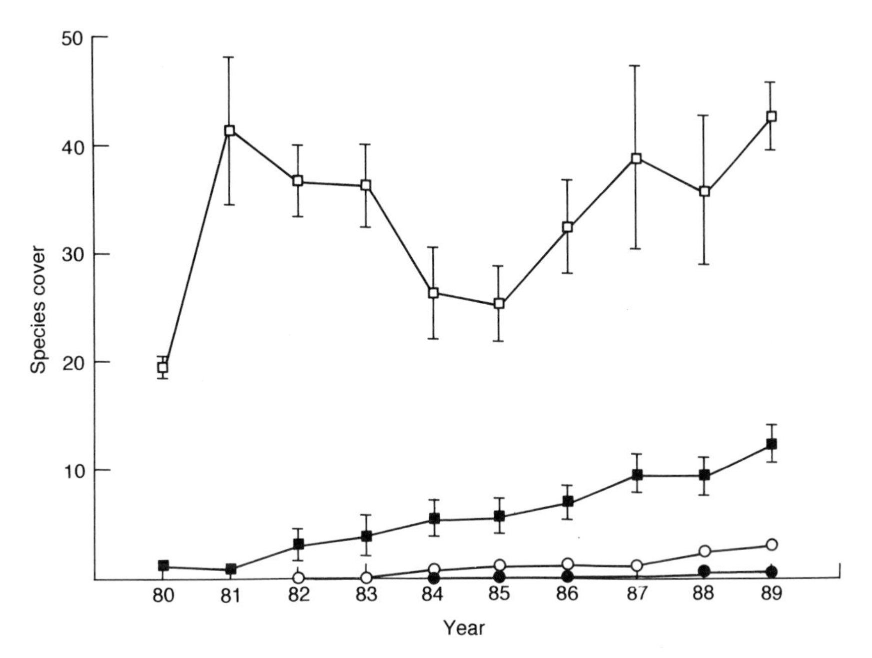

FIG. 7. Mean % cover determined annually in permanent plots. ■——■, scour; ○——○, lahar; ●——●, blasted ridge; □——□, tephra. Vertical lines are SEM.

cover since 0.1%, the minimum score, greatly overstates their abundance. The scoured ridge, with a legacy of survivors, has increased gradually each year to date. Comparisons with less severely scoured plots suggest that this development will continue for several years. Tephra cover recovered by 1981 and has fluctuated in response to highly variable summer precipitation (Pfitsch & Bliss 1988).

Dispersal

Even short distances reduce invasion in stressful habitats. The subalpine flora of Mount St Helens is depauperate, further limiting the pool of potential invaders. There are several cases of isolated establishment followed by local expansion. For example, on the Pumice Plains, several patches of *Lupinus lepidus* became established by 1981. These have expanded gradually, produced a few isolated individuals at substantial distances from the founder populations and have gone through two cycles of density increases and crashes. Distance *per se* is a major barrier to invasion, particularly at higher elevations where a limited flora with poor dispersal abilities occurs. Wood & del Moral (1987) measured invasion distances across a lahar–tephra boundary and demonstrated the limited dispersal ability of subalpine species. Mean cover across the ecotone during 1983–85 showed a sharp discontinuity that remained in 1989. Once a reproducing individual becomes established in an isolated, severe habitat, recruitment is far more likely to be due to this individual rather than from long-distance dispersal.

Permanent plots

At Pine Creek, distance is the sole dispersal barrier. Pre-existing conditions were similar along the ridge, yet species-richness declined rapidly with small elevation gain. In 1989, mean species-richness at 1300 m was 20·3, at 1370 m it was 15·2 and at 1520 m, 500 m uphill, it was 9·6. Species capable of growing under these conditions, such as *Achillea millefolium, Aster ledophyllus, Polygonum newberryi* and *Lomatium martindalei*, have been unable to migrate.

Grids

On the isolated and severely impacted Pumice Plains, there was scant vegetation after 10 growing seasons and wind-dispersed species dominated. Seed traps revealed that wind-blown seeds strongly dominated the invasion. Wood & del Moral (1988) found 32 species in 2475 plots at 1150 m in 1986. Most were confined to favourable microsites and 42% of the plots lacked plants. Species-richness averaged 1·31 species per plot. Readily dispersed weeds were common (22%), but *Anaphalis margaritacea* (37%) and *Epilobium angustifolium* (17%) plus poorly dispersed, but locally well-established *Lupinus lepidus* (23%), dominated the landscape. In 1989, a nearby grid on pumice at 1400 m had all plots occupied, but only 10% had lupins, while nearly all other individuals were of wind-dispersed species.

Two grids on lahars at 1350 m near intact vegetation were more diverse than those of the Pumice Plains or the Plains of Abraham, with high frequencies, though low cover (Table 2). These grids, sampled in 1987, 1988 and 1989 demonstrated rapid vegetation development. Recruitment was primarily from local parents, as determined from the pattern of colonization. In nearly every case, newly occupied plots were adjacent to plots in which flowering individuals of the

TABLE 2. Mean species-richness and calculated cover in two lahars during 1987–89 and the significance of changes. Lahar I is adjacent to intact vegetation; lahar II is more isolated, but exposed to up-valley winds

	1987	1988	Change	*P*	1989	Change	*P*
Lahar I ($n = 175$)							
Total species-richness	31	35	4		35	0	
Mean species-richness	8·40	9·75	1·35	<0·0001	11·27	1·52	<0·0001
Mean cover est.	1·57	2·69	1·12	<0·0001	3·22	0·43	<0·0001
Lahar II ($n = 317$)							
Total species-richness	31	33	2		37	4	
Mean species-richness	7·23	10·05	2·83	<0·0001	11·91	1·86	<0·0001
Mean cover est.	0·95	1·86	0·91	<0·0001	1·97	0·11	NS
Tephra (10 250 m^2 permanent plots)							
Total species-richness	34	34	0		34	0	
Mean species-richness	18·9	18·0	−0·9	NS	21·5	1·9	NS
Mean cover	38·9	35·8	−3·1	NS	42·4	4·4	NS

species occurred in the previous year. The mean species-richness of the two lahars was comparable with tephra, but the rank orders were uncorrelated. Cover was vastly lower on the lahars. On both lahars, mean number of species per 100-m^2 plot increased significantly between each year. Estimated cover increased significantly between 1987 and 1988 on both lahars, and on the less isolated lahar I between 1988 and 1989. Cover was lower on lahar II and the increment between 1988 and 1989 was significant only at the 6% level.

Table 3 summarizes several comparisons between these grids and tephra. All seven species found on lahars but absent from tephra are wind dispersed. While many more wind-dispersed disseminules must arrive on the lahar, large-seeded species have an establishment advantage. *Agrostis diegoensis*, the leading tephra dominant, neither large seeded nor well dispersed, was the 11th most abundant species on lahars. Grasses were poorly represented on lahars. Ten tephra species were absent from lahars, including: *Phlox diffusa, Fragaria virginiana* and *Castilleja miniata*.

Whereas dispersal to lahars is difficult for most species, establishment and growth is an even greater hurdle. Low cover and altered vegetation structure imply that lahars are very stressful. Much development remains before lahars begin to resemble adjacent sites.

Amelioration

Soils were tested for their growth-sustaining ability in a series of bioassays between 1980 and 1983. Materials were planted with seeds of 20 species into each of four 250 cm^3 pots per soil and grown in a growth chamber for 30 days. Control soil was from an infertile subalpine site. Shoot and root dry weights were used to assess growth.

The ability of pyroclastic material, lahars and tephra to support native seedling growth has previously been studied (del Moral & Clampitt 1985). Between 1980 and 1983, mean growth in substrates collected annually from plant-free locations

TABLE 3. Species and their % relative cover according to whether each species is primarily wind dispersed or has a large seed

	Lahars		Tephra	
	No.	% Relative cover	No.	% Relative cover
Wind dispersed	19	30·3	14	16·2
Large seeded	5	51·8	5	21·1
Bird dispersed	0	0	2	6·7
Grasses	4	3·5	7	41·4
Species found on only one site				
Wind dispersed	7	5·0	1	0·4
Small seeded	0	0	4	13·0
Grasses	0	0	3	1·2

increased from nil to 60% of control, a result of allogenic factors. Nutrients were low, but increased measurably during the study.

Much of the increase in cover on tephra between 1980 and 1981 was the result of erosion removing tephra and revealing plants that had survived burial for an entire season. On the scoured ridge at Pine Creek, silt had been deposited that smothered the vegetation. During the summer of 1980, rill erosion occurred to expose buried plants. In 1981, by which time winter rains had removed most of the remaining silt, the only survivors were found where rills had first formed. A pronounced pattern of long, narrow vegetation patches separated by barren terrain had been formed. These survivors provided the local seed pool for primary succession on adjacent sites.

Initial establishment

We have investigated the ability of species to invade barren habitats under different conditions (del Moral & Wood 1986). Classic succession theory suggests that primary succession in harsh habitats is initially confined to 'safe sites'. In order to test this hypothesis, the location of seedlings with respect to topographic features was investigated on lahars, at the Plains of Abraham and at the Pumice Plains.

At each location, on the grids described above, 0·25-m^2 quadrats were arrayed along randomly selected grid lines at 5 m intervals. On the lahar and the Pumice Plains, surface texture (very fine to rocks), texture at 3 cm (very fine to pebbles) and microtopography (flat, undulating, channel edge, drainage bottom or exposed) were determined. At the Plains of Abraham, due to restricted variation, a site was characterized by one of the following categories: pebbles-smooth (38·0%), pumice (about 2 cm diameter, 29·5%), rocky (11·5%), undulating (10·3%), drainage edge (4·1%) and drainage (7·3%). The presence of any species was recorded. The patterns were assessed for the more common species by a Chi-squared test for heterogeneity.

On the lahars, conditions are less stressful than other sites and the surfaces lie close to sources for colonists. Here the distribution of *Lupinus lepidus, Polygonum newberryi* and *Eriogonum pyrolifolium* was analysed. All species showed a preference for small rocks, 1–5 cm in diameter. All species showed a weak tendency for undulated surfaces.

On the Pumice Plains, *Anaphalis, Epilobium* and *Carex mertensiana* occurred at least 20 times in 240 quadrats. None showed any pattern with respect to surface or subsurface characteristics. Each responded to microtopography and all avoided flat areas. *Anaphalis* and *Carex* strongly preferred channels and undulating areas while *Epilobium* occurred primarily in undulating areas and in rocky sites.

At the Plains of Abraham, a similar pattern emerged. Here, *Anaphalis* and *Hypochoeris radicata* were sufficiently common for analysis, though they occurred in a combined total of less than 7% of the quadrats. *Anaphalis* occurred in no smooth, pumice or drainage habitats and was disproportionately represented in rocks, undulating and edge habitats. *Hypochoeris* occurred only next to rocks.

Facilitation

Facilitation is widely recognized as an important factor in controlling early primary succession (Connell & Slatyer 1977). Yet once established, plants may facilitate, inhibit or not affect a potential colonist. *Lupinus lepidus* dramatically ameliorates conditions by facilitating nitrogen fixation, trapping fine soil material, lowering surface temperatures and reducing evaporation. Wind-blown seeds are also trapped by the adult plant. These effects can markedly enhance seedling germination and survival on pyroclastic surfaces and lahars.

Further evidence of the importance of facilitation was obtained from nitrogen addition studies (del Moral & Wood 1986) that demonstrated a more than 400% increase in growth in grasses and composites, but only limited growth increases in lupins.

Permanent plots and grid patterns suggest that *Lupinus lepidus* can facilitate invasions. On lahars at Butte Camp, lupins were the first species to occupy the sites. On grids on the Pumice Plains, lupins were distributed widely (Wood & del Moral 1988) despite poor dispersal. Local population densities build quickly because lupins rapidly produce copious seeds *in situ*. Subsequently, species cover accelerates.

Early primary succession on volcanic substrates rarely yields biotic densities sufficient to inhibit invasion. Wood & del Moral (1987) suggested that nurse-plant effects were responsible for the anomaly that while virtually no seedlings of several common species emerged in garden trials, seedlings of these species were abundant in sample plots. *Aster*, *Penstemon* and *Lomatium* were strongly correlated with adult vegetation of other species.

Morris & Wood (1989) demonstrated the potentially contrary effects of nurse-plants early in primary succession. They planted seedlings of *Anaphalis margaritacea* and *Epilobium angustifolium* into dense and diffuse *Lupinus lepidus* patches and adjoining bare sites. Compared with the barren ground, survival was enhanced by diffuse lupins but eliminated by competition in dense patches. Even when survival was reduced by the lupins, surviving plants grew better. In a separate study of senescent lupins, I found that 82% of mounds formed by lupins now dead had been invaded by wind-blown species, while only 11% of mature lupins sheltered invaders. These short-lived lupins appear to facilitate invasion far better after their demise.

Results such as these emphasize that the dichotomy between facilitation and inhibition is arbitrary. The biological mechanisms operating at any given site, even one as simple as the Pumice Plains of Mount St Helens, will be complex. Whether the presence of a species facilitates or inhibits the invasion of another species depends on the density of the pioneer and on the identity and natures of the two species.

DISCUSSION

I will relate relevant results from Mount St Helens to the wider story of volcanic primary succession. The appendix summarizes the volcanoes discussed and the major substrates found.

Early recovery patterns

Vegetation structure

Primary succession is initially slow, then accelerates as soils develop and recruitment shifts from long-distance dispersal to locally produced seeds (Griggs 1933; Tagawa 1964; Hendrix 1981; Clarkson & Clarkson 1983; Hirose & Tateno 1984). Many factors influence the later course of succession, among them competition, dispersal vectors, chance and human disturbance.

Turnover and extinction

Any new substrate can be treated as an island on which establishment and extinction may be followed and mechanisms explored. Early in succession, net species-richness will increase, but local extinctions dampen species accumulation. Data documenting this process for volcanoes are scarce and existing data have yet to document any true equilibrium within a century, though plateaux have been noted on a large scale (Whittaker *et al.* 1989) and at local scale (del Moral & Wood 1988a).

Rejmanek *et al.* (1982) compared floristic records of El Paricutín over 20 years and found that the net rate of species increase slowed after several years. Of 52 recorded species, only 39 survived. Fridriksson (1987) presented vascular plant data for Surtsey over 21 years, recording 23 taxa, of which only 13 occurred in 1986. Omitting unknowns, his data imply 17 extinctions, of which seven reinvaded. The number of species has varied between 10 and 15 since 1973, a plateau far below equilibrium. Plants rarely establish. Therefore, turnover rates should stay high until many species begin to reproduce on the island. They should then decline as conditions become less stressful and the seed rain is enhanced. Whittaker *et al.* (1989) discussed minimum turnover rates on Krakatoa and found that losses are more common in forest than beach assemblages. They show that species-richness continues to increase, but that forests are species-poor and have not equilibrated. Turnover is a complex process concentrated on marginal species.

Local extinction and reinvasion is being documented on lahars and the Pumice Plains of Mount St Helens. Most species capable of growth on the substrate have arrived, but extinction continues to occur on a small scale. When taller plants become dominant, a new episode of colonization and habitat-wide extinction is expected.

The colonization pattern on volcanoes is that species accumulation is influenced strongly both by dispersal and establishment barriers, that extinctions fall heavily on rare, non-reproductive species and that these species reinvade frequently. Species that become reproductive on new substrates can expand their populations quickly.

Dispersal

Impact size influences the seed rain. Distant oceanic islands depend primarily on sea currents and bird dispersal (Fridriksson 1987), so that invasion is restricted to a few taxa and is diffuse. Subsequent succession is usually by expansion from these pioneering populations. On nearby islands, such as Wizard Island (Jackson & Faller 1973) or Motmot (Ball & Glucksman 1975), wind plays an increasingly important role. Wind dispersal is paramount on land-locked volcanoes. While sea birds are attracted to islands, devastated terrestrial volcanoes hold little fascination.

Several studies (Tezuka 1961; Smathers & Mueller-Dombois 1974; Timmins 1983) have shown that isolation slows colonization on volcanoes. The narrow Kautz Creek lahar on Mount Rainier was quickly invaded by *Alnus* and *Salix* (Frenzen *et al.* 1988), while the broad Muddy River lahar is scantily vegetated (Halpern & Harmon 1983; Braatne & Chapin 1986).

There are two invasion patterns. Along an ecotone with an impact discontinuity, there may be an advancing vegetation front, with outliers. However, if establishment probabilities are very low or if distance from sources is great, the invasion is more diffuse.

On Mount St Helens, invasion has been diffuse. Even along ecotones with recovered vegetation, conditions are so stringent that few seedlings can establish, and these only in favourable spots. In isolated habitats, the seed rain is scant. Any established reproductive plant greatly outproduces the general seed rain and thus is a locus for accelerated development. For example, on a lahar at Butte Camp, the wind-dispersed herb *Hieracium gracile* occurred in 97 quadrats in 1988 from which it was absent in 1987. Of these, half were isolated by more than 10 m and half were contiguous to existing plants. This species is still partially dependent on the general seed rain. In contrast, *Lupinus lepidus* increased in 209 plots, due to flowering parents, and established in 66 contiguous new quadrats. Local lupins produced all seedlings.

Establishment

The role of pioneers

There is no general rule about which species can pioneer volcanic habitats, though earlier workers were struck by the importance of mosses, liverworts and ferns. In drier climates, porous substrates such as tephra, and even lava beds, seed plants are frequent pioneers (Tagawa 1964; Ball & Glucksman 1975; Riviere 1982a). On lava, multiple concurrent successions occur, with rapid development among cracks, very slow development on rock faces.

Habitat stress

Recovery rates are related to physical stress and decrease with elevation and with latitude. Succession is also slow on substrates with either limited water-holding ability or low nutrients.

Suppressed timberlines on volcanoes have been noted frequently (Clements 1936; Lawrence 1938; Fosberg 1959; Jackson & Faller 1973; Timmins 1983; Nakamura 1985) and are symptoms of the difficulty of upward colonization. Studies of Mount Fuji (Ohsawa 1984) best document that treelines advance gradually to their climatic limit in such a way that invasion recapitulates succession. The treeline is severely depressed on the north side of Mount St Helens, though lowland conifers are invading the Pumice Plains. It will take many decades, if not centuries, of eruption-free recovery before a treeline even as high as that in 1979 is achieved.

Amelioration operates on volcanoes before the first dispersal unit falls on a barren site. Timmins (1983) reported that erosion on Mount Tarawera formed gullies within which vegetation established. Kadomura *et al.* (1983) demonstrated the importance of erosion to recovery on Mount Usu. On Mount St Helens, erosion removed large quantities of loose deposits from many unstable substrates, resulting in stabilization, creation of favourable establishment sites or the exposure of old substrates. The jump apparent in species-richness between 1980 and 1981 at Pine Creek was due to erosion that exposed surviving plants.

Edwards (1986a, b) showed that air-fall deposits significantly contributed to carbon and nitrogen accumulation on barren substrates. They estimated that during spring and summer, over 10 mg m^{-2} day^{-1} fell throughout the bare Pumice Plains. Nitrogen increases are due to symbiotic fixation and from fallout.

Facilitation

At high elevations, the entire flora is 'stress-tolerant' (*sensu* Grime 1977). Wood & del Moral (1987) found that species with good dispersal abilities tolerated stress poorly, while tolerant species dispersed poorly. Thus colonization rates were directly related to overall habitat stress. It is these stress-tolerant species, rather than 'pioneers', that possess the greatest ability to facilitate invasions, particularly by well-dispersed species unable to establish in barren substrates.

Types of plants that have been considered as 'nurse plants' include members of all multicellular groups: algae (Trueb 1888; Griggs 1918a; Booth 1941; Smathers & Mueller-Dombois 1974); lichens (Eggler 1941; Cooper & Rudolph 1953); mosses (Tagawa 1964; Clarkson & Clarkson 1983); liverworts (Griggs 1933); nitrogen-fixing species (Vitousek *et al.* 1987; Halvorson 1989), species that encourage free-living nitrogen-fixing bacteria (Hirose & Tateno 1984); conifers (Heath 1967; Jackson & Faller 1973; Clarkson & Clarkson 1983); and other flowering plants (Fridriksson 1987; Tsuyusaki 1987).

Factors other than the presence of pioneer species facilitate invasion. Allen

(1987; Allen & MacMahon 1988) suggested that the absence of mycorrhizae curtailed flowering plant invasion on Mount St Helens, but there is as yet little good evidence for this in subalpine habitats.

Unusual circumstances can markedly accelerate invasion rates. Pools of water on Surtsey attracted birds (Fridriksson 1987), while fumeroles and a spring on Mount St Helens provided moisture that permitted wind-blown seeds to establish in otherwise barren habitats (Wood & del Moral 1988). The presence of such 'safe sites' often permits primary succession to accelerate. The importance of such safe sites is apparent from the importance of microtopographic factors in determining the location of early pioneer species on Mount St Helens.

Seed size

The establishment of a newly arrived seed is far from assured. One major factor that enhances survival is its size. On Mount St Helens, drought is frequent and seedlings must establish quickly (Chapin & Bliss 1988). Wood & del Moral (1988) demonstrated a clear positive relationship between seed size and survival rates in field studies. The generality of this pattern awaits studies under other conditions.

The longer term

Floristic patterns

Frequent eruptions may produce species-poor and floristically unbalanced floras on volcanoes. Chronic disturbances often cause local extinctions (del Moral & Wood 1988b), eliminate suitable habitats and help maintain communities lacking a normal floristic composition. Previous eruptions also may have eliminated species suited to recolonization.

Isolation effects are obvious on oceanic volcanoes, but terrestrial ones are similarly, though more subtly, impacted. Large devastated areas were only slowly colonized. Mount St Helens is exceptional in that subalpine environments are very young and isolated from colonization sources. Immigration rates were thus low and many expected taxa were poorly represented (Kruckeberg 1987).

Novel communities

New communities are frequently assembled by vagaries of dispersal. Pioneers sometimes quickly establish dense vegetation and suppress further immigration (Tagawa *et al.* 1985). Whittaker *et al.* (1989) noted that seral forests on Krakatoa are stabilized at low species-richness levels and appear to be resisting further rapid invasion. At the higher elevations of Mount St Helens, pioneer communities, such as the common *Anaphalis–Epilobium* association, appear to have no analogue on other volcanoes, but because such associations are sparse, they probably will not arrest succession.

Future studies

Volcanoes produce many conditions that challenge pioneer species. How do populations become established? What are the roles of the first colonists and more distant seed donors? Are there genetic consequences due to the limited number of successful invading individuals? These substrates also offer opportunities to study constraints on primary succession that may instruct efforts to reclaim the products of human industrial activity or human-caused disasters.

Weeds

There have been few studies of the long-term effects of exotic species on primary succession, though arrested and deflected secondary successions are common. Introduced species may accelerate (Vitousek *et al.* 1987) or arrest succession through competition. Are exotics transient or can they seriously alter community structure? European weeds could delay succession on Mount St Helens, since they form a major component of the invaders at lower elevations. However, most of them are shade-intolerant and should decline as conifers become established. At higher elevations, weeds cannot survive in appreciable numbers and have had little impact. As more volcanoes erupt in biologically transformed landscapes, the effects of exotic species should become more clear.

Isolation

The importance of isolation acting as an immigration filter is well-established. However, interactions between isolation and habitat severity are poorly understood. On Mount St Helens, succession is often delayed because stress tolerance and dispersal ability are inversely related. Reclamation programmes that seek to mimic natural succession may be examined to determine if 'seral species' could be established early in the programme, their natural tardiness being a dispersal function rather than a facilitation requirement. Programmes to accelerate reclamation in this way would have to consider if additional fertilization would be required.

Facilitation and inhibition

The importance of facilitation and inhibition in early succession is only generally known. To what degree is physical amelioration required for invasion? What are the species-specific and density-related aspects of facilitation and inhibition? What is the best balance between site preparation and biotic facilitation? At what densities should facilitator species be established?

Long-term studies of primary succession on volcanoes such as Surtsey, Sakurajima, Mount Usu and Mount St Helens will contribute greatly to our specific and general understanding of primary succession.

ACKNOWLEDGMENTS

My colleagues D. M. Wood, L. C. Bliss, J. Braatne, D. M. Chapin, J. E. Edwards, W. A. Pfitsch and F. Ugolini have greatly assisted me in various aspects of this paper. J. E. Edwards provided useful information concerning Mount Tarawera. The paper was improved by the comments of E. A. Brosseau and D. M. Wood. Funding was provided by NSF Grants BSR 81–07042, BSR 84–07213 and BSR 89–06544.

APPENDIX 1. Volcanic habitats likely to undergo primary succession

Substrate	Volcano	References
Lava	Craters of Moon	Eggler (1941)
	Hawaii	Fosberg (1959); Smathers & Mueller-Dombois (1974); Vitousek *et al.* (1987)
	El Paricutín	Eggler (1948, 1963)
	Sakurajima	Tagawa (1964, 1965)
	Krakatoa	Tagawa *et al.* (1985); Whittaker *et al.* (1989)
	Crater Lake	Jackson & Faller (1973)
	Galapagos Islands	Hendrix (1981)
Pumice cinder	Kilauea Iki	Smathers & Mueller-Dombois (1974)
	El Paricutín	Eggler (1963); Rejmanek *et al.* (1982)
	Mount Fuji	Hirose & Tateno (1984); Ohsawa (1984); Mazusawa (1985); Nakamura (1985)
	Motmot	Ball & Glucksman (1975)
	Krakatoa	Tagawa *et al.* (1985); Whittaker *et al.* (1989)
	Mount Tarawera	Clarkson & Clarkson (1983); Timmins (1983)
	Surtesy	Fridricksson (1987)
	Crater Lake	Horn (1968); Jackson & Faller (1973)
	Mount St Helens	Wood & del Moral (1987); del Moral & Wood (1988a); Allen (1988)
	Mount Usu	Kadomura *et al.* (1983); Riviere (1982a, b); Tsuyuzaki (1987)
Lahars	Mount St Helens	Halpern & Harmon (1983); Dale (1986); del Moral & Wood (1986, 1988a)
	Mount Rainier	Frenzen *et al.* (1988)
	Mount Lassen	Heath (1967)
	Mount Shasta	Beardsley & Cannon (1930)
	Mount Usu	Kadomura *et al.* (1983); Riviere (1982a, b); Tsuyuzaki (1987)
Thick tephra	Motmot	Ball & Glucksman (1975)
	Katmai	Griggs (1918a, 1933)
	El Paricutín	Eggler (1963)

REFERENCES

Allen, M.F. (1987). Reestablishment of mycorrhizae on Mount St Helens: migration vectors. *Transactions of the British Mycological Society*, **88**, 413–7.

Allen, M.A. & MacMahon, J.A. (1988). Direct VA mycorrhizal inoculation of colonizing plants by pocket gophers (*Thomomys talpoides*) on Mount St Helens. *Mycologia*, **80**, 754–6.

Antos, A.J. & Zobel, D.B. (1985). Recovery of forest understories buried by tephra from Mount St Helens. *Vegetatio*, **64**, 105–14.

Antos, A.J. & Zobel, D.B. (1986). Seedling establishment in forests affected by tephra from Mount St Helens. *American Journal of Botany*, **73**, 495–9.

Ball, E. & Glucksman, J. (1975). Biological colonization of Motmot, a recently-created tropical island. *Proceedings of the Royal Society of London*, **190**, 421–42.

Beard, J.S. (1976). The progress of plant succession on the Soufriere of St. Vincent: Observations in 1972. *Vegetatio*, **31**, 69–77.

Beardsley, G.F. & Cannon, W.A. (1930). Notes on the effects of a mudflow at Mount Shasta. *Ecology*, **11**, 326–36.

Blong, R.J. (1984). *Volcanic Hazards*. Academic Press, Sydney.

Booth, W.E. (1941). Algae as pioneers in plant succession and their importance in erosion control. *Ecology*, **22**, 38–46.

Braatne, J. & Chapin, D.M. (1986). Comparative water relations of four subalpine plants at Mount St Helens. *Mount St Helens: Five Years Later* (Ed. by S.A.C. Keller), pp. 163–72. Eastern Washington State University Press, Cheney.

Bullard, F.M. (1984). *Volcanoes of the Earth*. University of Texas Press, Austin.

Chapin, D.M. & Bliss, L.C. (1988). Soil–plant water relations of two subalpine herbs from Mount St Helens. *Canadian Journal of Botany*, **66**, 809–18.

Chapin, D.M. & Bliss, L.C. (1989). Seedling growth, physiology and survivorship in a subalpine volcanic environment. *Ecology*, **70**, 1325–34.

Clarkson, B.R. & Clarkson, B.D. (1983). Mt. Tarawera : 2. Rates of change in the vegetation and flora of the high domes. *New Zealand Journal of Ecology*, **6**, 107–19.

Clements, F.E. (1936). Nature and structure of the climax. *Journal of Ecology*, **24**, 252–84.

Connell, J.H. & Slatyer, R.O. (1977). Mechanisms of succession in natural communities and their role in community stability and organization. *American Naturalist*, **111**, 1119–44.

Cooper, R. & Rudolph, E.D. (1953). The role of lichens in soil formation and plant succession. *Ecology*, **34**, 805–7.

Crocker, R.L. & Major, J. (1955). Soil development in relation to vegetation and surface age at Glacier Bay, Alaska. *Journal of Ecology*, **43**, 427–48.

Dale, V.D. (1986). Plant recovery on the debris avalanche at Mount St Helens. *Mount St Helens: Five Years Later* (Ed. by S.A.C. Keller), pp. 208–14. Eastern Washington State University Press, Cheney.

del Moral, R. (1983). Initial recovery of subalpine vegetation on Mount St Helens, Washington. *American Midland Naturalist*, **109**, 72–80.

del Moral, R. & Clampitt, C.A. (1985). Growth of native plant species on recent volcanic substrates from Mount St Helens. *American Midland Naturalist*, **114**, 374–83.

del Moral, R. & Wood, D.M. (1986). Subalpine vegetation recovery five years after the Mount St Helens eruptions. *Mount St Helens: Five Years Later* (Ed. by S.A.C. Keller), pp. 215–21. Eastern Washington State University Press, Cheney.

del Moral, R. & Wood, D.M. (1988a). Dynamics of herbaceous vegetation recovery on Mount St Helens, Washington, USA, after a volcanic eruption. *Vegetatio*, **74**, 11–27.

del Moral, R. & Wood, D.M. (1988b). The high elevation flora of Mount St Helens. *Madroño*, **35**, 309–19.

Edwards, J.S. (1986a). Arthropods as pioneers: recolonization of the blast zone on Mount St Helens. *Northwest Environmental Journal*, **2**, 263–73.

Edwards, J.S. (1986b). Life in the allobiosphere. *Trends in Ecology and Evolution*, **3**, 111–4.

Eggler, W.A. (1941). Primary succession on volcanic deposits in southern Idaho. *Ecological Monographs*, **11**, 277–98.

Eggler, W.A. (1948). Plant communities in the vicinity of the volcano El Paricutín, Mexico, after two and a half years of eruption. *Ecology*, **29**, 415–36.

Eggler, W.A. (1963). Plant life of Paricutín volcano, Mexico, eight years after activity ceased. *American Midland Naturalist*, **69**, 38–68.

Fosberg, F.R. (1959). Upper limits of vegetation on Mauna Loa, Hawaii. *Ecology*, **40**, 144–6.

Franklin, J.F., MacMahon, J.A., Swanson, F.J. & Sedell, J.R. (1985). Ecosystem responses to the eruption of Mount St Helens. *National Geographic Society Research*, **1**, 198–216.

Franklin, J.F., Frenzen, P.M. & Swanson, F.J. (1988). Recreation of ecosystem at Mount St Helens: contrasts in artificial and natural approaches. *Rehabilitating Damaged Ecosystems* (Ed. by J. Cairns), vol. II, pp. 1–37. CRC Press, Boca Raton, Florida.

Frenzen, P.M., Krasney, M.E. & Rigney, L.P. (1988). Thirty-three years of plant succession on the Kautz Creek mudflow, Mount Rainier National Park, Washington. *Canadian Journal of Botany*, **66**, 130–7.

Fridriksson, S. (1987). Plant colonization of a volcanic island, Surtsey, Iceland. *Arctic and Alpine Research*, **19**, 425–31.

Griggs, R.F. (1918a). The recovery of vegetation of Kodiak. *Ohio Journal of Science*, **19**, 1–57.

Griggs, R.F. (1918b). The great hot mudflow of the Valley of 10000 Smokes. *Ohio Journal of Science*, **19**, 117–42.

Griggs, R.F. (1919). The character of the eruption as indicated by its effects on nearby vegetation. *Ohio Journal of Science*, **19**, 173–209.

Griggs, R.F. (1933).. The colonization of the Katmai ash, a new and inorganic 'soil'. *American Journal of Botany*, **20**, 92–111.

Grime, J.P. (1977). Evidence for the existence of three primary strategies in plants and its relevance to ecological and evolutionary theory. *American Naturalist*, **111**, 1169–94.

Halpern, C.B. & Harmon, M.E. (1983). Early plant succession on the Muddy River mudflow, Mount St Helens. *American Midland Naturalist*, **110**, 97–106.

Halvorson, J.J. (1989). *Carbon and nitrogen contributions to Mount St Helens volcanic sites by lupines*. PhD thesis, Washington State University, Pullman.

Heath, J.P. (1967). Primary conifer succession, Lassen Volcanic National Park. *Ecology*, **48**, 270–5.

Hendrix, L.B. (1981). Post-eruption succession on Isla Fernandina, Galapagos. *Madroño*, **28**, 242–54.

Hirose, T. & Tateno, M. (1984). Soil nitrogen patterns induced by colonization of *Polygonum cuspidatum* on Mt. Fuji. *Oecologia*, **61**, 218–23.

Hitchcock, C.L. & Cronquist, A. (1973). *Flora of the Pacific Northwest*. University of Washington Press, Seattle.

Horn, E.M. (1968). Ecology of the pumice desert, Crater Lake National Park. *Northwest Science*, **42**, 141–9.

Jackson, M.T. & Faller, A. (1973). Structural analysis and dynamics of the plant communities of Wizard Island, Crater Lake National Park. *Ecological Monographs*, **43**, 441–61.

Kadomura, H., Imkagawa, T. & Yamamoto, K. (1983). Eruption-induced rapid erosion and mass movements on Usu Volcano, Hokkaido. *Zeitschrift für Geomorphologie*, **46**, 123–42.

Kruckeberg, A.R. (1987). Plant life on Mount St Helens before 1980. *Mount St Helens 1980, Botanical Consequences of the Explosive Eruptions* (Ed. by D.E. Bilderback), pp. 3–21. University of California Press, Berkeley.

Lawrence, D.B. (1938). Trees on the march. *Mazama*, **20**, 49–54.

Masuzawa, T. (1985). Ecological studies on the timberline of Mt. Fuji I. Structure of plant community and soil development on the timberline. *Botanical Magazine of Tokyo*, **98**, 15–28.

Morris, W.F. & Wood, D.M. (1989). The role of *Lupinus lepidus* in succession on Mount St Helens: facilitation or inhibition? *Ecology*, **70**, 697–703.

Nakamura, T. (1985). Forest succession in the subalpine region of Mt. Fuji, Japan. *Vegetatio*, **64**, 15–27.

Ohsawa, M. (1984). Differentiation of vegetation zones and species strategies in the subalpine region of Mt. Fuji. *Vegetatio*, **57**, 15–52.

Pfitsch, W.A. & Bliss, L.C. (1988). Recovery of net primary production in subalpine meadows of Mount St Helens following the 1980 eruption. *Canadian Journal of Botany*, **66**, 989–97.

Rejmanek, M., Haagerova, R. & Haager, J. (1982). Progress of plant succession on the Paricutín volcano: 25 years after activity ceased. *American Midland Naturalist*, **108**, 194–8.

Riviere, A. (1982a). Invasion and recovery of plants after the 1977–1978 eruption of Usu Volcano, Hokkaido, Japan: a preliminary note. *Environmental Science Hokkaido*, **5**, 197–209.

Riviere, A. (1982b). Plant recovery and seed invasion on a volcanic desert, the crater basin of USU-san, Hokkaido. *Ecological Congress, Sapporo, Seed Ecology*, **13**, 11–18.

Smathers, G.A. & Mueller-Dombois, D. (1974). Invasion and recovery of vegetation after a volcanic eruption in Hawaii. *Hawaii National Park Service Science Monograph*, No. 5.

Tagawa, H. (1964). A study of volcanic vegetation in Sakurajima, southwest Japan. I. Dynamics of vegetation. *Memoirs, Faculty of Science, Kyushu University, Ser. E. (Biology)* **3**, 165–228.

Tagawa, H. (1965). A study of volcanic vegetation in Sakurajima, southwest Japan. II. Distributional pattern and succession. *Japanese Journal of Botany*, **19**, 127–48.

Tagawa, H., Suzuki, E., Partomihardjo, T. & Suriadarma, A. (1985). Vegetation and succession on the Krakatau Islands, Indonesia. *Vegetatio*, **60**, 131–45.

Tezuka, Y. (1961). Development of vegetation in relation to soil formation in the volcanic island of Oshima, Izu, Japan. *Japanese Journal of Botany*, **17**, 371–402.

Timmins, S. (1983). Mt. Tarawera : 1. Vegetation types and successional trends. *New Zealand Journal of Ecology*, **6**, 99–105.

Trueb, M. (1888). Notice sur la nouvelle flora de Krakatau. *Annales du Jardin Botanique de Buitenzorg*, **7**, 213–23.

Tsuyuzaki, S. (1987). Origin of plants recovering on the volcano Usu, Northern Japan, since the eruptions of 1977 and 1978. *Vegetatio*, **73**, 53–8.

Vitousek, P.M., Walker, L.R., Whittaker, L.D., Mueller-Dombois, D. & Matson, P.A. (1987). Biological invasion by *Myrica faya* alters ecosystem development in Hawaii. *Science*, **238**, 802–4.

Walker, L.R. & Chapin, F.S. III (1987). Interactions among processes controlling successional change. *Oikos*, **50**, 131–6.

Whittaker, R.J., Bush, M.B. & Richards, K. (1989). Plant recolonization and vegetation succession on the Krakatau Islands, Indonesia. *Ecological Monographs*, **59**, 59–123.

Whittaker, R.J. & Bush, M.B. (1993). Dispersal and establishment of tropical forest assemblages, Krakatoa, Indonesia. *Primary Succession on Land* (Ed. by J. Miles & D.W.H. Walton), pp. 147–60. Blackwell Scientific Publications, Oxford.

Wood, D.M. & del Moral, R. (1987). Mechanisms of early primary succession in subalpine habitats on Mount St Helens. *Ecology*, **68**, 780–90.

Wood, D.M. & del Moral, R. (1988). Colonizing plants on the Pumice Plains, Mount St Helens, Washington. *American Journal of Botany*, **75**, 1228–37.

Primary succession on the cone of Vesuvius

S. MAZZOLENI AND M. RICCIARDI
Istituto Botanico, Facoltà di Agraria, Portici (NA) 80055, Italy

INTRODUCTION

This chapter reports a study of the vegetation patterns on the substrates deposited by the last eruption in 1944 and discusses the colonization process in relation to substrate type and age.

The eruption of Vesuvius of AD 79 (described by Pliny the younger to Tacitus) overwhelmed the ancient Roman towns of Pompeii, Erculaneum, Oplontis and Stabiae. These were kept in good preservation under ash, pumice and mud and their archeological excavations are nowadays visited by hundreds of thousands of tourists, which make this volcano known world-wide. Its shape has also characterized all the paintings of the Bay of Naples since the impressive explosive eruption of 1631.

Vesuvius had an active eruptive history during the period 1631–1944, with 18 eruptive cycles characterized by lava fountains, gases and vapour emissions from the crater. The quiet periods never exceeded 7 years until after the last eruption in 1944.

Its gentle slopes at lower altitudes have supported arable crops and vineyards since ancient times because of the high fertility of the volcanic soils. The lack of activity since the 1944 eruption coincided with a period of rapid expansion of the towns surrounding Naples. This has seen the growth of dense human settlements at the base of the volcanic cone without any care for the inherent danger of living there. Even at higher altitudes there is substantial human disturbance from agriculture, tourism and far too extensive and low-yielding forestry plantations. Very few areas have been left to develop naturally and most of the vegetation patterns on the older substrates are largely man-made. This is not the case, however, on the surfaces covered by lava and lapilli from the 1944 eruption and a few areas with 1906 and 1858 substrates whose inaccessibility and/or unsuitability for any practical and economic use have limited human influences.

THE ENVIRONMENT

The Somma–Vesuvius volcanic complex covers an area of about 165 km^2 in the eastern part of the Gulf of Naples (Fig. 1) centred on 40°49′N, 14°26′E.

It is a single mountain from the base at sea-level up to 800 m when it divides into the older volcano Monte Somma (1132 m a.s.l.) and the more recent Vesuvius (1281 m) which was built inside the original caldera after its summit collapsed

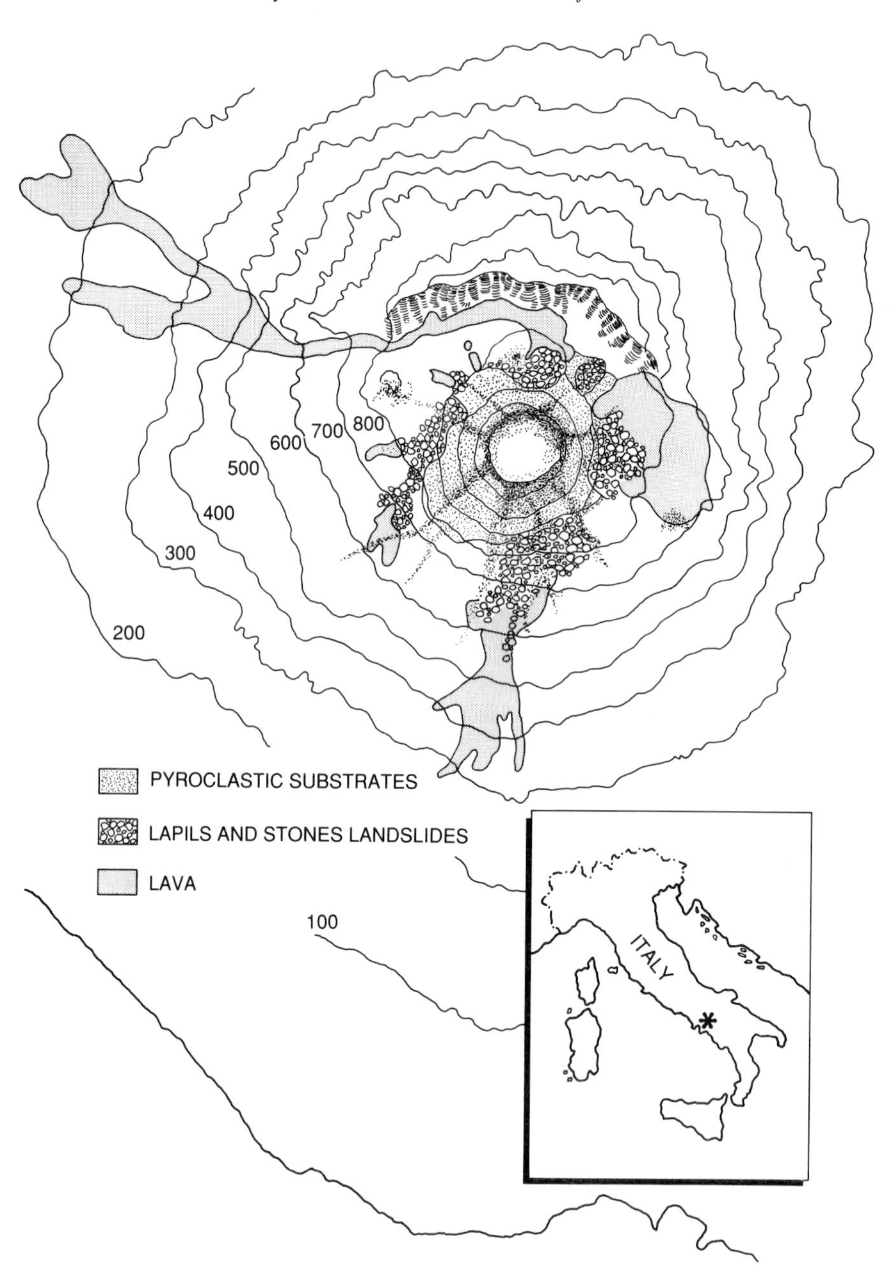

FIG. 1. Location of Vesuvius and map of the 1944 substrates.

during the AD 79 eruption. These two peaks are separated by a deep semicircular valley, Valle del Gigante, divided into Atrio del Cavallo westward and Valle dell'Inferno southward. The present crater of Vesuvius was formed at the end of the 1944 eruption and is about 450 m in diameter and 350 m deep. The cone has a regular shape with some landslide scars. Above 900 m the cone is covered by loose materials (sand, ash, lapilli and large blocks of pumice and lava), mainly produced during the explosive phase of the 1944 eruption. During this eruption there was substantial fallout of dense materials close to the crater from huge lava fountains over 2·0 km high (Principe *et al.* 1987). Landslides of the agglutinated hot tephra occurred on the sides of the cone. There was also noticeable emission of viscous lava which mainly flowed and accumulated in the Valle dell'Inferno giving the shape of a large lava 'river' surrounding the great cone.

Vesuvius has a mild Mediterranean-type climate with a dry period during summer. At the metereological station of the Osservatorio Vesuviano at 612 m, the average temperature varies from 22°C in July to 5–7°C in January. The mean yearly rainfall is 1000 mm with a mean monthly minimum in July (19 mm) and a maximum in November–December (137 mm). Snowfall is infrequent, occurring on only a few days in winter above 1000 m.

There is negligible surface water run-off because of the permeability of the volcanic soils and water-bearing strata occur close to the ground surface only where compact underlying layers of lava are present.

THE FLORA AND VEGETATION OF VESUVIUS

There is a long history of botanical survey on Vesuvius. The oldest reports of the flora are from AD 60–65 (Columella, *Libri rei rusticae*). Also, many wall paintings preserved in houses inundated by volcanic ash at Pompeii and Oplontis have shown easily recognized representations of plant species (Jashemski 1979). The discovery of carbonized plant remains buried during the AD 79 eruption has also given some information on the flora at that time (Ricciardi & Aprile 1979; Meyer 1980). There were few pre-Linnean floristic works related to the Vesuvian area (Imperato 1599; Colonna 1616) and the first extensive description of the flora of Vesuvius can be found in the *Flora Napolitana* (1811–38) and *Sylloge* (1831) of M. Tenore. Pasquale published two detailed *Floras of Vesuvius* (1840, 1869) while Licopoli (1873) reported on the cryptogams of the lavic soils of this volcano. Subsequent floristic studies by Baccarini (1881), Comes (1887), Martelli & Tanfani (1892), Migliorato (1896, 1897) and de Rosa (1906, 1907) added very little to the work of Pasquale.

Recently Ricciardi *et al.* (1986) published a new account of the flora of the Somma–Vesuvius. They found 610 species, but 293 taxa that had been recorded previously were not rediscovered. This loss of species probably reflects the extensive disturbance by humans in the last 50 years (Ricciardi *et al.* 1986). Therophyta predominate in the flora (42%), followed by hemicryptophyta (28%), phanerophyta (13%), geophyta (10%), chamaephyta (7%) and hydrophyta (0.2%). The flora is dominated by Mediterranean and widely-distributed species. Only 18 endemic species were found, perhaps reflecting the recent origin of this volcanic complex.

Almost no ecological research has previously been done in the area. The vegetation has received little attention compared with the flora and the only works available are a general description of the main vegetation types by Agostini (1975) and a phytosociological study of the pioneer stages by Mazzoleni *et al.* (1989).

The southern slopes and the valleys of Vesuvius have been largely afforested with *Pinus pinea*, with planting also of *Robinia pseudo-acacia*, *Genista aetnensis* and *Cytisus scoparius*. The northern slopes of Monte Somma have extensive planted stands of *Castanea sativa* below 900 m, and above this there is a mixed forest of *Ostrya carpinifolia* and *Acer neapolitanum*, with a few individuals of *Betula pendula*. On the steep southern cliffs of Monte Somma *Quercus ilex* is common, whereas *Cytisus scoparius* and *Spartium junceum* colonize the base and lower slopes of both Somma and Vesuvius. The volcanic substrates of recent origin support pioneer communities, which are described in this paper.

METHODS

The structure and composition of the vegetation on the 1944 substrates were studied during spring and summer 1988 and 1989, by stratified random sampling. Plots were randomly located within relatively uniform areas to avoid sampling overlapping sites with clearly different morphology or substrate types. A total of 39 sampling plots were located on the unwelded pyroclastic products of the volcanic cone between 800 and 1200 m, while another 10 plots were located on the lava flows at lower altitudes. Plant species occurring in each plot were recorded, with species cover rated on the five-point dominance–abundance scale of Braun-Blanquet. The altitude, slope, aspect, total vegetation cover, average plant height and substrate characters of each plot were described. Substrate depth and stability (stable, lightly mobile, mobile), stone and lapil dimension (small, 0.5–3 cm; medium, 3–10 cm; large, >10 cm) and presence of sand and/or developed soil was estimated by eye in the field. The relation of species abundance to the pyroclastic substrates was examined by numerical classification and principal component analysis. The cluster analysis was based on an average linkage agglomeration criterion of a correlation similarity matrix for the plant species and of a chord distance matrix for the sampling plots (Orloci 1978). Only species present more than twice were included in the analysis. Data from the plots on the lavas were not included in the main table because they were dominated by lichens with few phanerogams present. The corresponding table was recorded according to species frequencies and soil particle sizes.

RESULTS

The vegetation mosaic of the 1944 substrates showed a pattern strongly related to the character of the substrate, the main difference being between the lava flows and the unwelded pyroclastic products.

Table 1 shows the phytosociological relevées on the lava surfaces. These were mostly covered by lichens with dominant *Stereocaulon vesuvianum* and subdominant

TABLE 1. Phytosociological relevées on the lava flows of the 1944 eruption of Vesuvius. Species ordered according to frequencies, relevées ordered from left to right by increasing presence of small stones and finer soils

Species										
Stereocaulon vesuvianum Pers.	5	5	5	4	5	4	4	4	4	4
Candelariella xanthostigma (Pers.) Lett.	+	+	+	2	1	2	2	+	2	2
Parmelia conspersa (Ehrht.) Ach.	+	+	+	+	+	+	1	+	+	+
Moss spp.		+	+	+	+	+	+	+	+	+
Rumex scutatus L.			+		2	2	1	1		
Helichrysum litoreum Guss.					2	+		+	1	+
Centranthus ruber (L.) DC. subsp. *ruber*					2		+	1	1	1
Holcus lanatus L.							+		+	1
Arabis collina Ten.							1		1	1
Arabis turrita L.							+	+		+
Andryala integrifolia L.						2			+	1
Vulpia myuros (L.) C.C. Gmelin						+			1	2
Senecio vulgaris L.						+			+	+
Antirrhinum siculum Miller							+			1
Geranium purpureum Vill.									+	+
Asplenium trichomanes L. subsp. *quadrivalens* D.E. Mayer							+			+
Aira cariophyllea L. subsp. *caryophyllea*										2
Cynosurus echinatus L.										+
Rumex angiocarpus Murb.									3	
Linaria purpurea (L.) Miller										2
Bromus rigidus Roth subsp. *rigidus*								+		
Bromus tectorum L.					+					
Briza maxima L.										+
Dactylis glomerata L.										+
Reichardia picroides (L.) Roth									+	
Asplenium adiantum-nigrum L.									+	
Crepis lentodontoides All.									+	
Sedum cepea L.						+				
Trifolium arvense L.						+				

Candelariella xanthostigma. Mosses and mostly annual phanerogams were present only where some soil deposition had occurred in small depressions.

The loose soils of the upper part of the cone were much more variable than the lava surfaces because of varying proportions of ash, sand, lapilli and large stones between the different areas. Fig. 2 shows the vegetation table for these substrates structured according to the numerical clustering of both species and sampling plots. Two main groups are evident. The first is dominated by *Rumex scutatus* and represents the areas with coarse lapil substrates. This group can be subdivided into four clusters: the first representing the most pioneer stage and mainly occurring on the most mobile slopes with very gravelly soils; the second has codominant *Cytisus scoparius* and *Rumex scutatus* and corresponded to areas with more stable soils; the third and the fourth, slightly richer in species number, are intermediate in vegetation composition between this group and the other patches of the second main group.

In the second group two clusters could be distinguished. One tended to be

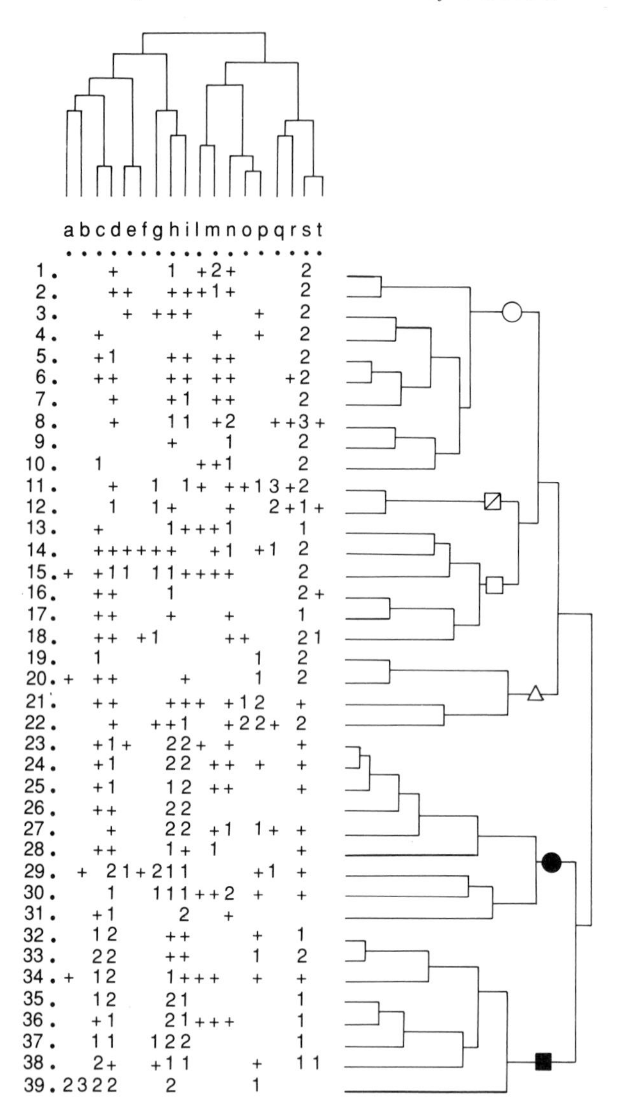

FIG. 2. Vegetation table and dendrograms produced by cluster analysis on sample plots and plant species from the Vesuvius cone. Symbols refer to groups of relevées discussed in the text: **a**: *Vulpia myuros* (L.) C.C. Gmelin; **b**: *Aira caryophyllea* L. subsp. *caryophyllea*; **c**: *Silene vulgaris* (Moench) Garcke v. *angustif.* (Miller) Hayek; **d**: *Rumex angiocarpus* Murbl; **e**: *Picris hieracioides* L.; **f**: *Daucus carota* L.; **g**: *Centranthus ruber* (L.) DC. subsp. *ruber*; **h**: *Scrophularia canina* L. subsp. *bicolor* (Sm.) W. Greuter; **i**: *Artemisia variabilis* Ten.; **l**: *Linaria purpurea* (L.) Miller; **m**: *Glaucium flavum* Crantz; **n**: *Solidago virgaurea* L.; **o**: *Chondrilla juncea* L.; **p**: *Bromus tectorum* L.; **q**: *Cytisus scoparius* L.; **r**: *Arabis hirsuta* L.; **s**: *Rumex scutatus* L.; **t**: *Helichrysum litoreum* Guss.

dominated by *Artemisia variabilis* and *Scrophularia canina*, which increased their dominance on relatively more stable soils with more fine fractions and sand. The second cluster had dominant *Rumex angiocarpus* and *Silene vulgaris* on areas with fine-textured and shallow soils mainly located over superficial rocks and compact lava layers.

The ordination of sample plots and the corresponding species weighting on the vectors by principal component analysis is shown in Fig. 3. The first axes (accounting for 18% of the total variability) reflected quite well the gradient from the lapilli and gravels with *Rumex scutatus* to the finer soil with *Rumex angiocarpus*, through increasing amounts of sand with *Artemisia variabilis* and *Scrophularia canina*. The second component (accounting for 14% of the remaining variability) can be related to soil stability, with positive scores reflecting higher substrate mobility, and negative values related to stable conditions (either consolidated soil or the presence of large rocks) with *Cytisus scoparius*, *Helichrysum litoreum* and *Centhrantus ruber*.

DISCUSSION

The succession since the 1944 eruption on Vesuvius has mostly been direct colonization of bare soil. Colonization depends on the available seed sources and the reinvasion is clearly related to dispersal distances and species capability (e.g.

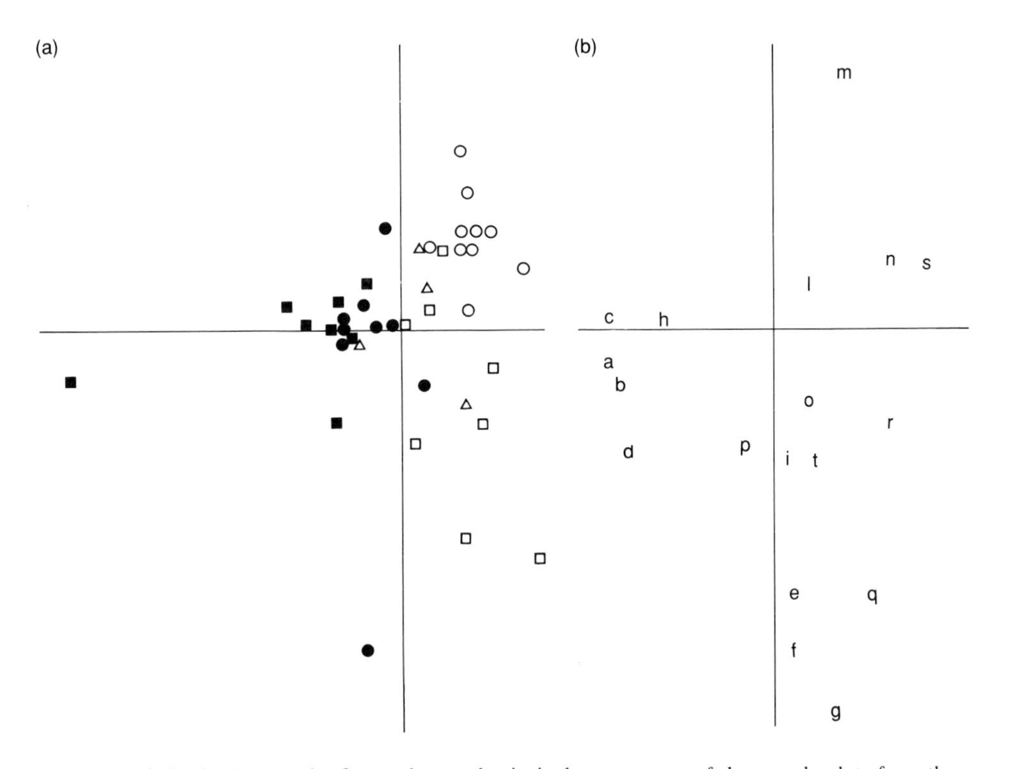

FIG. 3. (a) Ordination on the first and second principal components of the sample plots from the Vesuvius cone. (b) Weightings of species on the first and second vectors by principal component analysis. Symbols and letters as in Fig. 2.

see del Moral & Wood 1988). The main factor affecting the vegetation on Vesuvius seems to be soil particle size with soil stability second in importance, whereas altitude and aspect do not appear to have any major influence on the vegetation. Only one species, *Solidago virgaurea*, was not found on south-facing slopes, reflecting its relatively more mesic requirements. Great difference according to volcanic substrate types was found also in pioneer communities on Krakatoa islands (Tagawa *et al.* 1985).

Fig. 4 schematically represents these vegetation and soil mosaics to depict the strong linkage in their changing patterns. It does not show the condition of steep slopes with mobile sands, as such areas have no plants. However, as shown in Fig. 2, there are many transitional aspects. This vegetation can well-represent a didactical study case where it is shown that although the real vegetation 'can be viewed as a mosaic of distinct patches or types', it is in fact 'a pattern of intergrading populations' of the different component species (Miles 1979).

The first colonizers were bacteria and blue algae (Rossi & Riccardo 1926). Isolated individuals of *Stereocaulon vesuvianum* could be found soon after the lava had solidified and cooled. Agostini (1975) reported on the chronological sequence of appearance of new plant species after the 1944 eruption. Lichen colonization became evident on the scoriaceous lavas after 10–12 years and was characterized by the rapid spread and attainment of complete cover on these surfaces. Slow colonization by phanerogams occurred later. *Centranthus ruber* and *Helichrysum litoreum* were first recorded in 1958 and 1960. During 1960–66, a

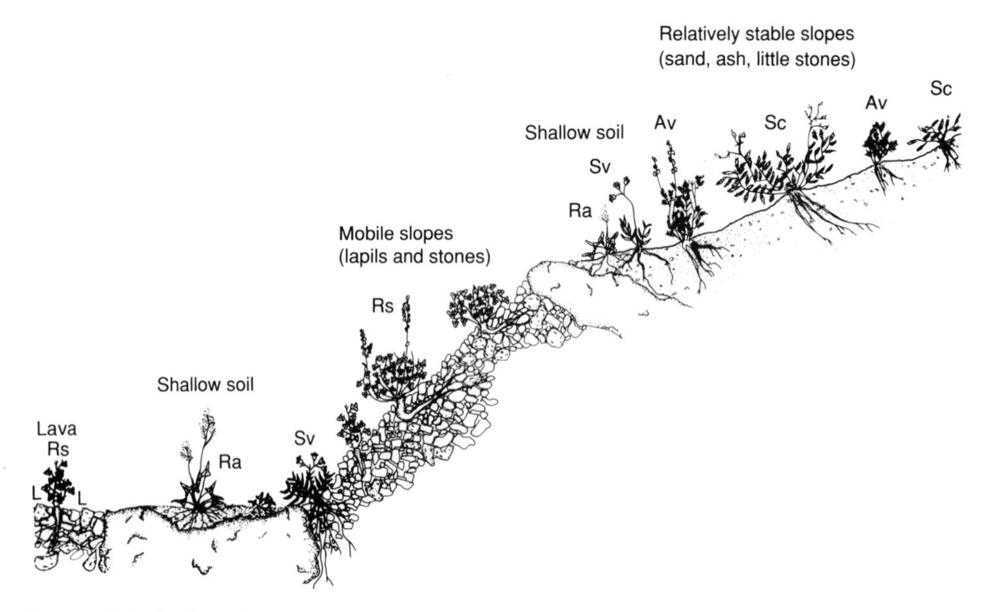

FIG. 4. Distribution of plants in relation to substrate characters on the Vesuvius 1944 eruption products. L: lichens; Rs: *Rumex scutatus*; Ra: *Rumex angiocarpus*; Sv: *Silene vulgaris*; Av: *Artemisia variabilis*; Sc: *Scrophularia canina*.

diffusion of *Aira caryophyllea*, *Reichardia picroides*, *Hypochoeris radicata* and *Vulpia myuros* was observed, while the first individuals of *Arabis hirsuta* were recorded in 1967. Current floristic composition on lava flows (see Table 1) is locally richer (sample plots 7–10), but most surfaces still lack abundant phanerogams. Weathering must be occurring, but these substrates seem to have hardly changed since 1944. A similar lichen cover was observed in 1926 on the 1906 and 1822 lavas, again showing the slow successional changes on this substrate type (Rossi & Riccardo 1926).

Colonization of the pyroclastic products was faster than on the lavas. A first appearance of *Artemisia variabilis*, *Scrophularia canina* and *Rumex angiocarpus* was recorded in 1957, with *Silene vulgaris* observed on the cone in 1958 (Agostini 1975). Since 1975 the patches dominated by these species have not changed in species composition. Changes in species cover probably occurred, but the lack of permanent sampling plots has not allowed us to evaluate this.

The spread of *Rumex scutatus* on Vesuvius in the last 10 years has been impressive. In 1956, Agostini reported this species as rare in the region of Campania with few individuals on a wall at the base of the volcano. In 1975, the species occurred only occasionally on the cone (Mazzoleni *et al.* 1989) whereas today it has become the dominant species. It has mainly colonized the mobile lapilli, showing less adaptation to the soils with more ash and sand, and its distribution only partially overlaps the areas with patches of *Artemisia variabilis* and *Scrophularia canina*.

Rumex scutatus is the first colonizer plant of the coarser mobile lapilli and pumices, and its presence seems to be facilitating the establishment of other species. For example, *Bromus erectus* was found growing inside the large plants of *Rumex scutatus* on the steeper and more mobile slopes, while it was growing alone on more stabilized sites. Moreover, many *Silene vulgaris* seedlings were observed downhill of large *Rumex scutatus* individuals, where they were presumably protected from soil erosion and burial by moving stones.

In a previous study (Mazzoleni *et al.* 1989), the pioneer communities of the substrates from the 1858, 1906 and 1944 eruptions were described in phytosociological terms and a marked difference in species composition was found in relation to substrate age. The analysis of these floristic data in terms of life-form spectra is shown in Table 2. A clear pattern can be observed, with a decrease in lichens, chamaephytes and hemicryptophytes according to age, and an increase of terophytes on the older lavas. Life-form spectra from adjacent forests shown an increase of phanerophytes and a reduction of terophytes with canopy closure. Perennial plants can show a new slight increase, but in any case due to different species than those of the first colonization stages. These results agree with Grubb's (1986) observations, the first colonizers during primary succession on coarse substrates on Vesuvius being woody species, with annuals being unimportant.

Many contrasting ecophysiological characteristics have been reported between early and late successional plants in seres from open fields to forests (Bazzaz 1979). On the other hand, studies on the autoecology of primary successional

TABLE 2. Raunkiaer life-forms average presence (%) in sampling plots on substrates of different ages (data from Mazzoleni *et al.* 1989)

	1944 lava	1944 cone	1906	1858
Chamaephytes	43	41	32	14
Hemicryptophytes	57	35	25	19
Terophytes	—	19	39	63
Phanerophytes	—	6	1	3
Geophytes	—	—	3	2

plants are rare. Grubb (1986) suggested nutrient supply was more important than water in the differentiation of plant pioneer types. However, the ability to grow on very loose substrates, lacking organic matter, implies an adaptation to limited water retention in the soil.

Recent studies by M. Amato (unpubl. data) have shown that during wetting the roots of *Rumex scutatus* absorbed water faster than those of other species such as *Rumex angiocarpus* and *Artemisia variabilis*, whereas flooding caused excessive root swelling with rupturing of the root periderm and cortex. During desiccation cycles a relatively quick diameter contraction was followed by a comparatively slower water loss in *Rumex scutatus* (M. Amato, unpubl. data). These observations match the distribution pattern of this common species which becomes dominant only on very permeable soils, as on Vesuvius, but also on calcareous and siliceous gravels in central Europe (Braun-Blanquet 1964) and on pyroclastites of other southern volcanos (Mazzoleni *et al.* 1989). Thus, we feel that comparative studies on water relations of these pioneer species may further our understanding of their adaptation to primary substrates.

APPENDIX 1. Site and substrate characters for the sampling plots on the Vesuvius cone. Numbers and symbols refer to relevées reported in Fig. 2.

	No.	Altitude	Aspect	Slope	Vegetation cover (%)	Stability*	Stone/lapil size†	Presence of sand/fine soil
○	1	1030	N	40	30	m	l(100)	
	2	1000	N	40	30	m	s(60)	x
	3	1070	W	35	25	l	s(100)	
	4	1100	NW	35	20	l	s(40) m(60)	
	5	1120	W	40	30	l	l(100)	
	6	1100	NW	35	15	l	m(100)	
	7	1120	W	40	45	s	l(100)	
	8	1040	N–NW	40	30	s	s(50) l(50)	
	9	1050	N–NW	45	30	l	s(50)	
	10	1050	N–NW	45	30	m	s(50) m(50)	
⧄	11	990	W			s	m(100)	
	12	990	W	30	40	s	m(100)	

APPENDIX 1. *Contd.*

	No.	Altitude	Aspect	Slope	Vegetation cover (%)	Stability*	Stone/lapil size†	Presence of sand/fine soil
□	13	1120	NE–E	40	30	l	s(70)	x
	14	1050	N	45	20	l	s(100)	
	15	1210	SE	40	10	l	s(60)	x
	16	1100	NE–E	40	30	l	m(60)	x
	17	1090	NE–E	40	40	s	s(50) m(20)	x
	18	1210	NW	15	15	l	s(70)	x
△	19	1200	NW–W	15	20	s	s(50)	x
	20	1190	NW–W	25	20	l	s(50)	x
	21	920	W	40	40	l	m(20)	x
	22	1180	W	20	15	s	s(40) m(40)	x
•	23	1070	NE	45	10	s	l(10)	x
	24	1065	N–NE	35	30	l	s(60) m(20)	x
	25	1065	N–NE	45	25	s	m(80)	x
	26	1080	NE	50	20	m	s(20)	x
	27	1070	NE	40	40	s	m(80)	x
	28	1040	S–SE	35	20	s	m(40)	x
	29	1160	S–SE	35	30	l	m(40) l(40)	
	30	1140	NE	45	35	s	s(40)	x
	31	1100	NE–E	40	10	s	s(60)	x
■	32	1020	SW	40	20	m	s(80)	x
	33	1020	SW	35	30	l	s(20) m(20) l(10)	x
	34	1020	S–SW	35	30	l	s(80)	x
	35	1050	N–NE	20	90	s	s(40)	x
	36	850	SW	25	35	s	s(20) m(20) l(5)	x
	37	1060	N–NE	30	20	s	s(10) m(40)	x
	38	1150	W–NW	40	30	s	s(30) m(10) l(5)	x
	39	1030	S–SW	35	30	s	s(30) l(5)	x

* s: stable; l: lightly mobile; m: mobile.
† s: small = 0.5 – 3 cm; m: medium = 3–10 cm; l: large = >10 cm. In parentheses: % values for each size class estimated by observations in the field.

REFERENCES

Agostini, R. (1975). Vegetazione pioniera del Monte Vesuvio: aspetti fitosociologici ed evolutivi. *Archivo Botanico e Biogeografico Italiano*, **51**, 11–34.

Baccarini, P. (1881). Studio comparativo sulla flora vesuviana e sulla etnea. *Nuovo Giornale Botanico Italiano*, **13**, 149–205.

Bazzaz, F.A. (1979). The physiological ecology of plant succession. *Annual Review of Ecology and Systematics*, **10**, 351–71.

Braun-Blanquet, J. (1964). *Pflanzensoziologie*. Audl, Vienna.

Colonna, F. (F. Columna) (1616). *Minus Cognitarum Rariorum nostro Coelo Orientium Stirpium Ecphrasis*. Pars altera. Jacobum Mascardum, Roma.

Comes, O. (1887). Le lave, il terreno vulcanico e la loro vegetazione. *Lo spettatore del Vesuvio e dei Campi Flegrei N. Ser.*, 35–51.

del Moral, R. & Wood, D.M. (1988). The high elevation flora of Mount St. Helens, Washington.

Madrono, **35**, 309–19.
de Rosa, F. (1906). La flora vesuviana e l'eruzione dell'Aprile 1906. *Bollettino della Societa dei Naturalisti in Napoli*, **20**, 132–9.
de Rosa, F. (1907). Poche osservazioni sulla flora vesuviana. *Atti Congresso Naturalisti Italiani*, pp. 413–8.
Imperato, F. (F. Imperatus) (1599). *Historia Naturale*. Combi e La Noce, Venezia.
Jashemski, W.F. (1979). The gardens of Pompeii, Herculaneum and the villas destroyed by Vesuvius. *Studia Pompeiana & Classica* (Ed. by R.I. Curtis), pp. 1–18. Orpheus Publications, New York.
Grubb, P.J. (1986). The ecology of establishment. *Ecology and Landscape Design* (Ed. by A.D. Bradshaw, D.A. Goode & E. Thorp), pp. 83–98. British Ecological Society Symposium, Vol. No. 24. Blackwell Scientific Publications, Oxford.
Licopoli, G. (1873). Storia naturale delle piante crittogame che crescono sulle lave vesuviane. *Atti della Accademia di Scienze Fisiche et Matematiche di Napoli*, **3**, 1–58.
Martelli, U. & Tanfani, E. (1892). Le Fanerogame e le Protallogame raccolte durante la riunione generale in Napoli della Società Botanicà Italiana nell'Agosto del 1891. *Nuovo Giornale Botanico Italiano*, **24**, 172–89.
Mazzoleni, S., Ricciardi, M. & Aprile, G.G. (1989). Aspetti pionieri della vegetazione del Vesuvio. *Annali di Botanica*, **XIVII** (Suppl. 6), 97–110.
Meyer, F.G. (1980). Carbinized food plants of Pompeii, Herculaneum and the Villa at Torre Annunziata. *Economic Botany*, **34**, 401–37.
Migliorato, E. (1896). Osservazioni relative alla flora napoletana. *Bollettino della Societa Italiana di Botanica*, **7**, 168–71.
Migliorato, E. (1897). Seconda nota di osservazioni relative alla flora napoletana. *Bollettino della Societa Italiana di Botanica*, **1**, 23–6.
Miles, J. (1979). *Vegetation Dynamics*. Chapman & Hall, London.
Orloci, L. (1978). *Multivariate Analysis in Vegetation Research*. W. Junk, The Hague.
Pasquale, G.A. (1840). La Flora del Vesuvio. *Esercitazioni Accedemia Aspiranti Naturalisti*, **2**, 25–66.
Pasquale, G.A. (1869). Flora vesuviana o catalogo ragionato delle piante del Vesuvio confrontate con quelle dell'Isola di Capri e di altri luoghi circostanti. *Atti della Accademia di Scienze Fisiche et Matematiche di Napoli*, **4**, 1–142.
Principe, C., Rosi, M., Santacroce, R. & Sbrana, A. (1987). Explanatory notes to the geological map. *Somma–Vesuvius* (Ed. by R. Santacroce), pp. 11–52. *Quaderni Ricerca Scientifica*, **114**, CNR, Rome.
Ricciardi, M., Aprile, G.G., La Valva, V. & Caputo, G. (1986). La Flora del Vesuvio. *Bolletino Società Naturalisti in Napoli. Officine Grafiche Napoletane*, **95**, 3–121.
Ricciardi, M. & Aprile, G.G. (1979). Identification of some carbonized plant remains from the archeological area of Oplontis. *Studia Pompeiana & Classica* (Ed. by R.I. Curtis), pp. 417–24. Orpheus, New York.
Rossi, G. & Riccardo, S. (1926). I terreni della regione del Vesuvio e la fissazione dell'azoto, Actes IV. *Conferenza Internazionale Pedologia*, Vol. III. International Agricultural Institute, Rome.
Tagawa, H., Suzuki, E., Partomihardjo, T. & Suriadarma, A. (1985). Vegetation and succession on the Krakatau islands, Indonesia. *Vegetatio*, **60**, 131–45.
Tenore, M. (1831). *Sylloge plantarium vascularium Florae Neapolitanae*. Tipografia del Fibreno, Napoli.
Tenore, M. (1811–38). *Flora napolitana*. I–V. I, LXXII + 324 pp., Stamperia Reale, Napoli; II, 398 pp., Tipografia del Giornale Enciclopedico, Napoli; III, XII + 412 pp., Stamperia Francese, Napoli; IV, 558 + XVIII pp., Stamperia Francese, Napoli; V, XIV + 379 pp.

The colonization of strandlines

A. J. DAVY* AND M. E. FIGUEROA†
* *School of Biological Sciences, University of East Anglia, Norwich NR4 7TJ, UK; and † Departamento de Biología Vegetal y Ecología, Universidad de Sevilla, Apartado 1095, 41080 Sevilla, Spain*

SUMMARY

1 Newly exposed land, at its interface with lakes or the sea, may be associated with the accretion of sediments or with changing land level relative to the water. Detritus and propagules are frequently washed ashore as 'strandlines' in this zone. Colonization by plants may be transient or it may represent the beginning of a true chronosequence.

2 The strandline environment is physically demanding; it is susceptible to disturbance by wind and wave action (arising from tidal oscillations, seasonal storms and changes in prevailing ocean currents); the porous sediments retain little water; there is little protection from solar irradiance and, in an essentially dry system, large diurnal temperature fluctuations may result; salt-spray is deposited on marine strandlines; there is scant nutrient capital, except from sea-spray and the local mineralization of detritus.

3 World-wide, strandline colonists show a diversity of life-histories. Annual or ephemeral species can complete their life cycles in the 5 months between successive periods of equinoctial spring (exceptionally high) tides. Many species are clonal perennials that may recolonize from fragments of rhizome or stolon after catastrophic disturbance. In the tropics, where disturbance is less frequent (tropical storms), vines and small shrubs may be prominent colonists. Disseminules of strandline species tend to be buoyant and long-lived in seawater; seed dimorphism and dormancy are important in some species; seeds tend to be large, allowing emergence from considerable depths of burial with sand and the rapid, deep anchorage of seedlings.

4 Both C3 and C4 plants are widespread, successful strandline colonists. Generally, they are not true halophytes but are able to cope with the ionic and osmotic consequences of moderately high soil salinities. Although many species are very tolerant of salt-spray, this is probably of secondary importance. Extensive root systems may exploit deep groundwater, and its upward movement, as 'internal dew'. Nitrogen availability, mainly from local mineralization of detritus, is a major determinant of establishment and success in the strandline.

5 Strandline colonization differs strikingly from primary colonization in other habitats. Dispersal by water is independent of propagule mass; much heavier propagules confer fitness in an unstable substrate. Seed and bud dormancy is associated with the environmental unpredictability of strandlines. Disturbance probably limits colonization by woody species, which are typical elsewhere.

Imported organic nitrogen, even though heterogeneously distributed, is a relatively unusual resource for primary colonization.

INTRODUCTION

> I do not know what I may appear to the world, but to myself I seem to have been only a boy playing on the seashore, and diverting myself in now and then finding a smoother pebble or a prettier shell than ordinary, whilst the great ocean of truth lay all undiscovered before me.
> [Sir Isaac Newton]

Strandlines along the shores of lakes, seas and oceans represent an essentially linear and ephemeral habitat at the interface of land with water where detritus and propagules are washed ashore. Wherever new land is claimed from the water, either by the accretion of sediments on prograding shores, or by changes in the level of land relative to that of the water, there is scope for primary succession. This review examines the processes by which plants colonize this harsh environment. We consider the peculiarities of the strandline environment, whether the colonizing species have distinctive life-history and physiological characteristics, and the possible role of these plants in the early stages of a primary succession.

The significance of strandline colonization for primary succession depends first and foremost on the physical environment: physiography and wave energy. Strandlines are probably of little consequence for the colonization of rocky (hard) shores by generally sparse terrestrial plants. On sedimentary (soft) shores it is usual for distinctive plant communities to develop at the strandline. But it is only in relatively high energy environments, where the sediments are coarse in texture (typically sands or shingle) that the current upper limit of the influence of wave action is also usually the most seaward zone that can be colonized by terrestrial plants. In contrast, low energy soft shores support the development of marsh communities on fine sediments (mainly silts and mud) within the sphere of influence of waves; primary colonization here is by aquatic or amphibious plants well below the strandline. Consequently, we will confine our attention to the colonization of strandlines on sandy and shingle shores.

The colonization of bare, mobile sands is a precursor to all sand-dune successions, probably the most celebrated of all primary successions. The earliest detailed analyses of autogenic processes in the development of vegetation were made on sand-dune systems associated with shorelines. The seminal monograph of Cowles (1899) on the physiographic and vegetational dynamics of the Lake Michigan Dunes was closely followed by Harshberger's (1900) study of Atlantic (New Jersey) coastal dunes. Similar processes have been described in many parts of the world since. The foreshore remains devoid of vegetation but represents the proximate source of sand which feeds dune development. Strandline colonists and associated litter reduce windflow locally and facilitate the deposition of sand. When the sand level surpasses the high-water mark of ordinary spring tides (HWMOST), perennial grasses or other plants with vertical and horizontal rhizome growth establish, if they are not already present. Their growth and continued trapping of sand leads to embryo dunes of a metre or more in height. The rapid and extensive rhizome growth of one of only a few species of dune-building grass can eventually produce high mobile dunes that progress landward in parallel ridges, with depressions (slacks) in between them. Eventually the mobile dunes may be stabilized by sand starvation and new colonists, or they may be destroyed at any stage by wind or storms. Although there are many variations world-wide, there is a general consensus as to the vegetational zonation and successional relationships represented by it (Doing 1985).

It would be wrong to believe that colonization events at the strandline necessarily represent the beginnings of a true chronosequence. Frequently the beach is too narrow to provide a sand supply sufficient to support dune development; alternatively, the prevailing wind direction may preclude an adequate sand supply. The strandline and foredune communities inevitably will be removed by tidal scour or storm surges if significant accretion of sand has not occurred, resulting in a cycle of colonization and destruction.

Shingle beaches tend to be less hospitable, because of the mechanical disturbance and the deposition of coarse sediments by high wave energies. As Fuller (1987) has pointed out, where the formation of successive shingle ridges provides a measure of stability and protection from wave action, strandlines are frequently colonized; elsewhere, colonization may be restricted to higher on the beach or absent altogether.

Any plant that can be washed into a lake or the sea and then survive for a few hours is a potential strandline colonist: an immense number and diversity of species can be found in strandlines. We have observed that species as different as the cactus *Opuntia ficus-indica* and the giant reed *Arundo donax*, for instance, will root adventitiously and survive for months in the strandlines of the Atlantic coast of south-west Spain. Here we define strandline colonizers as species that habitually colonize beaches near to the high-water line, often as components of typical communities. In practice, nearly all species that can be found frequently in the strandline can also be found higher on the beach or dune system, either at the same site or elsewhere. As the ephemeral nature of the habitat limits long-term survival, the question arises as to what extent these colonists are a cohesive group with particular characteristics that confer fitness in the severe strandline environment.

THE STRANDLINE ENVIRONMENT

The strandline environment is dominated by the physical forces of water and wind. Their underlying pattern is usually cyclic and predictable, but superimposed upon this are stochastic effects of the weather. Most familiar are the oscillations in tidal basins. The highest spring tides occur near to the equinoxes, with the result that the highest strandlines deposited at these times may persist for up to 6 months before they are reworked by tidal activity. In addition, in temperate regions, the probability and severity of storms are greater close to the equinoxes, and perhaps during winter. On a reasonably wide beach it is usually possible to see a series of parallel strandlines in midsummer, marking the successively lower spring high tides since the spring equinox; these strandlines are apt to be destroyed in reverse order towards the autumnal equinox. Tidal activity is not necessarily the driving force. The extensive, sandy coast of the extreme south of Brazil has a tidal range of only 50 cm but inundation is affected more by seasonal variation in wind direction (Costa *et al.* 1988). As a warm temperate transition zone, it is influenced by the warm Brazil current in summer and the cold Falkland current in winter (Cordazzo & Seeliger 1988). South-easterly winds in winter and spring tend to back up the water and inundate the beach, while north-easterly winds in summer tend to expose the beach for long periods. On the shores of Lake Huron the annual pattern of disturbance is associated simply with winter storms and fine weather in summer (Maun 1984). In tropical and subtropical areas violent seasonal storms can be the main agent of disturbance (Barbour *et al.* 1987). For whatever reason, the retreat of the water also often means exposure to potentially desiccating winds, abrasion and engulfment by blown sand, and an aerosol of salt-spray. Cycles of accretion and erosion usually depend substantially on variations in wind direction and velocity (Harris & Davy 1986a).

The sand or shingle is fairly typical of a previously uncolonized substrate: it has little nutrient capital and a negligible organic matter content, except for any litter that may be cast up by the water and probably remains unincorporated. Inorganic nutrient supply, especially for nitrogen, depends on the local mineralization of litter and the deposition of sea-spray. The high porosity allows little water retention against gravity or evaporation. There is generally no protection from the full glare of solar irradiance and, in an essentially dry system (i.e. with a high Bowen ratio), this produces large diurnal temperature fluctuations. The amplitude of temperature variations may be considerably reduced under tidal litter. It is in the context of these physical constraints that we must assess the life-history and physiological characteristics of strandline plants.

DIVERSITY, GROWTH FORM AND REPRODUCTIVE STRATEGY

A few of the most widespread and characteristic species are summer annuals that can complete their life-cycle between the spring and autumnal equinoxes in a wide range of temperate climates. Some behave as ephemerals in less seasonal climates. These represent only three genera: in the Brassicaceae there is *Cakile* (e.g. *C. maritima, C. edentula, C. lanceolata*); in the Chenopodiaceae there is *Atriplex* (e.g. *A. laciniata, A. glabriuscula, A. subcordata, A. gmelinii*), and *Salsola kali*.

The majority of species that colonize strandlines though are iteroparous, clonal perennials with extensive rhizomatous or stoloniferous growth. These include many of the grasses that are also able to form embryo or main dunes: in Europe typically *Elymus farctus*, less often *Ammophila arenaria* and in the northern part

Leymus arenarius; in North America, *Leymus mollis, Ammophila breviligulata, Uniola paniculata, Spartina patens, Sporobolus virginicus* and *Panicum amarum*; in South America, *Sporobolus* spp., *Panicum racemosum, Spartina ciliata* and *Paspalum vaginatum*. Various clonal forbs are prominent, depending on climate and geography: *Honkenya peploides*, (Caryophyllaceae), *Crambe maritima* (Brassicaceae), *Calystegia soldanella* (Convolvulaceae), *Eryngium maritimum* (Apiaceae), *Euphorbia paralias* (Euphorbiaceae), *Lathyrus japonicus* (Leguminosae), *Mertensia maritima* (Boraginaceae), *Polygonum maritimum* (Polygonaceae), *Blutaparon portucaloides* (Amaranthaceae) and *Hydrocotyle bonariensis* (Apiaceae) (Doing 1985; Fernandez-Palacios *et al.* 1987; Cordazzo & Seeliger 1988). In the tropics and subtropics the lowest beach vegetation may be creeping vines (*Ipomoea pes-caprae, I. stolonifera, Canavalia rosea*) or even small shrubs (*Croton punctatus*) (Doing 1985; Barbour *et al.* 1987; Moreno-Casasola 1988).

Life-history theory predicts that the annual habit will only be favoured when adult survival is low, relative to juvenile survival (Watkinson & Davy 1985). On most strandlines the disturbance is such that the chance of an adult surviving from one year to the next is extremely low. This is certainly consistent with the success of *Cakile, Salsola* and *Atriplex*. The question remains as to why there are so many perennials? Three possibilities emerge. First, disturbance may not necessarily kill adults: evidence is presented below that, although wave action disrupts clonal perennials, the resulting fragments of rhizome or stolon may survive and act as propagules. Second, strandline populations may be maintained by the immigration of propagules from more stable environments higher on the beach. Third (in the tropics), violent disturbances may be less frequent than annual, such that the probability of adult survival exceeds that of seeds and seedlings.

DISPERSAL

Dispersal is of paramount importance in a linear habitat. The cosmopolitan distribution of many species or species complexes around the shores of great oceans suggests that dispersal by water can be very effective, even allowing for introductions by humans. The *Cakile* complex appears to be excluded only from the tropics. Longshore drift and water currents can disperse propagules from one strandline to another, but storms and wind can also redistribute propagules between different levels on the beach.

Long-distance dispersal

We know remarkably little about long-distance transport of the propagules in water, although there are reports of strandline plants having travelled great distances (Guppy 1906; Ridley 1930). Darwin (1859) noted that the seeds of many different species could withstand prolonged immersion in seawater and that some would also retain buoyancy long enough to be transported considerable distances.

Ignaciuk & Lee (1980) investigated the buoyancy and survival of fruits of the annuals in an agitated solution of sodium chloride (600 mmol l^{-1}). *Atriplex glabriuscula* showed the greatest potential for range dispersal; most of its fruits remained buoyant and viable after 30 days' immersion. *Atriplex laciniata* remained similarly viable but sank within a few days, while *Salsola kali* retained moderate viability and also sank. Only a small proportion (*c*. 4%) of *Cakile maritima* fruits remained buoyant after 30 days but these few would probably have floated for much longer. *Cakile* has unusual fruits that consist of two unequal single-seeded joints (Wright 1927; Barbour 1970). The larger distal segment detaches readily, whereas the proximal one remains attached to the dying plant. This has been interpreted as bet-hedging in response to a high degree of disturbance: distal segments are free dispersers but the proximal ones are retained in or near the proven environment (Barbour 1972). The issue of buoyancy is complicated. Both distal and proximal segments of *Cakile edentula* var. *lacustris* fruits show high floatability (65 and 88%, respectively, after 20 days' agitation) once they have overwintered. Agitation to simulate wave action appears to prolong floatation and it is extremely rare for *Cakile* to germinate while actually afloat (Payne & Maun 1981). *Cakile maritima* seeds are large and well-endowed with nitrogen, phosphorus and micronutrients (Hocking 1982).

Further evidence for dispersal comes from completely new strandlines. The island of Surtsey was created by a volcanic eruption that started on 14 November 1963 and lasted for *c*. 3½ years. Seeds that had drifted there were discovered on its strandline as early as the spring of 1964. Subsequent colonization by the first four species to arrive, all strandline species, is shown in Fig. 1. Twenty-three plants of *Cakile maritima* were seen growing in the summer of 1965. It has re-invaded intermittently since but has not established a self-maintaining population. *Leymus arenarius* first arrived in 1966, and *Honkenya peploides* and *Mertensia maritima* in 1967; all three maintain thriving populations and *Honkenya* is by far the most abundant of the 23 species of vascular plant now present (Fridriksson 1987). The nearest seed sources are other small volcanic islands 5·1, 11 and 20 km away and the Icelandic mainland (35 km).

Local dispersal

There is also good evidence for dispersal between the strandline and higher levels of the beach and dunes. Keddy (1981, 1982) examined the population biology of *Cakile edulenta* on an environmental gradient from the open sandy beach to densely vegetated dunes to the landward. He found that there was a substantial flux of fruits, carried by wind and waves, from the large, fecund plants at the seaward end towards the landward end, where seedling survival and reproductive output declined dramatically. The population at the landward end could only exist because of the annual dispersal of seeds from seaward. The difference equation model of Watkinson (1985) shows that the abundance of *Cakile edulenta* plants on the dune system as a whole is regulated by the density-dependent control of

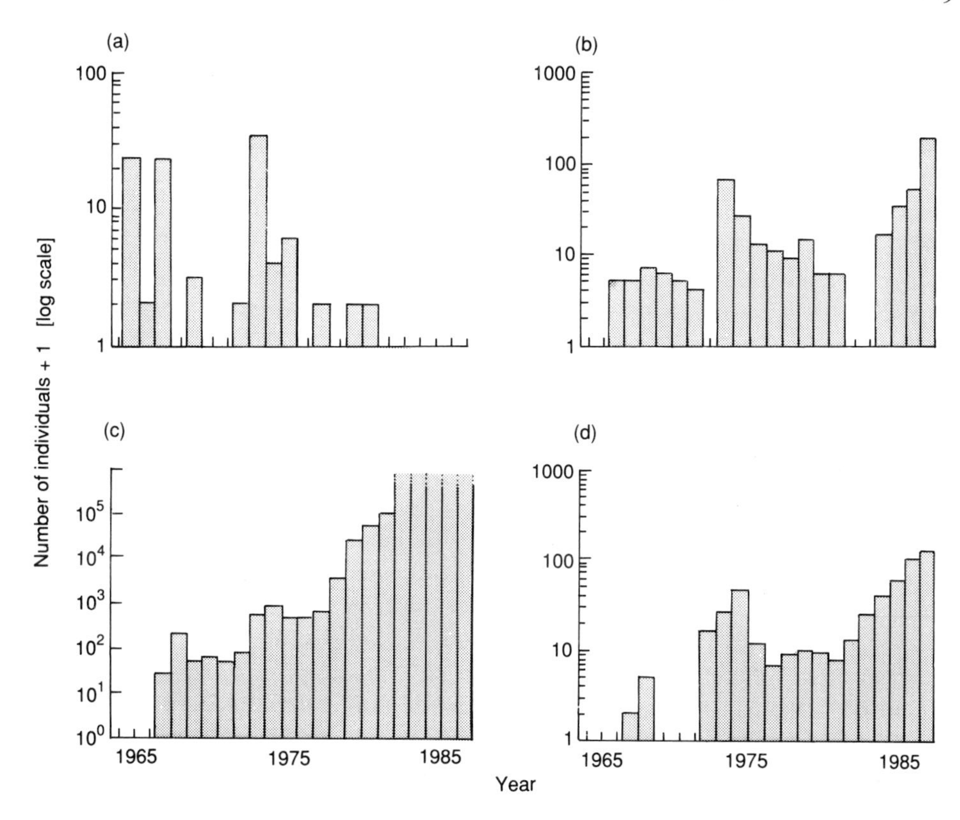

FIG. 1. Colonization of the newly formed volcanic island of Surtsey, Iceland (1965–86) by strandline species: (a) *Cakile maritima*; (b) *Leymus arenarius*; (c) *Honkenya peploides*; (d) *Mertensia maritima*. The total numbers of individuals present in the island in each year are represented. Data taken from Fridriksson (1987).

fecundity at the seaward end of the gradient; without the density-independent migration of seeds landward, populations could not persist in the areas where their densities are highest. Payne & Maun (1984) reported a very similar pattern of abundance of *Cakile edulenta* var. *lacustris* on the Lake Huron shoreline. A flux in the reverse direction may be important in clonal perennials. Both rhizome fragments and seeds of *Elymus farctus* are carried from the embryo dunes to strandlines lower on the shore by retreating storm tides on the north coast of Norfolk, Britain (Harris & Davy 1986a). There are similar reports for *Ammophila breviligulata* on the shore of Lake Huron (Maun 1984) and *Blutaparon portulacoides* on the coast of southern Brazil (Bernardi *et al.* 1987).

GERMINATION AND ESTABLISHMENT

Here we consider the germination and establishment of seeds and the establishment of vegetative disseminules in a broader sense.

Environment control of germination

Ignaciuk & Lee (1980) noted that the annuals *Atriplex glabriuscula, A. laciniata, Salsola kali* and *Cakile maritima* do not germinate in Britain until approximately a month after the spring equinox. They made a detailed analysis of the variety of factors that combine to prevent germination of mature seed in the autumn. *Cakile maritima* requires prolonged stratification to remove innate dormancy. Hocking (1982) concluded that the concentration of NaCl in the walls of recently matured fruits would be enough to inhibit germination until it was removed by leaching. Seed of *Atriplex* species is enclosed within persistent bracteoles and that of *Salsola kali* is similarly enclosed in a persistent perianth; both structures enforce dormancy in the autumn until they are weakened by decay and abrasion. Low temperature and salinity probably enforce dormancy in winter and early spring. *Atriplex* only germinates well in alternating temperatures and so the greater amplitudes of later spring would promote germination then; this response would also tend to prevent germination at depths beyond its capacity to emerge (Ignaciuk & Lee 1980). A number of other important beach species respond similarly to alternating temperatures (e.g. Barbour *et al.* 1985). Germination of *Salsola kali* is stimulated by moderate salinity, or short exposure to high salinity, both at low temperature. Barbour (1970) found that germination of *Cakile maritima* was inhibited by light. It is possible that this will prevent the germination of imbibed seed that have been exposed by sand movement but, as Barbour *et al.* (1985) point out, there are other North American beach plants whose germination is stimulated by light (*Physalis viscosa, Erigeron canadensis, Iva imbricata* and *Solidago sempervirens*) and most species are light neutral.

Response to burial

Burial by sand is one of the major hazards of the strandline, the more so as succession develops. The ability of seeds to germinate and emerge from depth is directly related to their size. Barbour *et al.* (1985) presented several relevant observations: seeds of California beach species were on average 73% heavier than their backdune counterparts; the heavier-seeded *Ipmomoea pes-caprae* tends to invade rapidly prograding beaches in preference to the *I. stolonifera* found elsewhere; most telling of all, there was a linear relationship between the logarithm of seed mass and the maximum depth from which successful establishment could take place for eight common beach species.

A small percentage of *Cakile maritima* seedlings can emerge after germinating at a depth of 160 mm, and *Atriplex laciniata* and *Salsola kali* can emerge well from 80 mm (Lee & Ignaciuk 1985). Harris & Davy (1986b) found that the maximum depth of burial that caryopses of *Elymus farctus* could emerge from was between 127 and 178 mm, with more than 10% able to emerge from the former depth (Fig. 2a). The capacity of buried rhizome fragments to produce emergent shoots was even more impressive, as would be expected from their greater reserves of

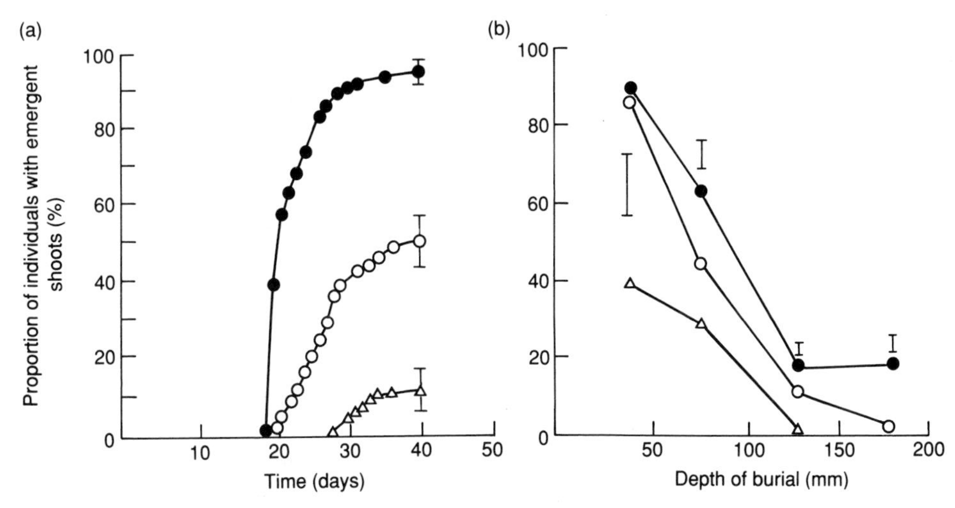

FIG. 2. Comparison of the ability of caryopses and rhizome fragments of *Elymus farctus* to produce emergent shoots after experimental burial with sand: (a) Caryopses: emergence over time from caryopses buried at depths of 38 mm (●), 76 mm (○) and 127 mm (△) (there was no emergence from 178 mm); (b) Rhizome fragments: final emergence from the same depths from three-node (●), two-node (○) and one-node (△) rhizome fragments, 120 days after burial. Vertical bars represent back-transformed SE derived from arc-sin transformed data. From Harris & Davy (1986b).

carbohydrates and nutrients. About 20% of three-node fragments were able to emerge from a depth of 178 mm within 120 days (Fig. 2b). Clearly, the larger fragments were better able to produce emergent shoots from greater depths. Greater vigour of multi-noded fragments was independent of initial fresh weight per node and suggests that dominance hierarchies were set between buds leading to more efficient use of resources: there was a strong tendency for the distal bud to develop preferentially, eventually inhibiting development of the others. Studies of single-node fragments collected over the course of a year and incubated at constant temperatures showed a peak of regenerative potential in autumn and winter (October–February). Careful excavation allows the origins of colonists to be ascertained in the field (Harris & Davy 1986a). In January 1978 a catastrophic storm surge on the north Norfolk coast of Britain obliterated strandline communities and eroded the embryo dunes. Six months later, colonizing clumps of *Elymus farctus* in the strandline were found to have originated in strikingly similar proportions from seeds and rhizome fragments: 49.3% of 367 clones in five sites, occupying a total of nearly 300 m^2, had been derived from rhizome fragments. The main difference between them was that clumps derived from rhizome fragments had, on average, nearly twice as many tillers as their seedling counterparts (see Table 1). Another important feature is the variation in size of the successful rhizome fragments: they ranged up to 600 mm in length but, with a skewed distribution, such that the mode was only 125 mm (Fig. 3). This is clearly much greater plasticity of biomass and meristem number than is seen in seeds, and the larger fragments would have had the potential to emerge from considerable depth

TABLE 1. Mode of recolonization of the strandline by *Elymus farctus* at Holkam National Nature Reserve in Norfolk, 6 months after the storm surge of January 1978. Values are mean (±SE) of six replicate samples of varying size, with a total area of 299 m^2. A total of 367 clones was excavated. Data from Harris & Davy (1986a)

Total clone density (m^{-2})	1.29 ± 0.35
Clone density from seed (m^{-2})	0.69 ± 0.30
Clone density from rhizome fragments (m^{-2})	0.60 ± 0.25
Total tiller density (m^{-2})	2.44 ± 0.60
Tiller density from seed (m^{-2})	1.02 ± 0.46
Tiller density from rhizome fragments (m^{-2})	1.42 ± 0.31
Rhizome bud density (m^{-2})	4.10 ± 1.11
Number of tillers per seed	1.39 ± 0.09
Number of tillers per rhizome fragment	2.41 ± 0.37
Number of tillers per rhizome bud	0.39 ± 0.08

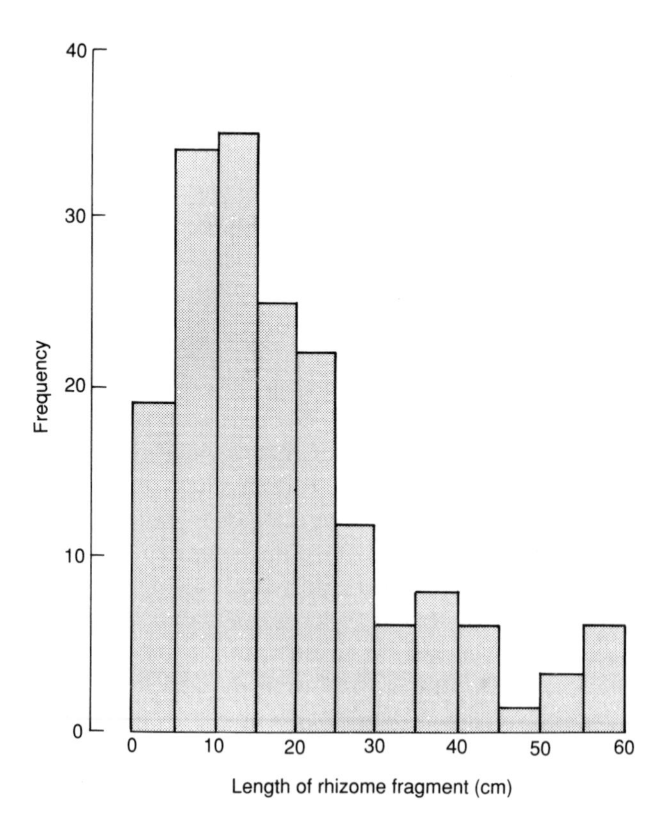

FIG. 3. Frequency distribution of rhizome fragment lengths of *Elymus farctus* excavated from Holkam National Nature Reserve, Norfolk in July 1978. These fragments had successfully recolonized the strandline during the 6 months after the catastrophic storm surge of 11 January 1978. From Harris & Davy (1986a).

of burial. Fragments carried on average seven buds, the majority of which had remained dormant as a 'bud bank' during establishment (Harris & Davy 1986a). Plants from fragments presumably represent locally selected genotypes, whereas seedlings offer at least the potential for greater genetic diversity.

Risks from burial do not end with emergence. Repeated episodes of burial and re-emergence are common, as the wind oscillates in direction and velocity. Yet we have little insight in the physiological responses to such burial. Newly established seedlings that have committed their reserves of nutrients and carbon are more vulnerable to temporary burial than either ungerminated seeds or well-established plants. Two week-old seedlings of *Elymus farctus* were able to survive 1 week of experimental burial to a depth of 60 mm with sand, but all those buried for 2 weeks perished. Although there was respiratory loss of dry mass during the week of burial, relative growth rates rapidly increased after uncovering to match those of unburied control plants. Seedlings appear to maintain leaf mass and function at the expense of roots and stems (Harris & Davy 1987) and the deleterious effects of burial are greater than those of simply keeping plants in darkness. Leaves had much higher concentrations of water-soluble carbohydrate than those of controls during burial and up to 20 days after uncovering. Equivalent losses from roots and stems suggest a reversal of the normal source–sink relationships in order to protect the leaves. Photosynthetic competence, measured by gas-exchange, returned in full within 24 h of uncovering, whereas the ability to translocate ^{14}C to stems, roots and newly-expanding leaves recovered more slowly. Proportional allocation of nitrogen, phosphorus and potassium to the leaves was also increased after burial, relative to controls (Harris & Davy 1988).

Effects of substrate texture

Strandline colonists are found on wave-sorted sediments ranging from fine sand to course gravel and shingle. We could predict that coarse-textured sediments also select for large-seeded species, partly because small seeds can be easily washed to greater depths in the interstices than they can later emerge from, and partly because of adverse water relations (large reserves allow rapid, deep root growth during establishment). Typical colonists of coastal shingle — *Lathyrus japonicus, Crambe maritima* and *Honkenya peploides* — are all exceptionally large seeded. Keddy & Constabel (1986) examined the germination responses of 10 lakeshore species of different seed size to sediments ranging from 0·13 to 16 mm diameter. There were strong responses to particle size when the water-table was at least 4 cm below the surface, with more recruitment on the fine sediments. Large-seeded species had the broadest tolerances for variation in sediment particle size.

GROWTH, PHYSIOLOGY AND SURVIVAL

Information on the survival of strandline plants after establishment is sparse and indicates large variations from place to place and from year to year. The physiological properties and adaptations that affect the ability to survive in such a physically adverse environment have attracted much work (Barbour *et al.* 1985; Lee & Ignaciuk 1985; Rozema *et al.* 1985).

Responses to salt-spray and soil salinity

It has been argued frequently that differential susceptibility of species to salt-spray, coupled with a sharp gradient of decreasing influence of salt-spray away from the sea, are the main determinants of coastal zonation and succession (Oosting 1945; Boyce 1954; Barbour & De Jong 1977; Barbour 1978; Rozema *et al.* 1983; Sykes & Wilson 1988). Van der Valk (1974) concluded that salt-spray played a secondary role, with burial by sand as the primary mechanism excluding non-coastal species from foredunes. The main toxic effect is of accumulating Cl^- ions absorbed through the epidermis or stomata (Barbour *et al.* 1985). Coastal strandline species are indisputably very tolerant of salt-spray in comparison with inland species. *Elymus farctus*, which occupies the zone where salt-spray deposition is highest, is among the most tolerant of species (Rozema *et al.* 1983; Sykes & Wilson 1988), probably as a result of its cuticular structure which resists entry of Cl^-. *Cakile maritima, Salsola kali* and *Atriplex* spp. are similarly tolerant but in this case at least partly because they respond to foliar entry of Cl^- with hypertrophy and become succulent; under conditions of low sand fertility, salt-spray actually stimulates dry matter production (Rozema *et al.* 1982). It is easy to forget that seawater is a dilute nutrient solution, notwithstanding the dominating influence of NaCl.

The role of soil salinity is rather different. Kearney (1904) concluded that beach plants were not generally true halophytes and a substantial body of more recent work has borne this out (see Barbour *et al.* (1985) for a review). Except in arid climates, salt is readily leached from sand or shingle strandlines. Nevertheless, strandline colonizers do cope with ionic and osmotic consequences of moderate salinities in the soil solution, generally in the range 10–130 mmol l^{-1} NaCl (Ignaciuk & Lee 1980; Rozema *et al.* 1983; Pakeman 1990). Barbour & De Jong (1977) periodically inundated beach species with seawater and then removed the seawater by leaching. All were more tolerant than a control glycophyte and those species closest to the shoreline were most tolerant. Root growth in seedlings of *Cakile maritima* is not inhibited by salinities up to 28% those of seawater (Barbour 1970). Benecke (1930) was the first to record the relatively high salt tolerance of *Elymus farctus*. Sykes & Wilson (1989) reported that *Elymus farctus*, an introduced species, was the most tolerant of 29 New Zealand dune species tested. On the other hand, Rozema *et al.* (1983) found that *Elymus farctus* was less tolerant of soil salinity than its foredune congener *Elymus pycnanthus* in The Netherlands, and they consider salt-spray to be the more important factor in the relative distributions of the two taxa.

Water relations

Investigation of water availability on sandy beaches (De Jong 1979) has yielded some unexpected results. Water is generally available at potentials of > −1.5 MPa at depths below 30 cm and plants show little sign of water stress, as indicated by dawn xylem sap water potentials. Such low stress can only be explained if there is upward movement of water from considerably deeper layers, possibly from the relatively non-saline water-table. Capillary rise would be too slow to account for the water content of dune sand, but movement could be as water vapour along a temperature gradient, as 'internal dew' (Olsen-Seffer 1909), or as liquid along a gradient of water potential. As De Jong (1979) points out, the water-holding capacity of sand is so low that relatively small fluxes could maintain high water potentials. He suggests several distinct strategies for the Californian beach: tap-rooted species (such as *Ambrosia chamissonis*) can exploit the water-table; species with large diurnal fluctuations of water potential and spreading, shallower root systems (such as *Atriplex leucophylla*) use precipitation and upward water movement; ephemerals (such as *Cakile maritima*) can complete their life-cycle before the dry season.

Both C3 and C4 plants are widespread and successful colonists in strandline communities in a wide range of climates (De Jong 1978; Lee & Ignaciuk 1985; Barbour *et al.* 1987). Two of the four British annuals (*Salsola kali* and *Atriplex laciniata*) have the C4 pathway, which is disproportionately high for the British flora, in an admittedly very small sample. The greater water-use and nitrogen-use efficiency of the C4 system seem only to be decisive at high temperatures. This is supported by a survey of the Gulf of Mexico beach vegetation showing a much higher proportion of C4 taxa than along temperate Atlantic and Pacific coasts (Barbour *et al.* 1987). Beach plants generally have leaf structures and photosynthetic light response curves that allow them to exploit high solar irradiances (Barbour *et al.* 1985).

Nitrogen nutrition

The sand and shingle cast up by the sea is practically devoid of organic matter and contains only the nitrogen present in films of seawater on the surface of the particles. A certain amount of algal foam may also be blown onshore. This lack of nutrient capital is characteristic of the starting point of primary succession; but where strandlines differ from most others is in the subsequent import and deposition of floating organic matter by tides or storms. On many beaches, especially where rocky headlands are nearby, this litter takes the form of detached thalli of sublittoral algae, which are rapidly decomposed to release nitrogen and other nutrients. Presumably alginates released from them assist in sand stabilization and water retention, although no work appears to have been done on this. Lee & Ignaciuk (1985) found that the peak soluble nitrogen content of litter bags containing algal material occurred in late May, coinciding with the peak growth rates of strandline annuals. Very large variations in individual plant size depend on the distribution of the litter. Lee *et al.* (1983) show that the standing crop of *Salsola kali* was vastly greater within the drift zone than above or below it, and the growth of individual plants was 100-fold greater in the drift zone. *Salsola kali, Cakile maritima, Atriplex glabriuscula* and *A. laciniata* all showed responses that were primarily due to nitrogen in nutrient-addition experiments. This corresponds well with the very high activities of nitrate reductase that are inducible in *Cakile maritima*; those of the perennial *Leymus arenarius* are substantially lower (Garcia Novo 1976). Single-node fragments of *Elymus farctus* rhizome responded to nitrogen consistently throughout the year; the ability to produce shoots and roots was substantially enhanced by incubation with 15 mmol l^{-1} KNO_3 rather than distilled water (Harris & Davy 1986b).

Lathyrus japonicus supports symbiotic di-nitrogen fixation in its root nodules and presumably augments the nitrogen capital of the strandline communities it is a component of. Judging from its vigour and colour, *Leymus arenarius* in Iceland can benefit from this nitrogen (S. Greipsson, pers. comm.). We might also speculate that di-nitrogen fixation by free-living bacteria at the root/sand interface occurs in strandlines. Mycotrophy may be significant, particularly after the initial stages of colonization. *Salsola kali* is definitely non-mycotrophic: inoculation with vesicular-arbuscular (VA) mycorrhizal fungi induces an incompatible response in the roots, which leads to reduced growth and survival of the host (Allen *et al.* 1989). Strandline grasses, like most others, support VA mycorrhiza when inoculum is available.

CONCLUSIONS

The strandline is sandwiched between the foreshore and the foredune. It is a transient habitat, distinct from its neighbouring habitats, but equally capable of reverting to one or developing into the other in a very short time, depending on the impact and frequency of storms and tides. In this respect it must be admitted that strandlines do not match the normal concept of primary succession: repeated colonizations of virgin territory can occur at one place. Neither do strandline colonizers conform to the general model of a primary successional colonizer, which has a profusion of light, wind-dispersed seeds with minimal dormancy; inevitably the small seedlings are rather vulnerable (Huston & Smith 1987; Chapin 1993). Their propagules have a similar capacity for long-range dispersal, but the usual association of this with a small mass is uncoupled because water is the agent of transport. Without this constraint, it may be assumed that selection pressures associated with substrate instability and nutrient poverty have favoured large seeds that produce well-endowed seedlings. The frequently observed seed dormancy on strandlines allows dispersal in time as well as in space and is a typical response to extreme environmental unpredictability in other habitats (e.g. Venable & Lawlor 1980). Strandline species also are typically not woody (except perhaps in the tropics), unlike the primary colonizers of many other habitats

(Chapin 1993). Lichens and bryophytes are conspicuously absent from strandlines.

What makes a successful strandline colonizer? An appropriate life-history is crucial, but this may be one of at least two contrasting types: summer annual/ephemeral or clonal perennial. Strandline species with these two different life-histories do nevertheless have identifiable determinants of fitness in common. Perhaps foremost of these is extreme flexibility of growth and reproduction, in response to the unpredictability of the abiotic environment in time and space. An individual *Cakile edentula* plant may produce a few seeds or nearly 2000 at the end of its 5 month life, depending on nutrient availability and the proximity of neighbours (Keddy 1981; Payne & Maun 1984). Such plasticity is evident also in the enormous growth responses of *Salsola kali* and *Cakile maritima* to variations in their nitrogen supply (Lee *et al.* 1983; Pakeman 1990). In clonal perennials the rate and pattern of development of rhizome or stolon systems, with their banks of dormant buds, are similarly responsive to local conditions. The apparent ability of *Elymus farctus* to reverse the usual source–sink relationships, so as to protect its leaves during burial, indicates flexibility at a physiological level also (Harris & Davy 1988). Both life-history types have relatively large propagules (seeds or clonal fragments with buds) that are endowed with nutrient reserves sufficient to permit emergence from 100 to 200 mm depth and then establishment in the face of a minimal external nutrient supply. These same propagules are generally capable of floating long enough to have a moderate probability of being washed up on a strandline alive; some can clearly travel great distances in water and found new colonies.

Strandline communities are relatively simple and are dominated by the fluctuating forces of the physical environment. The distribution of individuals tends to be irregular, often sparse and to a large measure fortuitous. By implication, biotic interactions may be weak. In such a harsh environment the primary colonists would be expected to facilitate the establishment of subsequent invaders. For instance, on Surtsey, *Leymus arenarius* seedlings become established mainly in patches already stabilized by *Honkenya peploides* (Fridriksson 1987). *Salsola kali* seems to facilitate the establishment of perennial grasses even on newly regraded subsoil at a semi-arid inland site, where presumably there is no sand accretion (Allen & Allen 1988). Overall, there is little evidence that competition between species is a major determinant of the structure of strandline communities. Nevertheless, the striking spread and productivity of *Ammophila arenaria* along the Pacific coast of North America, since its introduction from Europe in 1869, appears to have been partly at the expense of the native perennial grass, *Leymus mollis* (Pavlik 1983a, b, c). Strandline plants persist well in embryo dunes but it is not entirely clear why they tend to be eliminated from the later stages of sand-dune succession. The most probable explanation is that they are engulfed: they cannot cope with the sand accretion rates associated with the rapid vertical and horizontal rhizome growth of the main dune building grasses. To the extent that the invading dune-building grasses promote high sand accretion rates, they could be regarded as exerting a biotic effect on the strandline colonizers.

There is more evidence of competition within species on the strandline. The only species to receive thorough demographic study (*Cakile edentula*) shows strong negative density-dependence in its seed production under certain circumstances (Keddy 1981, 1982; Payne & Maun 1984). Indeed, such density-dependence at the seaward limit is capable of interacting with migration and mortality rates to regulate abundance throughout the beach and foredunes (Watkinson 1985). It is probable that such interactions are significant only where the deposition of organic litter during winter, or near the spring equinox, provides a large pulse of mineralized nitrogen early in the growing season. Elsewhere, the nutrient-poverty of sand or shingle would be unlikely to allow plants to attain sufficient size for interaction. This highlights an important difference between the strandline and the usual conception of the initial stages of primary succession: an initial pulse of nutrients, even though often heterogeneously distributed, may provide a critical capital of nitrogen and phosphorus that can be acquired only very slowly from precipitation, weathering and nitrogen fixation in other ecosystems. It is necessary to know much more about the nutrient budgets of strandlines and their influence on subsequent succession.

The parallel patterns of colonization on lakeshores and maritime strandlines suggest that tolerance of salt-spray and soil salinity may be of secondary importance. A good case in point arises from comparison of *Cakile edentula* along the Pacific coast and shores of the Great Lakes; this suggests that the divergence in tolerance of salt-spray between maritime and lacustrine forms has evolved only in the last 9000 years (Boyd & Barbour 1986). Salt tolerance is an attribute of maritime plants in a wide variety of habitats. The corollary of this is perhaps controversial: we suggest that the traditional emphasis placed on investigating static gradients of salinity as determinants of zonation in coastal vegetation is likely to contribute little to our understanding of primary succession. The dynamics of colonization process, from dispersal through establishment and growth to reproduction, may offer a more profitable approach for the future.

ACKNOWLEDGMENTS

We thank Dr C. S. B. Costa and Dr Sigurdur Greipsson for information and stimulating discussion, and Dr M. G. Barbour for comments on a draft of the manuscript.

REFERENCES

Allen, E.B. & Allen, M.F. (1988). Facilitation of succession by the nonmycotrophic colonizer *Salsola kali* (Chenopodiaceae) on a harsh site: effects of mycorrhizal fungi. *American Journal of Botany*, **75**, 257–66.

Allen, M.F., Allen, E.B. & Friese, C.F. (1989). Responses of the non-mycotrophic plant *Salsola kali* to invasion by vesicular-arbuscular mycorrhizal fungi. *New Phytologist*, **111**, 45–9.

Barbour, M.G. (1970). Germination and early growth of the strand plant *Cakile maritima*. *Bulletin of the Torrey Botanical Club*, **97**, 13–22.

Barbour, M.G. (1972). Seedling establishment of *Cakile maritima* at Bodega Head, California. *Bulletin of the Torrey Botanical Club*, **99**, 11–6.

Barbour, M.G. (1978). Salt spray as a microenvironmental factor in the distribution of beach plants at Point Reyes, California. *Oecologia*, **32**, 213–24.

Barbour, M.G. & De Jong, T.M. (1977). Response of West Coast beach taxa to salt spray, seawater inundation, and soil salinity. *Bulletin of the Torrey Botanical Club*, **104**, 29–34.

Barbour, M.G., De Jong, T.M. & Pavlik, B.M. (1985). Marine beach and dune plant communities. *Physiological Ecology of North American Plant Communities* (Ed. by B.F. Chabot & H.A. Mooney), pp. 296–322. Chapman & Hall, New York.

Barbour, M.G., Rejmanek, M., Johnson, A.F. & Pavlik, B.M. (1987). Beach vegetation and plant distribution patterns along the northern Gulf of Mexico. *Phytocoenologia*, **15**, 201–33.

Benecke, W. (1930). Zur biologie der strand- und dunenflora, 1. Vergleichende versuche uber die salztoleranz von *Ammophila arenaria* Link, *Elymus arenarius* L. und *Agriopyrum junceum* L. *Berichte der Deutschen Botanischen Gesellschaft*, **148**, 128–39.

Bernardi, H., Cordazzo, C.V. & Costa, C.S.B. (1987). Efeito de ressacas sobre *Blutaparon portulacoides* (St Hill.) Mears, nas dunas costeiras do sul do Brasil. *Ciencia e Cultura*, **39**, 545–7.

Boyce, S.G. (1954). The salt spray community. *Ecological Monographs*, **24**, 29–67.

Boyd, R.S. & Barbour, M.G. (1986). Relative salt spray tolerance of *Cakile edentula* from lacustrine and marine beaches. *American Journal of Botany*, **73**, 236–41.

Chapin III, F.S. (1993). Physiological controls over plant establishment in primary succession. *Primary Succession on Land* (Ed. by J. Miles & D.W.H. Walton), pp. 161–78. Blackwell Scientific Publications, Oxford.

Cordazzo, C.V. & Seeliger, U. (1988). Phenological and biogeographical aspects of coastal dune plant communities in southern Brazil. *Vegetatio*, **75**, 169–73.

Costa, C.S.B., Seeliger, U. & Kinas, P.G. (1988). The effect of wind velocity and direction on the salinity regime of the Lower Patos Lagoon estuary. *Ciencia e Cultura*, **40**, 909–12.

Cowles, H.C. (1899). The ecological relations of the vegetation on the sand dunes of Lake Michigan. *Botanical Gazette*, **27**, 95–117, 167–202, 281–308, 361–91.

Darwin, C. (1859). *The Origin of Species*. John Murray, London.

De Jong, T.M. (1978). Comparative gas exchange of four California beach taxa. *Oecologia*, **34**, 343–51.

De Jong, T.M. (1979). Water and salinity relations of Californian beach species. *Journal of Ecology*, **67**, 647–63.

Doing, H. (1985). Coastal fore-dune zonation and succession in various parts of the world. *Vegetatio*, **61**, 65–75.

Fernandez-Palacios, J.M., Martos, M.J. & Figueroa, M.E. (1987). Estructura de la vegetacion dunar en la flecha litoral del Rompido (Huelva). *Actas de la Real Sociedad Espanola de Historia Natural*, **8**, 375–82.

Fridriksson, S. (1987). Plant colonization of a volcanic island, Surtsey, Iceland. *Arctic and Alpine Research*, **19**, 425–31.

Fuller, R.M. (1987). Vegetation establishment on shingle beaches. *Journal of Ecology*, **75**, 1077–89.

Garcia-Novo, F. (1976). Ecophysiological aspects of the distribution of *Elymus arenarius* and *Cakile maritima* on the dunes at Tents-Muir point (Scotland). *Oecologia Plantarum*, **11**, 13–24.

Guppy, H.B. (1906). *Observations of a Naturalist in the Pacific between 1869–1899. Plant Dispersal*. Macmillan, London.

Harris, D. & Davy, A.J. (1986a). Strandline colonization by *Elymus farctus* in relation to sand mobility and rabbit grazing. *Journal of Ecology*, **74**, 1045–56.

Harris, D. & Davy, A.J. (1986b). The regenerative potential of *Elymus farctus* from rhizome fragments and seed. *Journal of Ecology*, **74**, 1057–67.

Harris, D. & Davy, A.J. (1987). Seedling growth in *Elymus farctus* after episodes of burial with sand. *Annals of Botany*, **60**, 587–93.

Harris, D. & Davy, A.J. (1988). Carbon and nutrient allocation in *Elymus farctus* seedlings after burial with sand. *Annals of Botany*, **61**, 147–57.

Harshberger, J.W. (1900). An ecological study of the New Jersey strand flora. *Proceedings of the Academy of Natural Sciences of Philadelphia*, **52**, 623–71.

Hocking, P.J. (1982). Salt and mineral nutrient levels in fruits of two strand species, *Cakile maritima* and *Arctotheca populifolia*, with special reference to the effect of salt on the germination. *Annals of Botany*, **50**, 335–43.

Huston, M. & Smith, T. (1987). Plant succession: life history and competition. *American Naturalist*, **130**, 168–98.

Ignaciuk, R. & Lee, J.A. (1980). The germination of four annual strand-line species. *New Phytologist*, **84**, 581–91.

Kearney, T.H. (1904). Are plants of sea beaches and dunes true halophytes? *Botanical Gazette*, **37**, 424–36.

Keddy, P.A. (1981). Experimental demography of the sand-dune annual *Cakile edentula*, growing along an environmental gradient in Nova Scotia. *Journal of Ecology*, **69**, 615–30.

Keddy, P.A. (1982). Population ecology on an environmental gradient: *Cakile edentula* on a sand dune. *Oecologia*, **52**, 348–55.

Keddy, P.A. & Constabel, P. (1986). Germination of ten shore-line plants in relation to seed size, soil particle size and water level: an experimental study. *Journal of Ecology*, **74**, 133–41.

Lee, J.A. & Ignaciuk, R. (1985). The physiological ecology of strandline plants. *Vegetatio*, **62**, 319–26.

Lee, J.A., Harmer, R. & Ignaciuk, R. (1983). Nitrogen as a limiting factor in plant communities. *Nitrogen as an Ecological Factor* (Ed. by J.A. Lee, S. McNeil & I.H. Rorison), pp. 94–112. Blackwell Scientific Publications, Oxford.

Maun, M.A. (1984). Colonizing ability of *Ammophila breviligulata* through vegetative regeneration. *Journal of Ecology*, **72**, 565–74.

Moreno-Casasola, P. (1988). Patterns of plant species distribution on coastal dunes along the Gulf of Mexico. *Journal of Biogeography*, **15**, 787–806.

Olsson-Seffer, P. (1909). Hydrodynamic factors influencing plant life on sandy sea shores. *New Phytologist*, **8**, 37–49.

Oosting, H.J. (1945). Tolerance to salt spray of plants of coastal dunes. *Ecology*, **26**, 85–9.

Pakeman, R.J. (1990). *Mineral nutrition of strandline annuals*. PhD thesis, University of Manchester.

Pavlik, B.M. (1983a). Nutrient and productivity relations of the dune grasses *Ammophila arenaria* and *Elymus mollis*. I. Blade photosynthesis and nitrogen use efficiency in the laboratory and field. *Oecologia*, **57**, 227–32.

Pavlik, B.M. (1983b). Nutrient and productivity relations of the dune grasses *Ammophila arenaria* and *Elymus mollis*. II. Growth and patterns of dry matter and nitrogen allocation as influenced by nitrogen supply. *Oecologia*, **57**, 233–8.

Pavlik, B.M. (1983c). Nutrient and productivity relations of the dune grasses *Ammophila arenaria* and *Elymus mollis*. III. Spatial aspects of clonal expansion with reference to rhizome growth and the dispersal of buds. *Bulletin of the Torrey Botanical Club*, **110**, 271–9.

Payne, A.M. & Maun, M.A. (1981). Dispersal and floating ability of dimorphic fruit segments of *Cakile edentula* var. *lacustris*. *Canadian Journal of Botany*, **59**, 2595–602.

Payne, A.M. & Maun, M.A. (1984). Reproduction and survivorship of *Cakile edentula* var. *lacustris* along the lake Huron shorline. *American Midland Naturalist*, **111**, 86–95.

Ridley, H.N. (1930). *The Dispersal of Plants throughout the World*. Reeve & Co., Kent.

Rozema, J., Bijl, F., Dueck, T. & Wesselman, H. (1982). Salt-spray stimulated growth in strand-line species. *Physiologia Plantarum*, **56**, 204–10.

Rozema, J., Manen, Y.van, Vugts, H.F. & Leusink, A. (1983). Airborne and soilborne salinity and the distribution of coastal and inland species of the genus *Elytrigia*. *Acta Botanica Neerlandica*, **32**, 447–56.

Rozema, J., Bijwaard, P., Prast, G. & Broekman, R. (1985). Ecophysiological adaptations of coastal halophytes from foredunes and salt marshes. *Vegetatio*, **62**, 499–521.

Sykes, M.T. & Wilson, J.B. (1988). An experimental investigation into the response of some New Zealand sand dune species to salt spray. *Annals of Botany*, **62**, 159–66.

Sykes, M.T. & Wilson, J.B. (1989). The effect of salinity on the growth of some New Zealand sand dune species. *Acta Botanica Neerlandica*, **38**, 173–82.

van der Valk, A.G. (1974). Environmental factors controlling the distribution of forbs on coastal foredunes in Cape Hatteras National Seashores. *Canadian Journal of Botany*, **52**, 1057–73.

Venable, D.L. & Lawlor, L. (1980). Delayed germination and dispersal in desert annuals: escape in time and space. *Oecologia*, **46**, 271–82.

Watkinson, A.R. (1985). On the abundance of plants along an environmental gradient. *Journal of Ecology*, **73**, 569–78.

Watkinson, A.R. & Davy, A.J. (1985). Population biology of salt marsh and sand dune annuals. *Vegetatio*, **62**, 487–97.

Wright, J. (1927). Notes on strand plants. II. *Cakile maritima*, Scop. *Transactions of the Botanical Society of Edinburgh*, **29**, 389–401.

Plant distribution patterns and primary succession on a glacier foreland: A comparative study of cryptogams and higher plants

H. J. CROUCH*

School of Environmental Sciences, University of Greenwich, Rachel McMillan Building, Creek Road, Deptford, London SE8 3DU, UK

SUMMARY

1 Species composition was recorded at 190 sites located by an objective sampling design on and outside the glacier foreland of Storbreen, southern Norway. Cryptogam and higher plant data were analysed separately and collectively, using mapping, classification and ordination techniques.

2 Time-dependent distribution patterns were common, and a successional sequence of species groups defined by two-way indicator species analysis (TWINSPAN) was evident for both cryptogams and higher plants. In each case, species groups on the oldest terrain were relatively distinct. Results support the contention that there are three major successional stages at Storbreen.

3 Cryptogam and higher plant species groups show little correspondence when plotted on a common detrended correspondence analysis ordination diagram. Furthermore, maps of community types defined by TWINSPAN for cryptogam, phanerogam and combined data sets only partially correspond. Percentage correspondence decreases as the level of division increases.

4 There is evidence for much individualistic behaviour of species on the Storbreen glacier foreland, supporting the validity of a population approach. Yet many species exhibit similarity in their distribution patterns, which suggests that a community approach is not inappropriate. The study of separate cryptogam and higher plant communities is considered expedient.

INTRODUCTION

Glacier forelands – the areas exposed by glacial retreat in historic times – provide an excellent opportunity for the study of primary plant succession (e.g. Cooper 1923; Crocker & Major 1955; Viereck, 1966; Persson 1964; Archer 1973; Richard 1973; Elven 1978; Birks 1980; Matthews & Whittaker 1987). Most studies of foreland succession have employed a chronosequence approach, in which vegetation

* Present address: The Library, University of Reading, Whiteknights, Reading, Berkshire RG6 2AE, UK.

is sampled along a transect perpendicular to the glacier. Variation in vegetation with terrain age is taken to represent a successional sequence through time. This approach is limited by the assumption that there is no effective environmental variation between sites of the same age. Matthews (1976, 1978a, 1979a, b), Whittaker (1985, 1987, 1989) and Matthews & Whittaker (1987) developed a spatial approach to vegetation chronosequences on the Storbreen glacier foreland, southern Norway, which avoids this limitation. The spatial approach involves examination of vegetation at sites distributed two-dimensionally. This allows greater separation of the effects of time and environmental factors on vegetation than the one-dimensional chronosequence approach. A spatial approach has been employed in the present study.

The degree to which cryptogams have been included in such studies varies. At Storbreen, which is also the site of the present study, Matthews' and Whittaker's investigations have been confined to higher plants. Other authors have included cryptogams in the discussion of glacier foreland succession, without making any distinction between cryptogam and phanerogam successions. Relatively few have studied succession of the two groups separately. Bryophytes, lichens, blue-green algae and other algae are the main constituents of the 'black crust' phenomenon investigated by Worley (1973) in the early stages of succession at Glacier Bay, Alaska. Faegri (1933) described nine 'Kryptogamengesellschaften' (cryptogam communities) on the foreland of Nigardsbreen, southern Norway. Stork (1963) found that in front of glaciers in the Tarfala valley, northern Sweden, moss and lichen communities 'constitute special cryptogamic successions'.

That cryptogams form discrete communities which may only partially correspond with higher plant communities has been demonstrated by Alpert & Oechel (1982). Studying cryptogam distributions along a snow-accumulation gradient in Alaska, they found that bryophyte species and communities display distributions which are partially independent of vascular vegetation. They suggest that analysis of cryptogam and vascular vegetation 'cannot always be successful if one attempts to fit both into a single system'.

This chapter examines the extent to which cryptogams and phanerogams differ in their distribution patterns, and hence their successional sequences, on the glacier foreland of Storbreen, southern Norway.

STUDY SITE

Storbreen is a valley glacier situated on the eastern side of the Smørstabb massif in Leirdalen, western Jotunheimen, central southern Norway (61°35′N, 8°11′E) (Fig. 1). The study area, which includes the glacier foreland up to an altitude of 1400 m, has a generally north-easterly aspect and an altitudinal range of 300 m. Available climatic data indicate a mean annual temperature ranging from about −0·6°C at 1100 m to −2·2°C at 1400 m. Mellor (1985) estimated mean annual precipitation to be approximately 1550 mm. The parent material of the foreland consists chiefly of pyroxene-granulite gneiss and mylonite.

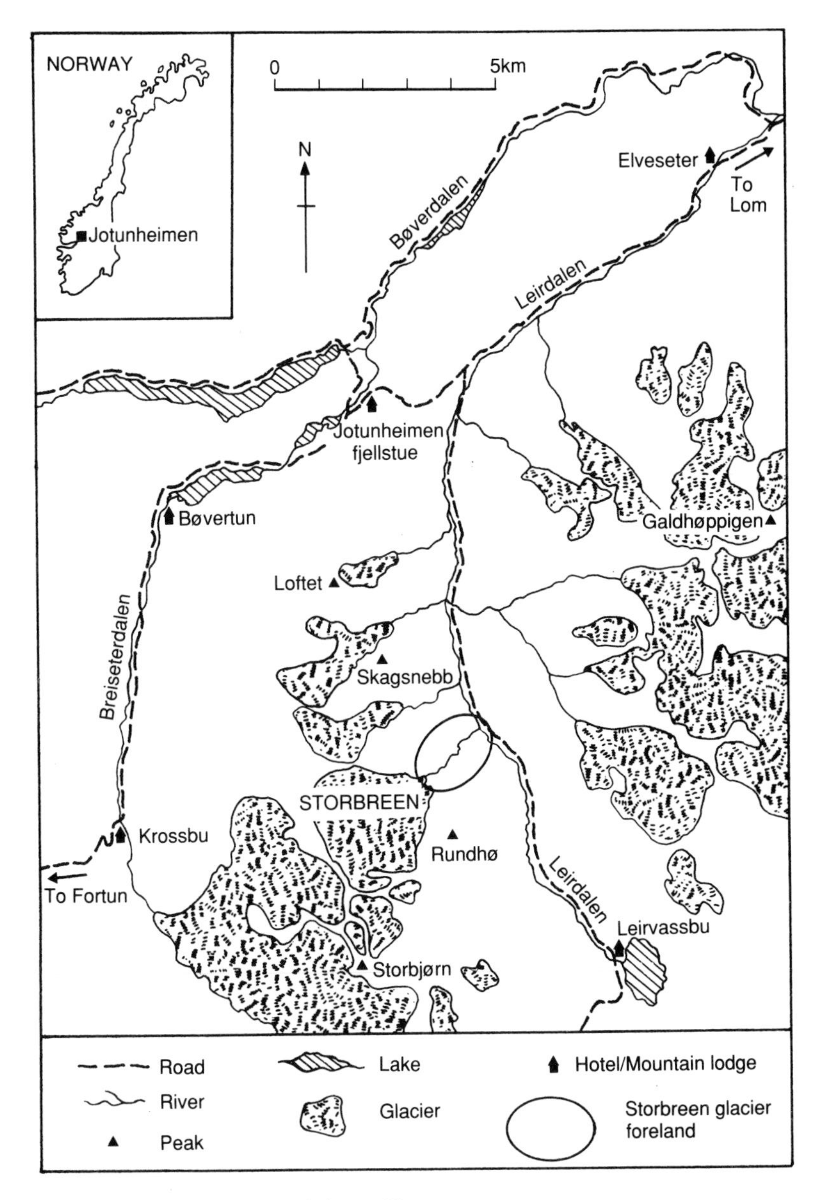

FIG. 1. Location of Storbreen, Jotunheimen, Norway.

The study area is situated above the treeline in the low- and mid-Alpine belts (Nordhagen 1943; Dahl 1956). It is approximately rectangular with dimensions $1{\cdot}5 \times 1{\cdot}0$ km (Fig. 2). The glacier foreland is delimited by the outermost of a sequence of arcuate 'Little Ice Age' end moraines. A detailed areal chronology has been established for the foreland by Liestøl (1967) and Matthews (1974, 1975, 1976, 1977) using lichenometry, direct observations and historical evidence. Thus,

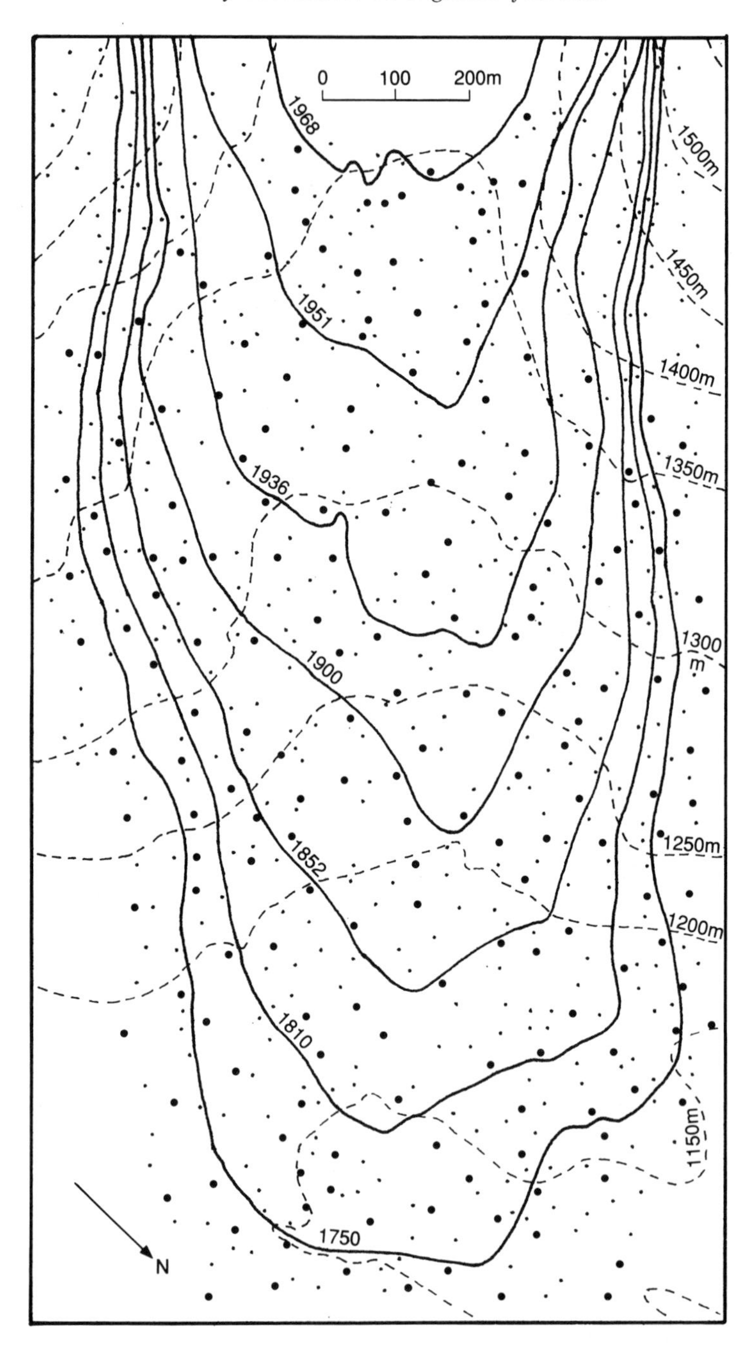

FIG. 2. Map of the Storbreen glacier foreland, showing the distribution of study sites. Symbols: isochrones (———); contours (– – – –); sites used for the present study (●); other sites established by Matthews (1976; •). Areal chronology is after Matthews (1978a, 1979a).

the foreland represents an accurately dated two-dimensional chronosequence from 0 to 239 years (Fig. 2). Terrain outside the foreland is believed to have been undisturbed by glacial ice since about 9000 BP (Matthews *et al.* 1986).

FIELD AND ANALYTICAL METHODS

Percentage cover was recorded for all species of vascular plants and terricolous bryophytes and lichens in 190 sites (2 × 2 m), which represent a one-in-three subsample of Matthews' (1976) original systematically stratified random design (Fig. 2). The cover value for each site was the mean for four 1 m^2 quadrats. Distribution maps of individual species were produced using the program GEFMAP (G. P. Ibbett, unpubl. data). Multivariate analyses were carried out on three different data sets: cryptogams only, phanerogams only and all species together. Data were ordinated by detrended correspondence analysis (DCA), using this option within the program CANOCO (ter Braak 1987). Two-way indicator species analysis (TWINSPAN) was used for classification (Hill 1979). Maps of assemblage groups were produced using BASMAP, within the VESPAN II package (Malloch 1988).

RESULTS

Mapping of individual species and the analysis of species groups

All species were mapped, particularly as an aid to the interpretation of multivariate analyses. Quantitative distribution maps for 11 species are shown in Figs 3 and 4. These illustrate different types of distribution, representative of some of the TWINSPAN species groups defined (see below), which can be interpreted to a considerable extent in terms of terrain age, altitude and aspect. The pioneers, *Saxifraga cespitosa* and *Pohlia bulbifera*, both occur almost exclusively in the proglacial area. *Oxyria digyna* was found at many relatively recently deglaciated sites, but also at older sites, particularly at high altitude on the north-facing side of the foreland. *Pannaria pezizoides* also occurs mostly on younger ground, but also at some older sites. Some species, for example *Salix lanata* and *Racomitrium canescens*, exhibit no apparent time-dependence and occur throughout the altitudinal range of the study area. *Empetrum hermaphroditum* and *Cetraria islandica* are both widespread across the foreland, but have greatest cover on terrain outside the AD 1852 moraine. *Betula nana* is restricted to older ground, with greatest cover outside the AD 1750 moraine at low altitude. *Hylocomium splendens* and *Pleurozium schreberi* are similarly distributed, but the latter is found in greater quantities on the north-west side of the study area.

Separate TWINSPAN classifications of cryptogams and phanerogams are compared on a common basis in Fig. 5a and b. The axes are derived from a DCA analysis of the combined data set. Classifications are interpreted at two levels: two divisions of the data set produce the 'four-group level', while a third division

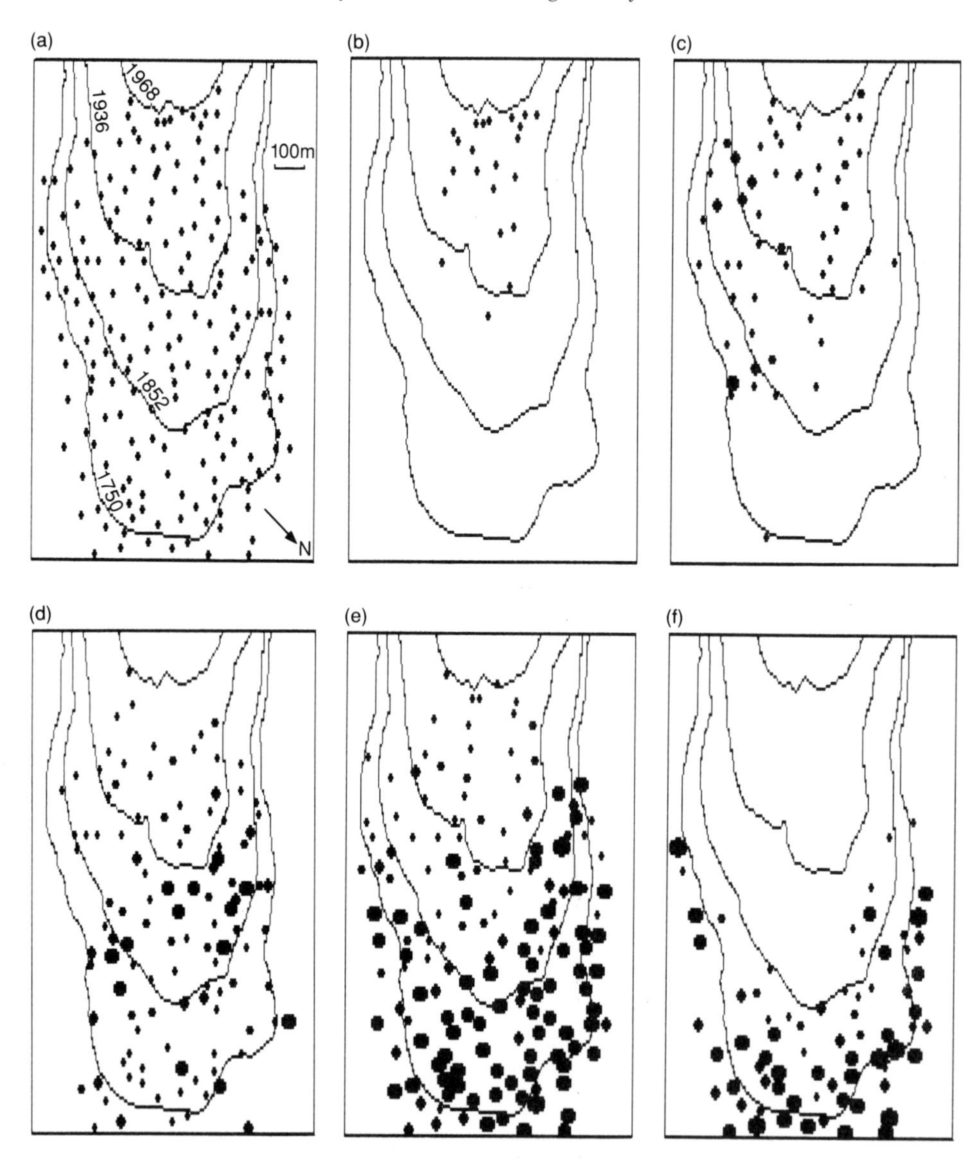

FIG. 3. Distribution of selected phanerogam species in relation to terrain age: (a) location of all sites, (b) *Saxifraga cespitosa*, (c) *Oxyria digyna*, (d) *Salix lanata*, (e) *Empetrum hermaphroditum*, (f) *Betula nana*. The size of each symbol in (b) to (f) represents the percentage cover at that site, on a scale with seven intervals, as follows: 1, 2–3, 4–7, 8–15, 16–31, 32–63, 64–100.

produces the 'eight-group level'. For each TWINSPAN group at Fig. 5a the four-group and Fig. 5b the eight-group level of division, the mean (centroid) of the DCA scores of all species in that group is plotted, together with its 95% confidence limits. Axis 1 is interpreted as a time axis, since the correlation coefficient (r) between this axis and site age is −0·76. Axis 2 is most strongly correlated with a

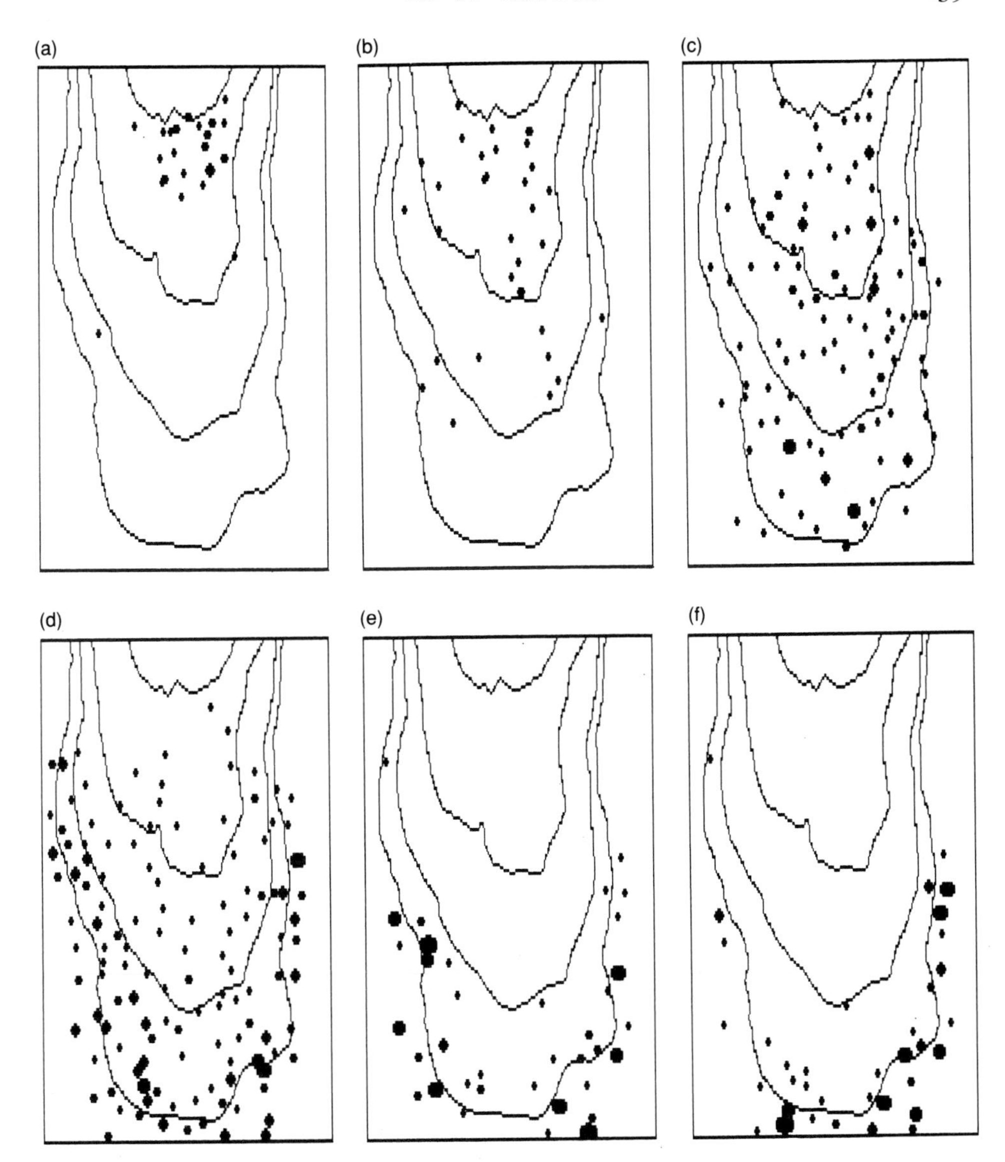

FIG. 4. Distribution of selected cryptogam species in relation to terrain age: (a) *Pohlia bulbifera*, (b) *Pannaria pezizoides*, (c) *Racomitrium canescens*, (d) *Cetraria islandica*, (e) *Hylocomium splendens*, (f) *Pleurozium schreberi*. The scale of symbol sizes is as for Fig. 3.

soil moisture index ($r = -0{\cdot}33$). Environment–vegetation relationships of cryptogams at Storbreen will be discussed in detail elsewhere.

At the four-group level, pioneer groups (1c and 1p) can be recognized for both cryptogams and phanerogams. The confidence limits indicate that these two groups, represented in Figs 3 and 4 by *Saxifraga cespitosa*, *Pohlia bulbifera*, *Oxyria digyna* and *Pannaria pezizoides*, are relatively distinct. At the other end of

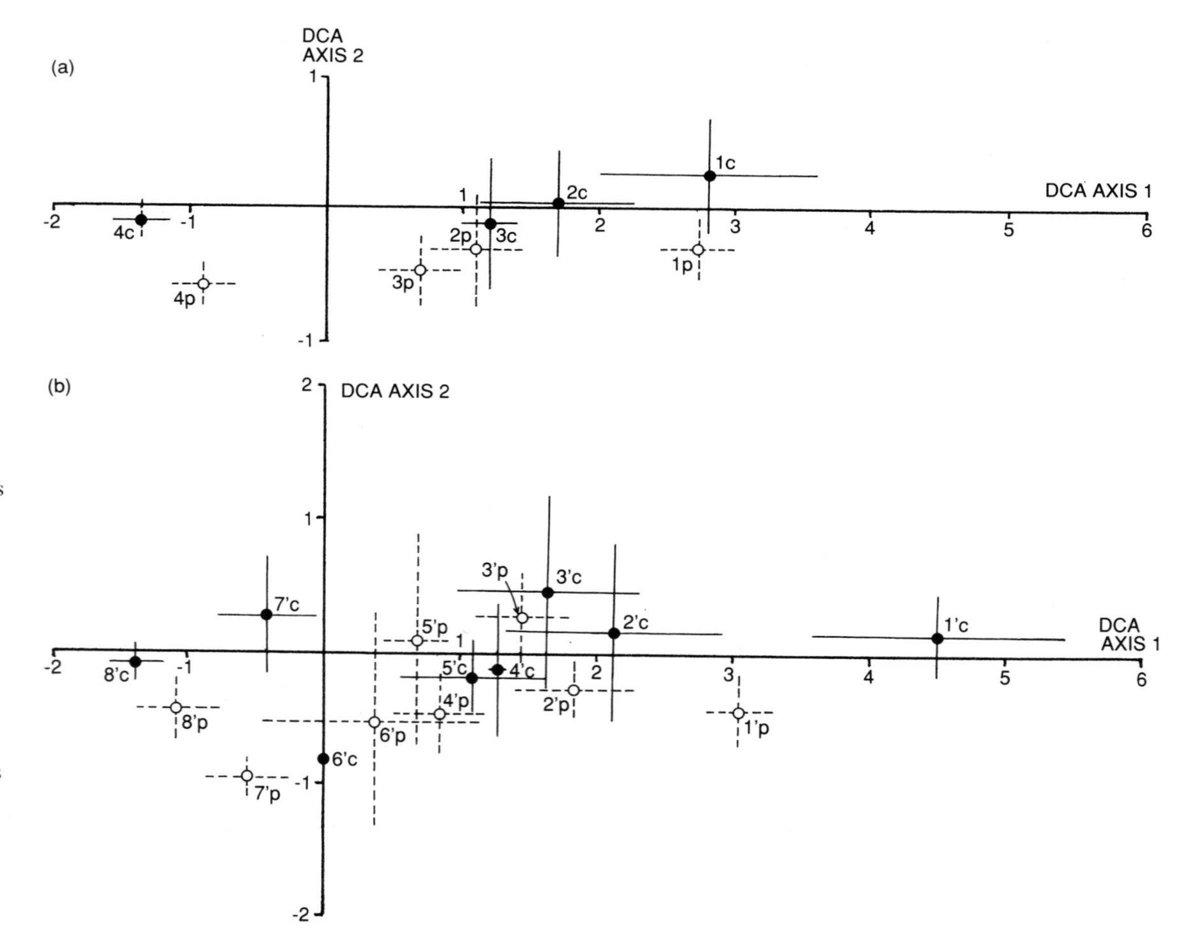

FIG. 5. Cryptogam and phanerogam species groups represented in relation to common DCA axes. Centroids of the respective TWINSPAN groups are represented as mean DCA scores on axes 1 and 2 at (a) four-group and (b) eight-group levels. Filled circles are centroids of cryptogam groups; open circles are centroids of phanerogam groups. Confidence limits of 95% of the centroids are shown. Axis 1 has eigenvalue 0·52 and is correlated with terrain age ($r = -0{\cdot}76$). Axis 2 has eigenvalue 0·38 and is correlated with the soil moisture index ($r = -0{\cdot}33$).

axis I are two even more distinct groups, one of cryptogams (4c) and one of phanerogams (4p), characteristic of older ground. These groups are represented by *Betula nana*, *Empetrum hermaphroditum*, *Hylocomium splendens* and *Pleurozium schreberi*. Between these extremes, relatively indistinct groups are characteristic of ground of intermediate age or no preferred age. Only one pair of centroids (3c, represented by *Cetraria islandica*; and 2p, represented by *Salix lanata*) corresponds closely in this area of the ordination diagram.

At the eight-group level (Fig. 5b), there is also a pioneer group of cryptogams and of phanerogams (1′c and 1′p), distinct from each other and all other groups. At the other end of axis I are two cryptogam groups (7′c and 8′c) and two phanerogam groups (7′p and 8′p) with non-overlapping confidence limits. The remaining groups between these extremes are again less distinct, each overlapping with at least one other group; no pair of centroids corresponds very closely.

Classification of sites

Fig. 6 shows the TWINSPAN classification of sites for the three data sets (all species, cryptogams and phanerogams) at (a) the four-group level and (b) the eight-group level of division. The percentage correspondence between these classifications is given in Table 1. Classifications based on cryptogams and phanerogams separately are more dissimilar to each other than either is to the classification based on all species. The degree of similarity decreases drastically between the four- and eight-group levels, from 62 to 39%.

At the four-group level the cryptogam classification (Fig. 6a (ii)) results in a more restricted pioneer group than the phanerogam classification (Fig. 6a (iii)). On the oldest terrain, the cryptogam group is again more restricted than the respective phanerogam group (Fig. 6a (ii and iii): ■). Apart from two sites (Fig. 6a (ii): ▲), the remaining sites in the cryptogam classification form a single group (▽). In the phanerogam classification, however, the corresponding intermediate-aged sites are separated as two groups with less unequal numbers of sites in each (Fig. 6a (iii): ▽ and ▲).

At the eight-group level (Fig. 6b), despite a much lower level of agreement between the classifications, maps for both cryptogam and phanerogam classifications demonstrate strong age-dependence. As at the four-group level, cryptogam communities on the youngest and oldest ground have more restricted ranges than the respective phanerogam communities.

TABLE 1. Percentage correspondence between site classifications using the three data sets: all species, cryptogams only and phanerogams only, at two levels of division

Level of division	Cryptogams and all species	Phanerogams and all species	Cryptogams and phanerogams
4	78	74	62
8	60	61	39

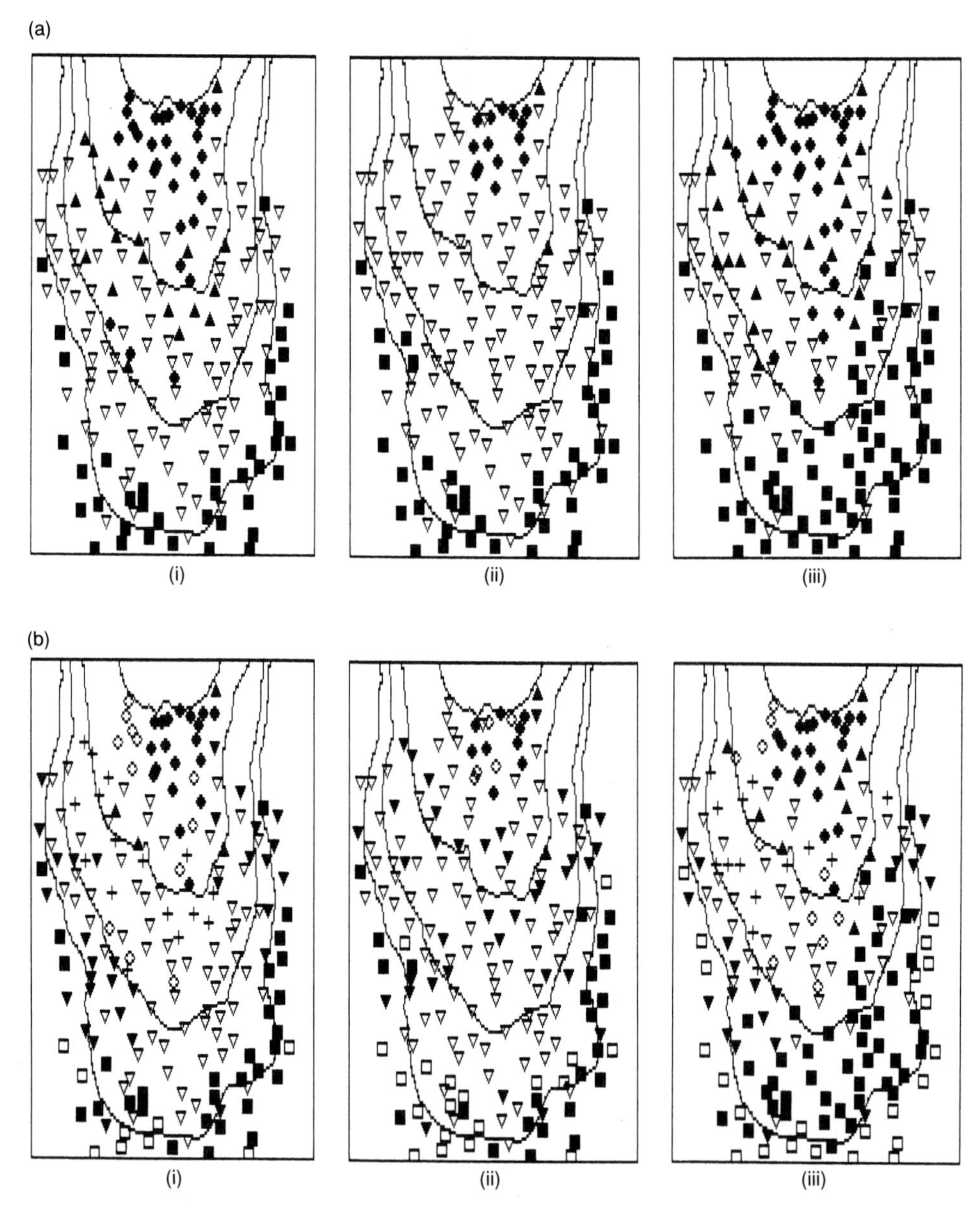

FIG. 6. Maps of community types (TWINSPAN site classifications) based on (i) all species, (ii) cryptogams only and (iii) phanerogams only, at (a) the four-group level and (b) the eight-group level. Community types (site groups) are represented by distinctive symbols.

DISCUSSION

Plotting the TWINSPAN species groups on common DCA axes indicates a successional sequence of communities for both cryptogams and higher plants. Species groups on the oldest and youngest ground are distinct. On intermediate-aged ground, there is some overlapping of phanerogam species groups, and of cryptogam species groups, indicating that within each classification these groups are less distinct from each other than from the groups characteristic of older or younger ground. This suggests that there are in fact three broad distribution types that correspond to three major successional stages for both cryptogams and phanerogams at Storbreen, as has been demonstrated by Matthews (1978a, b) for vascular plants.

All lines of evidence employed indicate only partial correspondence between the distributions of cryptogam and higher plant species and species groups. Pioneer cryptogam and phanerogam species groups are distinct from each other, as are cryptogam and phanerogam species groups on older ground. On intermediate-aged terrain, confidence limits of cryptogam centroids mostly overlap with those of phanerogam centroids, but this is only ever partial. Partial correspondence between cryptogam and phanerogam distributions is also reflected in the maps of TWINSPAN site classifications (Fig. 6).

This lack of correspondence between the distribution of cryptogam and higher plant species and communities seems to reflect the differential sensitivity of the two groups of species to environmental factors. Attributes responsible for the differences may include the lack of a well-developed root system or conducting tissues in most of the cryptogam species involved and the generally higher tolerance of cryptogams to desiccation. Differences in their capacity for dispersal and establishment may also be of significance, particularly in early stages of succession. Spence (1981), studying cryptogams in front of glaciers in the Teton Range, Wyoming, found their distributions to be controlled by a combination of site age, substrate type and aspect. In the site classifications above, differences in the way the four groups are split into eight are evident. For example, the group of sites on the oldest terrain (Fig. 6a: ■) appears to have been split largely according to site age for the phanerogams, but altitude for the cryptogams (Fig. 6b: ■ and □). This supports the notion that, relative to the time factor, the cryptogam communities may be more sensitive than the phanerogam communities to environmental variation. Other authors have suggested that bryophyte and lichen distributions are controlled by microenvironmental factors to a greater extent than phanerogam distribution (e.g. Faegri 1933; Flock 1978). A thorough analysis of relevant environmental variables at Storbreen is in progress.

The separate analyses of phanerogam and cryptogam data have revealed differences in their respective distribution patterns and successional sequences not detectable in an analysis of vegetation as a whole. Results suggest that analysis of the whole plant community is not always the most informative approach. There is evidence for much individualistic behaviour in succession at Storbreen. The distribution maps of no two species are identical. Even species which are very close together in the species ordination and are classified in the same group, for example *Hylocomium splendens* and *Pleurozium schreberi*, exhibit distinct distribution patterns (Fig. 4). This could be taken to support Gleason's (1927) view that succession is 'no more than the mass-effect of the action or behaviour of individual plants'. It certainly suggests that a population ecology approach is valuable. Yet the similarities exhibited by many species distributions indicate that a community approach to vegetation is realistic, at least initially, for the study of successional sequences.

ACKNOWLEDGMENTS

I thank Dr J. A. Matthews for support and encouragement, both in fieldwork and in the preparation of this manuscript; J. P. Crouch, D. Barnes and K. MacLean for assistance with fieldwork, which was undertaken on the Jotunheimen Research Expeditions, 1987 and 1988; Dr P. D. Coker, J. P. Crouch and particularly G. P. Ibbett for help with computing. Funding was provided by a Science and Engineering Research Council Studentship. This chapter is Jotunheimen Research Expeditions Contribution No. 86.

REFERENCES

Alpert, P. & Oechel, W.C. (1982). Bryophyte vegetation and ecology along a topographic gradient in montane tundra in Alaska. *Holarctic Ecology*, **5**, 99–108.

Archer, A.C. (1973). Plant succession in relation to a sequence of hydromorphic soils formed on glacio-fluvial sediments in the alpine zone of the Ben Ohau Range. *New Zealand Journal of Botany*, **11**, 331–48.

Birks, H.J.B. (1980). The present flora and vegetation of the moraines of the Klutlan glacier, Yukon territory, Canada: a study in plant succession. *Quaternary Research*, **14**, 60–86.

Cooper, W.S. (1923). The recent ecological history of Glacier Bay, Alaska: II. The present vegetation cycle. *Ecology*, **4**, 223–46.

Crocker, R.L. & Major, J. (1955). Soil development in relation to vegetation and surface age at Glacier Bay, Alaska. *Journal of Ecology*, **43**, 427–48.

Dahl, E. (1956). Rondane: mountain vegetation in South Norway and its relation to environment. *Skrifter utgitt av det Norske Videnskaps-academi i Oslo (Matematisk-naturvidenskapelig klasse)*, **3**, 1–374.

Elven, R. (1978). Association analysis of moraine vegetation at the glacier Hardangerjøkulen, Finse, South Norway. *Norwegian Journal of Botany*, **25**, 171–91.

Faegri, K. (1933). Über die Längenvariationen einiger Gletscher des Jostedalsbre und die dadurch bedingten Pflanzensukzessionen. *Bergens Museums Årbok*, **7**, 1–255.

Flock, J.W. (1978). Lichen–bryophyte distribution along a snowcover–soil–moisture gradient, Niwot Ridge, Colorado. *Arctic and Alpine Research*, **10**, 31–47.

Gleason, H.A. (1927). Further views on the succession concept. *Ecology*, **8**, 299–326.

Hill, M.O. (1979). *TWINSPAN – a FORTRAN Program for Arranging Multivariate Data in an Ordered Two-way Table by Classification of the Individuals and Attributes*. Cornell University, Ithaca, NY.

Liestøl, O. (1967). Storbreen glacier in Jotunheimen, Norway. *Norsk Polarinstitutt Skrifter*, **141**, 1–63.

Malloch, A.J.C. (1988). *VESPAN II. A Computer Package to Handle and Analyse Multivariate Species Data and Handle and Display Species Distribution Data*. University of Lancaster, Lancaster.

Matthews, J.A. (1974). Families of lichenometric dating curves from the Storbreen gletschervorfeld, Jotunheimen, Norway. *Norsk Geografisk Tidsskrift*, **28**, 215–35.

Matthews, J.A. (1975). Experiments on the reproducibility and reliability of lichenometric dates, Storbreen gletschervorfeld, Jotunheimen, Norway. *Norsk Geografisk Tidsskrift*, **29**, 97–109.

Matthews, J.A. (1976). *A phytogeography of a gletschervorfeld: Storbreen-i-Leirdalen, Jotunheimen, Norway*. PhD thesis, University of London (King's College).

Matthews, J.A. (1977). A lichenometric test of the 1950 end-moraine hypothesis: Storbreen gletschervorfeld, southern Norway. *Norsk geografisk Tidsskrift*, **31**, 129–36.

Matthews, J.A. (1978a). Plant colonization patterns on a gletschervorfeld, southern Norway: a meso-scale geographical approach to vegetation change and phytometric dating. *Boreas*, **7**, 155–78.

Matthews, J.A. (1978b). An application of non-metric multidimensional scaling to the construction of an improved species plexus. *Journal of Ecology*, **66**, 157–73.

Matthews, J.A. (1979a). The vegetation of the Storbreen gletschervorfeld, Jotunheimen, Norway. I. Introduction and approaches involving classification. *Journal of Biogeography*, **6**, 17–47.

Matthews, J.A. (1979b). The vegetation of the Storbreen gletschervorfeld, Jotunheimen, Norway. II. Approaches involving ordination and general conclusions. *Journal of Biogeography*, **6**, 133–67.

Matthews, J.A. & Whittaker, R.J. (1987). Vegetation succession on the Storbreen glacier foreland, Jotunheimen, Norway: a review. *Arctic and Alpine Research*, **19**, 385–95.

Matthews, J.A., Harris, C. & Ballantyne, C.K. (1986). Studies on a gelifluction lobe, Jotunheimen, Norway: ^{14}C chronology, stratigraphy, sedimentology and palaeoenvironment. *Geografiska Annaler*, **68A**, 345–60.

Mellor, A. (1985). Soil chronosequences on Neoglacial moraine ridges, Jostedalsbreen and Jotunheimen, southern Norway: a quantitative pedogenic approach. *Geomorphology and Soils* (Ed. by K.S. Richards, P.R. Arnett & S. Ellis), pp. 289–308. Allen & Unwin, London.

Nordhagen, R. (1943). Sikilsdalen og Norges fjellbeiter. En plantesosiologisk monografi. *Bergens museums skrifter*, **22**, 1–607.

Persson, A. (1964). The vegetation at the margin of the receding glacier Skaftafellsjökull, south eastern Iceland. *Botaniska Notiser*, **117**, 323–54.

Richard, J.L. (1973). Dynamique de la vegetation au bord du grand glacier d'Aletsch (Alpes Suisses). *Berichte der Scheizerischen Botanischen Gesellschaft*, **83**, 159–74.

Spence, J.R. (1981). Comments on the cryptogam vegetation in front of glaciers in the Teton Range. *The Bryologist*, **84**, 564–58.

Stork, A. (1963). Plant immigration in front of retreating glaciers, with examples from the Kebnekajse area, Northern Sweden. *Geografiska Annaler*, **45**, 1–22.

ter Braak, C.J.F. (1987). *CANOCO – a FORTRAN Program for Canonical Community Ordination by [Partial] [Detrended] [Canonical] Correspondence Analysis, Principal Components Analysis and Redundancy Analysis (Version 2.1)*. ITI-TNO, Wageningen.

Viereck, L.A. (1966). Plant succession and soil development on gravel outwash of the Muldrow glacier, Alaska. *Ecological Monographs*, **36**, 181–99.

Whittaker, R.J. (1985). *Plant community and population studies of a successional sequence: Storbreen Glacier foreland, Jotunheimen, Norway*. PhD thesis, University of Wales (University College Cardiff).

Whittaker, R.J. (1987). An application of detrended correspondence analysis and non-metric multidimensional scaling to the identification and analysis of environmental factor complexes and vegetation structures. *Journal of Ecology*, **75**, 363–76.

Whittaker, R.J. (1989). The vegetation of the Storbreen gletschervorfeld, Jotunheimen, Norway. III. Vegetation–environment relationships. *Journal of Biogeography*, **16**, 413–33.

Worley, I.A. (1973). The 'black crust' phenomenon in upper Glacier Bay, Alaska. *Northwest Science*, **47**, 20–9.

Dispersal and establishment of tropical forest assemblages, Krakatoa, Indonesia

R. J. WHITTAKER AND M. B. BUSH*
*School of Geography, University of Oxford, Mansfield Road, Oxford OX1 3TB, UK; and * Department of Botany, Duke University, Durham, North Carolina 27706, USA*

SUMMARY

1 The redevelopment of vegetation on the Krakatoa Islands following sterilization in 1883 is examined by analysis of two data sources:

(a) a numerical classification of vegetation survey data collected in 1979–83, which is presented in the context of the detailed documentary evidence for the preceding century;

(b) the complete floristic database for 1883–1983, which provides data on the contributions of flowering plants of different growth forms and dispersal mechanisms to the developing vegetation communities.

2 Several alternative successional pathways are recognized, involving both habitat determinants and deflection of succession due to renewed disturbance.

3 The development of differing vegetation communities is found to be underlain by differences in patterns of dispersal. Later successional forest species, as exemplified by certain large trees (including new families and genera) and forest orchids, have continued to accrue over the last 50 years, and yet it is noted that:

(a) the islands remain depaupaurate in inland tree species; and

(b) a minority of species provide the majority of both canopy cover and of individuals within the forests, which thus remain early successional in nature.

INTRODUCTION

Primary succession from bare ground through to relatively stable forest assemblages is typically a lengthy process, which ideally requires a duration of study in excess of the generation times of the dominant species. Few successional studies can call upon a 100-year span of empirical data as is the case with the redevelopment of vegetation of the Krakatoa Islands (Fig. 1). Despite the heated debate on the issue (Backer 1929; Docters van Leeuwen 1936) there is little doubt that the islands were completely sterilized in the eruptions of 20th May to 27th August 1883, which culminated in a massive reshaping of the island group and the deposition of great thicknesses of volcanic ash substratum (reviewed in Whittaker *et al.* 1989). The exceptional opportunity that this provided for studying primary succession was realized at an early stage, in particular by Treub (1888) who did

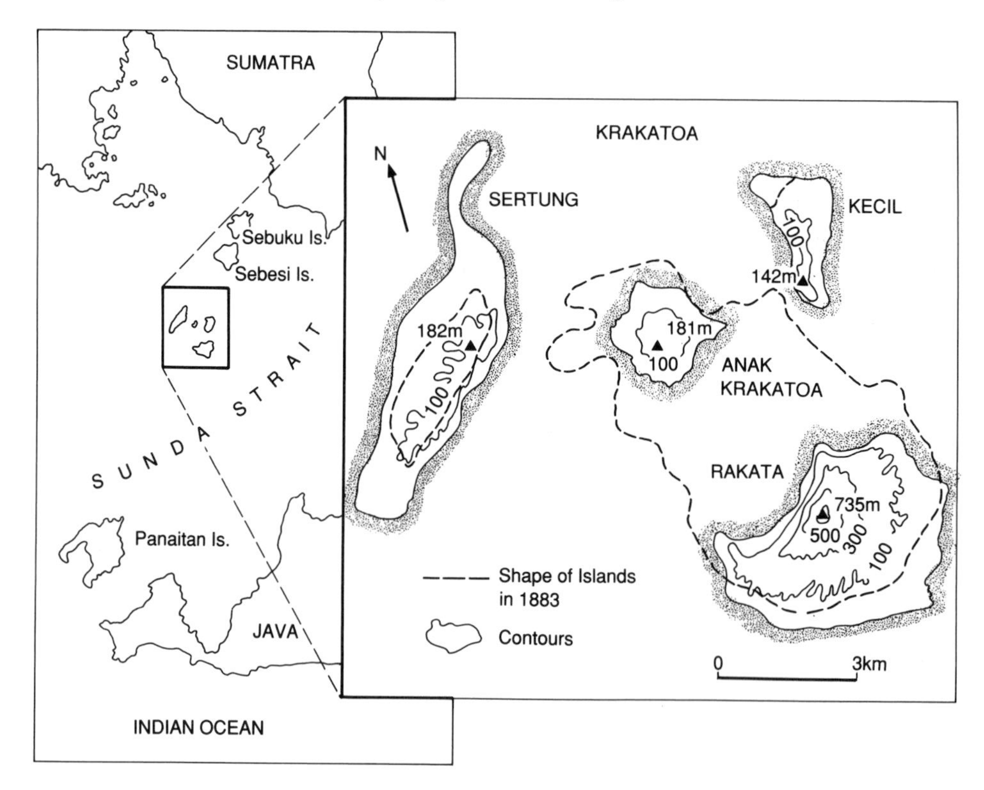

FIG. 1. Location and configuration of the Krakatoa islands, Sunda Straits, Indonesia.

the first survey in 1886, by Ernst (1908) and by Docters van Leeuwen (1936). The length of the study is unusual enough, but the humid tropical location of the islands, 6° south of the equator, provides for a unique combination of circumstances, making it the longest running 'natural experiment' on tropical forest regeneration. The floral survey data for the islands, grouped into 13 separate collection periods, have been compiled into a database (Whittaker *et al.* 1989, appendix 1), to which certain autoecological data have now been added. Although these unstandardized survey data provide an incomplete account, they nonetheless constitute a rich source of empirical data.

The group lies approximately equidistant 44 km from mainland Java and Sumatra, although 'connected' to the latter by two large stepping stone islands, Sebesi and Sebuku (Fig. 1). This geographical location has undoubtedly been of fundamental importance in shaping the succession on the islands, as certain groups of organisms may be unable to reach the site, while chance early arrivals may have been able to steal a lead over other species. Whittaker *et al.* (1989) found that not only was an understanding of dispersal mechanisms crucial to the pattern of development but that differing habitats, crudely — sub-montane, lowland and coastal — effectively sampled different species pools. Although the isolation of the islands may lead to an overemphasis on dispersal limitations in comparison with mainland situations, these factors do operate within mainlands, albeit to lesser degrees and with a different balance of mechanisms. The principal aim of this paper is therefore to examine the manner in which the forests of Krakatoa have reassembled, by analysis of plant growth form and dispersal mechanisms. First, however, the broad pattern of 'community' succession will be examined. The interrupted succession on the new island, Anak Krakatoa, will not be discussed in this chapter (but see Whittaker *et al.* 1989; Partomichardjo *et al.* 1992).

MATERIALS AND METHODS

Two forms of analysis are presented here, first a community classification of 1979–83 vegetation survey data and second, database analyses of the floral surveys of 1883–1983.

Classification by TWINSPAN

Two-way indicator species analysis, as performed by the TWINSPAN programme (Hill 1979) is a polythetic hierarchical technique commonly used in ecology for classifying groups and attributes (Gauch 1982). The analysis is based on 35 sites, located to maximize altitudinal and habitat coverage and varying in size from 400–2500 m^2. For each site, diameter at breast height of trees over 10 cm d.b.h. was recorded, with the data standardized to per cent total cross-sectional area prior to analysis.

Floral database analyses

These analyses are based upon the complete floral lists for Rakata, Sertung and Kecil (Whittaker *et al.* 1989, appendix 1, which includes all taxonomic authorities for this paper). We have subsequently added certain unpublished autoecological data to the published database. In this chapter, two categories are used, dispersal mechanism and growth form (Table 1), but the treatment is restricted to flowering plants because of limitations of space.

Dispersal mechanisms

The categories used (Table 1) refer to the most probable mechanism of dispersal to the island group. Docters van Leeuwen (1936) pays particular attention to the dispersal mechanisms of the Krakatoa flora of 1883–1936, citing many observations, e.g. of viable seeds recovered from bird droppings or sea drift. In several cases, two or more means of dispersal are noted in the sources cited in Table 1. Generally we have then assumed that sea dispersal provides the long distance transport, with wind or animals providing subsequent spread within the islands.

TABLE 1. Dispersal mechanisms and growth form categories used in the analyses of the floral data. The data were compiled from the following principal sources, in order of significance: Docters van Leeuwen (1936), Backer & Bakhuizen van den Brink (1963–68), Ridley (1930), Whitmore & Tantra (1986) and Corner (1988)

*Dispersal categories**

Endozoical	= animal dispersal by defecation of ingested seeds
Exozoical	= animal dispersal by external attachment of seeds
Humans	= species planted or accidentally introduced by humans
Sea	= seeds dispersed by sea and cast upon the beach
Wind	= seeds dispersed by wind

Growth form (and habit) categories

Large trees	= trees of roughly 30 m or over at maturity
Medium trees	= trees of *c.* 15–30 m at maturity
Small trees	= <15 m, one or few trunks
Shrubs[†]	= woody plants, of bushy form
Herbs[†]	= mostly non-woody plants of small size
Large liana	= woody climbers or scramblers

Habit

Terrestrial	= terrestrial orchids, including two saprophytes
Epiphytic	= mostly orchids, but a few other flowering plants are epiphytic, e.g. in the Gesneriaceae and Moraceae

* A few species could not be assigned with confidence to one category and have been excluded from the analysis.
[†] These categories include climbing or trailing herbs and shrubs, although the distinction between some of these and lianas is rather arbitrary.

Growth form

Table 1 gives the guidelines followed in these analyses, with some discretion allowed for different growth form types (some trees may be quite tall but of narrow crown and bole diameters, and might therefore be classed as of medium size rather than large). It can be a rather subjective matter to determine which category a particular species falls into, as many species are highly variable throughout their range. However, the broad pattern is probably fairly reliable. The Orchidaceae were separated into epiphytic and terrestrial species, strictly habit rather than growth form, but have been included here for convenience.

RESULTS AND DISCUSSION

Community development

Community succession on each of the three main islands followed a similar and fairly well-documented pattern over the first 50 years (Docters van Leeuwen 1936; Whittaker *et al.* 1989). The coastal communities established rapidly, with the typical dominant strandline species among the first to colonize, producing a one or two phase succession (Fig. 2). In the interior of the islands, an early phase of

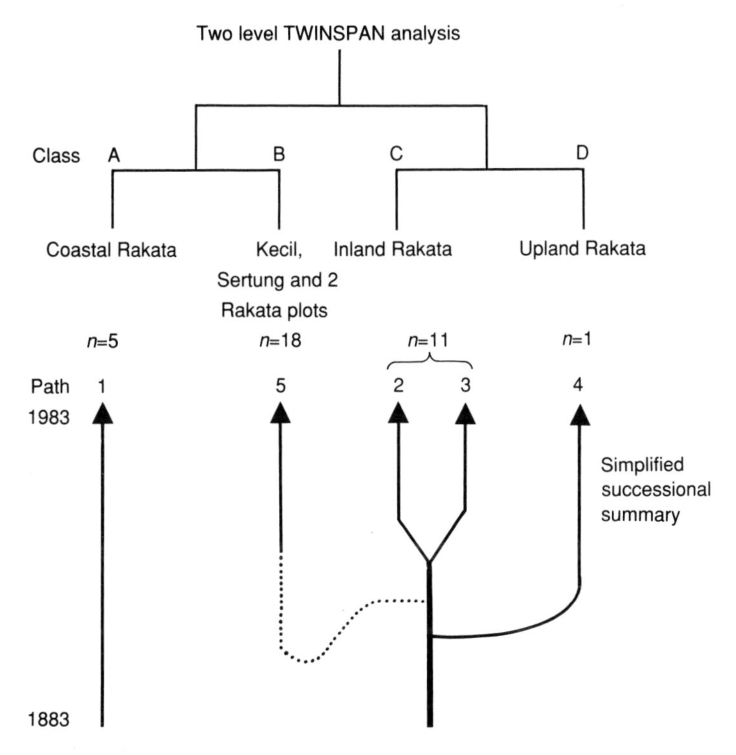

FIG. 2. Simplified successional model, incorporating a two-level TWINSPAN classification of 35 sample sites recorded in 1979–83 and a simplified successional scheme, for the period 1883–1983. The positions of the TWINSPAN classes A–D correspond to paths 1–5 in the manner shown. Typical species of class A sites: *Terminalia catappa*, *Barringtonia asiatica* (sometimes *Casuarina equisetifolia*); class B sites: *Timonius compressicaulis* (sometimes *Dysoxylum gaudichaudianum*); class C sites: *Neonauclea calycina*, *Ficus pubinervis*; class D site: *Saurauia nudiflora*, *Villebrunea rubescens* (*Schefflera polybotrya* in understorey). Path 1: coastal communities, either (a) rapid establishment of *Terminalia catappa*/*Barringtonia asiatica* woodland; or (b) the same, after a stage of *Casuarina equisetifolia* woodland. Path 2: ferns→grass savannah→*Macaranga tanarius*/*Ficus* spp. forest→ *Neonauclea calycina* forest. Path 3: as path 2, but current stage of *Ficus pubinervis*/*Neonauclea calycina* forests. Path 4: ferns→grass savannah→*Cyrtandra sulcata* scrub→*Schefflera polybotrya*/ *Saurauia nudiflora*/*Ficus ribes*, submontane forest. Path 5: — ferns→grass savannah→*Macaranga tanarius*/*Ficus* spp., cut-back by volcanism→grass savannah→*Timonius compressicaulis*→ *Dysoxylum gaudichaudianum* forest.

highly dispersive ferns, grasses (anemo- and zoochorous) and a few Compositae, gradually diversified, gaining by 1897 such species as the heliophilous terrestrial orchids *Arundina graminifolia*, *Phaius tankervilliae* and *Spathoglottis plicata*. A scattering of shrubs and trees, interspersed among the savanna vegetation, quickly spread and forest closure on Rakata (Fig. 1) was achieved by the end of the 1920s, with the summit the last area to gain forest. A relatively high proportion of early colonists were obligate open-habitat species, many of which were lost with forest closure (Docters van Leeuwen 1936). In addition, as open areas were reduced to relatively small gaps, spatially and temporally disconnected, any fauna reliant on open areas were also seriously disadvantaged. Losses included several species of Hesperiidae (skippers), and birds such as the red-vented bulbul (*Pycnonotus cafer*) and the long-tailed shrike (*Lanius schach bentet*) (Dammerman 1948; Hoogerwerf 1953).

The early forests of the interior were mostly species-poor *Ficus–Macaranga* forest. However, an anemochorous early successional species, *Neonauclea calycina*, became dominant over much of the south-east side of Rakata in the early 1930s and remains so today (Fig. 2). On Kecil and Sertung (Fig. 1) the course of redevelopment was interrupted at this time and subsequently by renewed volcanic activity.

The detailed descriptive accounts of the first 50 years (Docters van Leeuwen 1936) have been supplemented by a brief survey in 1951 (Borssum Waalkes 1960) and by the quantitative vegetation surveys of 1979–83. The results of the TWINSPAN analyses at the second level of the division are displayed in Fig. 2, in conjunction with the simplified scheme for the successional pathways within the island group. Typical strandline sites were sampled only on Rakata, explaining the anomalous absence of Sertung and Kecil sites from this class (class A). The remaining three classes (B, C, D) are of inland or near-coastal vegetation. The 16 plots on Sertung and Kecil remain as one group (class B), together with two near-coastal Rakata plots (Fig. 2). They are species-poor forests in comparison with inland Rakata and have different species compositions (Bush & Whittaker 1986). The forests of Rakata have continued to diversify, dominating the analysis in Fig. 2; although large areas are dominated by *N. calycina* and/or one or two species of *Ficus*, especially *F. pubinervis*. The highest altitude plot on Rakata splits off at the second level of the classification, indicative of a further upland successional pathway (class D, path 4). Perhaps the most interesting development of the last 50 years, however, is the spread of *Timonius compressicaulis* and to a lesser extent *Dysoxylum gaudichaudianum* as canopy dominants on Sertung and Kecil. Whittaker *et al.* (1989) attributed this deflection from the previous successional pathway to rapid invasion from within the group upon the creation of large gaps by ash-fall since *c.* 1930 (Rakata has not been affected). Significantly, neither species was observed in the early years of forest development (up to the mid-1920s). Both are endozoically dispersed.

Analysis of growth form and dispersal data

From 1883 to 1983, *c.* 88 species of trees have been recorded on the Krakatoa islands, of which 28 are sea-dispersed. Few of these have spread to any distance inland. One notable exception is *Terminalia catappa*, which appears to be carried inland by crabs and which can be found at altitudes as high as 80 m. More typical are *Guettarda speciosa*, *Calophyllum inophyllum* and *Hernandia peltata*, whose secondary dispersal agent appears to be bats (Docters van Leeuwen 1936) but which nonetheless remain coastal and near-coastal. The remaining 60 species are made up of five anemochorous and 55 endozoically dispersed species, together incorporating the true inland tree flora. Thus in terms of the successional development of tropical forest, certain species groups represent noise and others signal. Noise elements include weed species introduced by humans and most strandline plants. The signal might be expected to lie with zoochorous tree species, especially larger ones, and with groups sensitive to forest maturity, such as the Orchidaceae. The strandline assemblages are nonetheless of interest in terms of their own succession and are therefore included in the discussion.

Whittaker *et al.* (1989) noted two patterns within the thalassochorous (sea-dispersed) species: (a) an early set of arrivals, which have been species stable; and (b) a later set, which have shown a higher rate of turnover. The latter included ephemeral species (due to 'over-sampling' of strandline habitats) and species lost as a result of habitat destruction due to coastal erosion. Those thalassochorous species which established on all three islands within the first 50 years have suffered very few losses (six lost of 48 species). The data summarized in Table 2 exemplify this pattern in the early arrival and establishment of the typical sea-dispersed strandline species, including *Barringtonia asiatica*, *Cocos nucifera*, *Erythrina orientalis*, *Hernandia peltata*, *Hibiscus tiliaceus*, *Morinda citrifolia* and *Terminalia catappa*, all of which were found as early as 1897. Very few of the sea-dispersed tree and shrub species had not been found by 1908 (Table 2, Fig. 3). By contrast, the build-up in numbers of true inland trees (i.e. animal- and wind-dispersed species) was slow to begin but has continued to the present day. This has proved to be a remarkably species-stable assemblage (Table 3), particularly when the difficulty of carrying out a thorough survey in the interior of these islands is considered. Of the medium and large animal- and wind-dispersed trees found in the first 50 years, all were also recorded in the 1979–83 period; compared with a figure of 56% for all remaining flowering plants.

Table 4 gives the cumulative species pattern for six of the numerically larger families of flowering plants of the Krakatoa Islands. The Moraceae of Krakatoa consists of 22 species of figs (*Ficus* spp.) plus *Artocarpus elasticus*, all of which are endozoically dispersed. Comparison of Tables 2 and 4 reveals the importance of this family in the species-richness of these forests. Of the 23 species of Moraceae, 17 were present within the first 40 years, at which point they represented 63% of the zoochorous tree species, including all six large species. Since then the forest of the interior has diversified somewhat, while remaining species-poor for the region (see Whitmore 1984).

The contribution of inward-migrating thalassochorous species has probably not been particularly great at any distance from the coast (above), while only five anemochorous tree species have been recorded altogether (three Rubiaceae, *Vernonia arborea* (Compositae) and *Bombax ceiba* (Bombacaceae)). The principal contribution to diversification has thus been in the addition of genera and families of endozoically dispersed tree species. Table 5 details the nine families of flowering plant first recorded after 1934, of which six are endozoically dispersed trees or shrubs, and four are large tree species of the interior. At the genus level, considering only the 1979–83 period, eight new genera were recorded which were

TABLE 2. Cumulative numbers of tree species on the three main islands since 1883. For definition of growth forms, see Table 1.

Date:	1886	1897	1908	1920	1922	1924	1929	1932	1934	1951	1979	1982	1983
Years elapsed:	3	14	25	37	39	41	46	49	51	68	96	99	100
Animal dispersed (all endozoically)													
Trees													
Large	0	0	1	5	6	6	6	8	8	11	12	13	13
Medium	0	0	0	2	2	2	3	4	4	6	9	11	11
Small	0	5	12	18	19	20	22	23	23	23	29	30	31
Total	0	5	13	25	27	28	31	35	35	40	50	54	55
Shrubs	0	1	8	12	17	19	21	23	23	25	31	33	33
Wind dispersed													
Trees													
Large	0	0	0	0	0	0	0	0	0	0	0	0	1
Medium	0	0	1	2	3	3	3	3	3	3	4	4	4
Small	0	0	0	0	0	0	0	0	0	0	0	0	0
Total	0	0	1	2	3	3	3	3	3	3	4	4	5
Shrubs	0	1	1	2	2	2	2	2	2	2	4	4	4
Sea dispersed													
Trees													
Large	1	1	1	1	1	1	1	1	1	1	1	3	3
Medium	0	5	7	7	7	7	7	8	8	8	8	8	8
Small	3	7	14	15	15	15	16	17	17	17	17	18	18
Total	4	13	22	23	23	23	24	26	26	26	26	28	28
Shrubs	2	5	13	14	14	14	14	15	15	15	16	16	16

TABLE 3. Persistence of the early assemblages. In each case column 1 (T1) provides the number of species found in period 1, and column 2 (T2) the number of these which were found in the 1979–83 collection period. In addition to the data given below, five small trees have been introduced by humans, two of which (*Mangifera indica*, *Gnetum gnemon*) were present in 1979–83

	1886–1908		1886–1934	
Period 1	T1	T2	T1	T2
Animal dispersed				
Trees				
Large	1	1	8	8
Medium	—	—	4	4
Small	12	9	23	15
Total	13	10	35	27
Wind dispersed				
Total	1	1	3	3
Sea dispersed				
Trees				
Large	1	1	1	1
Medium	7	4	8	5
Small	14	8	17	8
Total	22	13	26	14

represented by endozoically dispersed tree species. The character of these successful colonists indicates that access may be assumed to be more important in limiting establishment of other such species than is stage of habitat development.

It must, however, be remembered that species-richness is only one facet of forest maturity. The Krakatoa forests remain remarkably inequitable in the contributions to the canopy. Although Rakata has the most diverse forests, large areas of the island remain dominated by one or two species. Of 10 plots recorded in 1979–83 from between 30 and 500 m altitude, *Neonauclea calycina* accounted for >50% of the total basal area in seven (Bush & Whittaker 1986). The early regrowth forests of Kecil and Sertung are even more depauperate, as exemplified, respectively, by plots in which 153 of 163 and 112 of 120 trees (of >10 cm d.b.h.) were *Timonius compressicaulis*.

The Euphorbiaceae of Krakatoa also consist mostly of endozoically dispersed species, 14 of the 16 species (Table 4) being small trees or shrubs dispersed in this manner. In contrast, all but two of the Compositae are anemochorous, most are herbs and all but four of the species recorded had arrived within the first 50 years. Indeed, together with the Gramineae, this constitutes one of the swiftest groups of flowering plants to have colonized the islands; a pattern entirely consistent with their early successional role. Another important anemochorous family is the Orchidaceae, which is divided into terrestrial and epiphytic types in Table 4. The three early terrestrial orchids were heliophilous species (above), and there has been something of a progression among the terrestrial species towards those

TABLE 4. Cumulative numbers of species in particular families, on the three main islands, since 1883. The table includes all known species in each family, except those believed to have been introduced by humans. Shrubs and herbs include climbing or trailing forms but not true lianas. The exception to this is the Leguminosae, in which small lianas have been included as herbs, and large lianas have been entered seperately

Date:	1886	1897	1908	1920	1922	1924	1929	1932	1934	1951	1979	1982	1983
Years elapsed:	3	14	25	37	39	41	46	49	51	68	96	99	100
Compositae													
Trees, medium	0	0	0	0	1	1	1	1	1	1	1	1	1
Shrubs	0	1	1	1	1	1	1	1	1	1	2	2	2
Herbs	2	7	9	12	12	14	14	14	14	14	16	17	17
Total	2	8	10	13	14	16	16	16	16	16	19	20	20
Euphorbiaceae													
Trees, small	0	3	5	6	7	7	8	9	9	9	12	12	12
Shrubs	0	0	0	1	1	2	2	2	2	2	3	3	3
Herbs	0	1	1	1	1	1	1	1	1	1	1	1	1
Total	0	4	6	8	9	10	11	12	12	12	16	16	16
Gramineae	2	7	7	11	11	11	14	15	15	15	15	16	16

Leguminosae													
Trees													
Large	0	0	0	0	0	0	0	0	0	0	0	1	1
Medium	0	1	3	3	3	3	3	3	4	4	4	4	4
Small	1	3	5	6	6	6	6	6	6	6	6	7	7
Shrubs	0	3	3	5	5	5	5	5	5	5	5	5	5
Large lianas	0	1	3	4	4	4	4	4	4	4	4	4	4
Herbs	0	2	4	6	6	7	7	7	7	7	11	11	11
Total	1	10	18	24	24	25	25	25	26	26	30	31	31
Moraceae													
Trees													
Large	0	0	1	5	6	6	6	6	6	7	7	7	7
Medium	0	0	0	1	1	1	1	1	1	2	2	4	4
Small	0	4	5	6	6	6	6	6	6	6	6	7	7
Shrubs	0	0	2	3	4	4	4	4	4	4	5	5	5
Total	0	4	8	15	17	17	17	17	17	19	20	23	23
Orchidaceae													
Terrestrial	0	3	5	12	12	14	15	16	19	19	23	26	28
Ephiphytic	0	1	1	6	14	14	15	16	16	17	20	22	24
Total	0	4	6	18	26	28	30	32	35	36	43	48	52

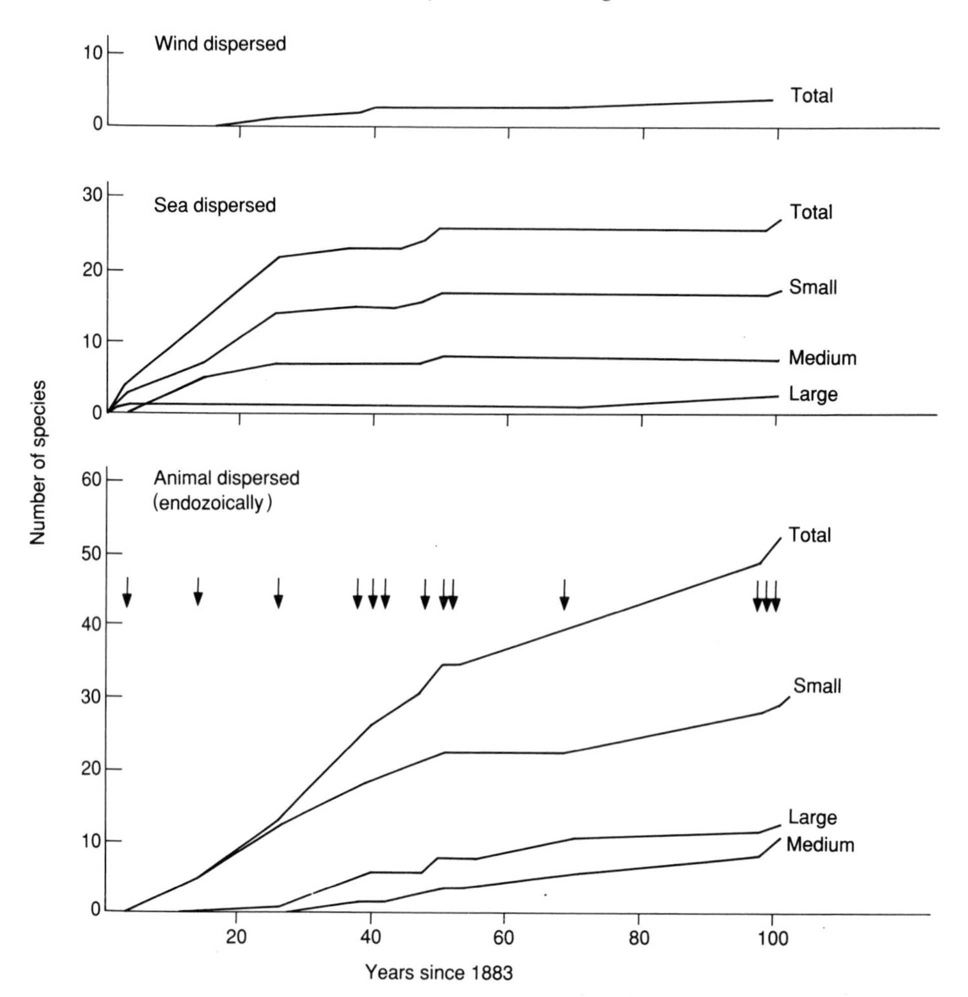

FIG. 3. Cumulative tree species totals through time on Krakatoa. Note the generally steady increase in zoochorous (animal-dispersed) species in comparison to the thalassochorous (sea-dispersed) species, in which an initial steep rise in number is followed by a long period of little change. Data from Table 2.

typical of shady, humid or damp habitats, e.g. *Tropidia curculigoides* and *Peristylis grandis* (both finds of 1979). Only two saprophytic species have been recorded, *Eulophia zollingeri* and *Galeola kuhlii* (a holosaprophyte, that lives on decaying tree trunks), both found in 1934. Interestingly the rate of increase of epiphytic and terrestrial species is very similar throughout the 100 years; however, one notable detail is that many of the epiphytes appear to be restricted to the humid submontane forest of upland Rakata (about 10 of the 24 species — according to Docters van Leeuwen, 1936).

The Leguminosae of Krakatoa are all thalassochorous, most are therefore

TABLE 5. Families of flowering plant first recorded after 1934. In each case only one species has been found, except for Commelinaceae, in which one other species was introduced by humans, but has since disappeared. All are endozoically dispersed species except *Clematis smilacifolia*, which is anemochorous. For each island, the first record only is given

Family	Only representative	Growth form	Year and island recorded
Burseraceae	*Canarium hirsutum*	Large tree	Kecil 1982
Elaeocarpaceae	*Elaeocarpus glaber*	Large tree	Rakata 1979
Myrtaceae	*Syzygium polyanthum*	Small tree	Rakata 1979 + Kecil 1982
Rosaceae	*Maranthes corymbosum*	Large tree	Rakata 1951
Sapotaceae	*Planchonella duclitan*	Large tree	Rakata 1951 Kecil 1982
Elaeagnaceae	*Elaeagnus latifolia*	Shrub	Rakata 1979 + Kecil 1982
Ranunculaceae	*Clematis smilacifolia*	Climber	Rakata 1979 + Kecil 1982
Commelinaceae	*Pollia secundiflora*	Herb	Rakata 1979
Begoniaceae	*Begonia isoptera*	Herb	Rakata 1983

strandline or very near-coastal in their distribution. Interestingly, several species occur in a much wider range of habitats on the mainland, e.g. *Albizzia chinensis* can occur up to 1300 m altitude on mainland Java, but at least up to 1934 was only found in the vicinity of the beach on Krakatoa. At that time, only one species, *Mucuna acuminata*, appeared to have successfully colonized to any distance inland, reaching *c*. 300 m altitude, presumably aided by bats or crabs. Our limited data do not suggest a marked changed in this situation to date (perhaps key secondary dispersal agents are lacking?), so that despite their numerical importance (Table 4), the Leguminosae remains significant only to the coastal assemblages.

CONCLUSIONS

The two independent forms of analysis presented in this paper have been shown to be mutually supportive in returning information on the manner in which the forest assemblages of the Krakatoa Islands are developing. These results detail the importance of dispersal mechanisms in determining patterns of succession, e.g. in the deflection of succession on Kecil and Sertung and in the gradual increase in inland tree species. The inland forests remain at an early successional stage, depaupaurate in tree species. The islands have continued to gain flowering plant genera and some families over the last 50 years, including several large endozoically dispersed tree species. In time, such colonists may be expected to be particularly important in enhancing the overall biological diversity of the islands (Bush & Whittaker 1991), although significantly, the difficulties of colonization continue to increase as the source areas of Java and Sumatra are depleted.

ACKNOWLEDGMENTS

We thank all those who contributed to data collection and, in particular, Keith Richards for his help in compiling the database. R.J.W. thanks Angela Whittaker for her support during the completion of this work and for comments on the text. The figures were drawn by Gillian Hardman. Krakatoa Research Project Publication No. 30.

REFERENCES

Backer, C.A. (1929). *The Problem of Krakatoa as seen by a Botanist* (translation from Dutch by A. Fitz). Published by the author, Surabaya.

Backer, C.A. & Backhuizen van den Brink, R.C. Jr (1963–68). *Flora of Java*, Three volumes. N.V.P. Noordhoff, Groningen.

Borrsum Waalkes, J. van. (1960). Botanical observations on the Krakatoa Islands in 1951 and 1952. *Annales Bogoriensis*, **4**, 1–64.

Bush, M.B. & Whittaker, R.J. (1986). The vegetation communities of Sertung, Rakata Kecil and Rakata. *The Krakatoa Centenary Expedition 1983: Final Report* (Ed. by M.B. Bush, P. Jones & K. Richards), pp. 14–55. Miscellaneous Series Number 33. Department of Geography, University of Hull, Hull.

Bush, M.B. & Whittaker, R.J. (1991). Krakatoa: Colonization patterns and hierarchies. *Journal of Biogeography*, **18**, 341–56.

Corner, E.J.H. (1988). *Wayside Trees of Malaya*, 3rd edn. United Selangor Press, Kuala Lumpur, Malaysia.

Dammerman, K.W. (1948). The fauna of Krakatoa 1883–1933. *Verhandelingen der Koninklijke Nederlandse Akademie van Wetenschappen Afdeling Natuurkunde (Tweede Sectie)*, **44**, 1–594.

Docters van Leeuwen, W.M. (1936). Krakatoa 1883–1983. *Annales du Jardin Botanique de Buitenzorg*, **46,47**, 1–506.

Ernst, A. (1908). *The New Flora of the Volcanic Island of Krakatoa*. Translation by A.C. Seward. Cambridge University Press, Cambridge.

Gauch, H.G. Jr (1982). *Multivariate Analysis in Community Ecology*. Cambridge Studies in Ecology 1. Cambridge University Press, Cambridge.

Hill, M.O. (1979). *TWINSPAN – a FORTRAN Program for Arranging Multivariate Data in an Ordered Two-way Table by Classification of the Individuals and the Attributes*. Cornell University, Ithaca, NY.

Hoogerwerf, A. (1953). Notes on the vertebrate fauna of the Krakatoa Islands, with special reference to the birds. *Treubia*, **22**, 319–48.

Partomihardjo, T.P., Mirmanto, E., Whittaker, R.J. (1992). Anak Krakatoa's vegetation and flora circa 1991, with observations on a decade of development and change. *Geojournal*, **28**, 233–48.

Ridley, H.N. (1930). *The Dispersal of Plants throughout the World*. Reeve, Ashford.

Treub, M. (1988). Notice sur la novelle flore de Krakatoa. *Annales du Jardin Botanique de Buitenzorg*, **7**, 213–22.

Whitmore, T.C. (1984). *Tropical Rain Forests of the Far East*, 2nd edition. Clarendon Press, Oxford.

Whitmore, T.C. & Tantra, I.G.M. (Eds) (1986). *Tree Flora of Indonesia: Check List for Sumatra*. Forest Research and Development Centre, Bogor, Indonesia.

Whittaker, R.J., Bush, M.B. & Richards, K. (1989). Plant recolonization and vegetation succession on the Krakatoa Islands, Indonesia. *Ecological Monographs*, **59**, 59–123.

Physiological controls over plant establishment in primary succession

F. S. CHAPIN III

Department of Integrative Biology, University of California, Berkeley, California 94720, USA

SUMMARY

1 The two main obstacles that confront a colonizer of primary succession are:
  (a) dispersal to a site that lacks a buried seed pool; and
  (b) establishment in a site that is infertile and often dry.

2 Primary successional colonizers commonly have traits that maximize dispersal but permit growth in dry, infertile sites. Most have light, wind-dispersed seeds with minimal dormancy requirements, germinating shortly after arrival rather than forming a buried seed pool. Woody species are more important as colonizers of primary succession than of secondary succession.

3 These primary colonizers have a low absolute growth rate because of small seed size rendering their small seedlings vulnerable to desiccation in soils that characteristically have low water-holding capacity. Their relative growth rate is higher than that of late successional species but lower than that of secondary successional colonizers.

4 Species with symbiotic N fixation are not among the initial colonizers because large seed size limits dispersal, but they quickly become part of the primary successional community because of their rapid growth on low-N soils.

5 Primary successional colonizers have high potential for both photosynthesis and nutrient uptake.

INTRODUCTION

There are well-accepted generalizations about differences between early and late successional species in terms of their life-history and physiological traits (e.g. Pianka 1978; Bazzaz 1979; Huston & Smith 1987). However, these generalizations are drawn almost entirely from studies of secondary plant succession. The purpose of this review is to describe the physiological traits of colonizers of *primary* succession and to suggest that these traits differ in important ways from traits of both secondary successional colonizers and late successional species. There has been much less ecophysiological research done in primary than in secondary succession, so that most of the generalizations made in this review should be treated as hypotheses that deserve more thorough examination in a range of successional sequences.

NATURE OF THE HABITAT

Primary succession is distinguished from secondary succession by its occurrence on soils with no previous history of plant growth. Surfaces that support primary succession include both natural features (e.g. glacial moraines, sand dunes, volcanic substrates and riparian point bars) and anthropogenically disturbed sites (e.g. mine spoils and gravel quarries). The lack of previous plant growth ensures that primary successional soils initially have a low organic content and therefore low-N availability, generally a low water-holding capacity and low cation exchange capacity. Successional patterns in these latter characteristics generally parallel the pattern shown for total soil N (Fig. 1; Roberts *et al.* 1981; Vitousek & Walker 1987). The high radiation input at the soil surface causes extreme fluctuations in soil temperature and rapid drying of surface soils in early succession (Bazzaz 1979), even in wet climates such as coastal rain forest (Cooper 1923). Thus, we expect primary successional colonizers to exhibit traits that maximize success in dry infertile soils.

Although primary successional soils initially have low-N availability, they can accumulate N quickly, if pH, mineral toxicity and climatic factors are not inhibiting and if colonizers are available to disperse to the site. Thus, many primary successional sites accumulate as much N within 50 years as secondary successional sites accumulate in 25 years (Fig. 1). Subsequently, the general patterns of N accumulation and of succession do not differ strikingly between primary and secondary succession. Only in extreme environments such as polar climates, deserts or toxic mine spoils is primary succession markedly slower than secondary succession.

Although early primary successional soils are usually less fertile than those of secondary succession (Fig. 1), there are numerous exceptions (Miles 1979). For example, abandoned agricultural soils that are highly degraded or which occur on infertile sand plains (Tilman 1987) often have low N availability, whereas riparian primary successional soils may have a substantial initial pool of soil N and organic matter, because the soils that are initially present are produced in part by upstream erosion of soils from established communities (Walker 1989). The initial N and organic contents of soils are probably more important than previous presence or absence of plants on the site in determining many physiological traits that are adaptive in early succession. Therefore, a strict dichotomy of traits between primary and secondary successional colonizers may not be particularly useful.

SEED SIZE AND GERMINATION

Primary successional soils initially lack a buried seed pool, requiring colonizers to disperse to the site. Species that colonize unvegetated sites in primary succession generally produce numerous small, wind-dispersed propagules. Thus, glacial moraines are typically colonized by genera such as *Salix, Epilobium* and *Dryas* (Cooper 1923) and river floodplains by *Salix, Populus* and *Equisetum*. Propagules

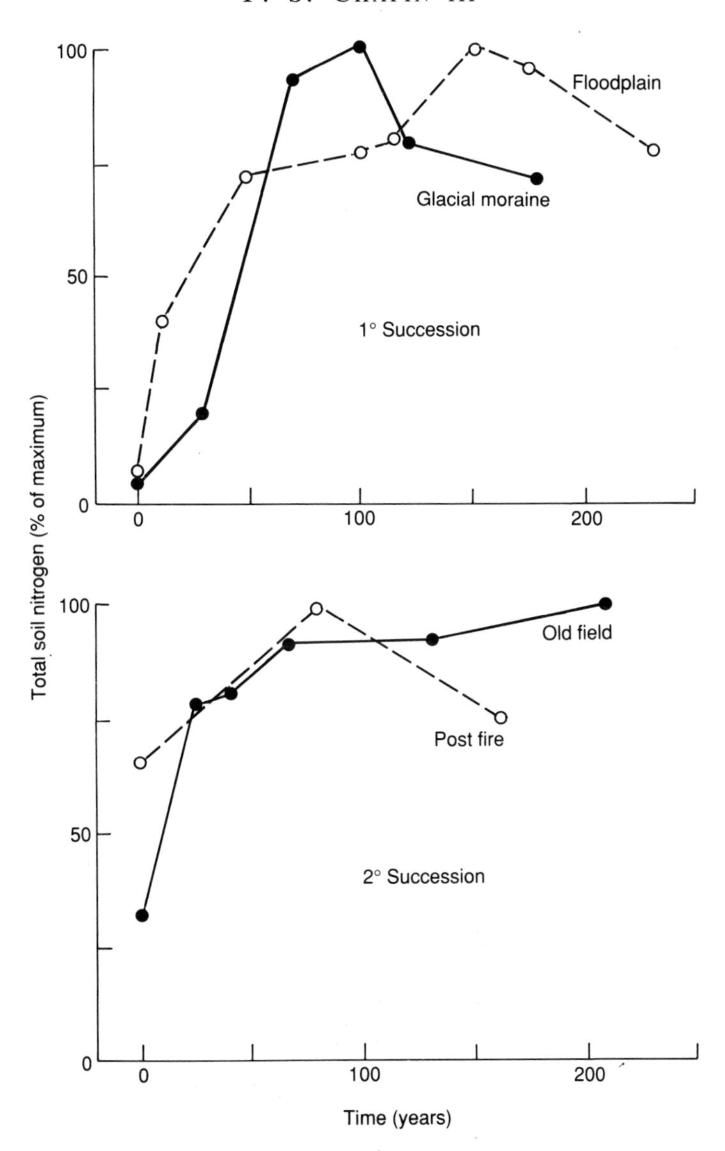

FIG. 1. Successional changes in total soil N (kg ha^{-1}) in two primary and two secondary successional ecosystems. The floodplain (Van Cleve *et al.* 1980), glacial moraine (Crocker & Major 1955) and post-fire (Van Cleve & Dyrness 1985) successions are in Alaska. The old-field succession (Switzer *et al.* 1979) is in Mississippi.

of these species are as small as, or smaller than, those of species which colonize in secondary succession (Table 1; Fig. 2), perhaps because colonizers in many primary successional environments have further to travel and, as discussed below, cannot accumulate in a buried seed pool. By contrast, late successional species of both primary and secondary succession produce large seeds (Salisbury 1942; Baker 1972) (Fig. 2).

TABLE 1. Seed mass and germination cues of some major plants of three Alaskan successional sequences. Data are mean ± SE, n = five groups of 50 seeds

	Successional status*	Seed mass (μg seed^{-1})	Germination cue
Riparian primary succession			
Salix alaxensis	Early	140 ± 3	None
Populus balsamifera	Early	315 ± 6	None
Alnus tenuifolia	Mid	514 ± 13	None
Picea glauca	Late	2565 ± 24	Cold
Glacial primary succession			
Epilobium latifolium	Early	72 ± 5	None
Dryas drummondii	Early	97 ± 18	None
Salix alaxensis†	Early	140 ± 3	None
Alnus sinuata	Mid	494 ± 23	None
Picea sitchensis	Late	2694 ± 26	Cold
Old-field secondary succession			
Chenopodium album	Early	1002 ± 9	Light, temperature
Hordeum jubatum	Early	1132 ± 65	Light, temperature
Rosa acicularis	Mid	11 668 ± 1482	Scarification
Betula papyrifera	Mid	276 ± 10	None
Picea glauca	Late	2565 ± 24	Cold

* Time of colonization.
† From Tanana River.

There are three general exceptions to the pattern of light, wind-dispersed seeds predominating in early primary succession. First, legumes are among the major colonizers of the banks of Arctic streams and rivers (Bliss & Cantlon 1957). These legumes have large seeds that are readily dispersed by water but are dispersed slowly to disturbed sites away from rivers (Bishop & Chapin 1989a). Legumes are also important colonizers of Alpine, glacial and volcanic disturbances where they are often dispersed by wind over the snow surface (Cooper 1923; Viereck 1967; Wood & del Moral 1987). Their large seed size ensures that N-fixing species are seldom among the first colonizers in primary succession (Grubb 1986), even though they form a prominent part of the community within a few years (Wood & del Moral 1987).

A second situation in which colonizing species have large seeds is in sites characterized by predictable physical disturbance (e.g. coastal strands, dunes and mud flats; Davy & Figueroa 1993). Here the need to rapidly establish a large root system apparently outweighs the advantage of effective dispersal.

Third, in some riparian successions vegetative material that is deposited during floods can be important as 'propagules' (Krasny *et al.* 1988).

Most seeds of wind-dispersed colonizers in primary succession have no dormancy requirement (Table 1). For example, *Salix* and *Populus* seeds generally die if they fail to land on a moist surface and germinate within 24 hours (Krasny *et al.* 1988). Because such species lack a dormancy requirement, they do not become incorpor-

small-seeded species have high root : shoot ratios immediately after germination (Fenner 1983). Early successional sites have low N availability and the species colonizing these sites have small seed reserves, thus the absolute growth rate of seedlings in the field is slow (Baker 1972; Fenner 1983), and seedlings remain vulnerable to drought for an extended period of time. As mentioned above, primary successional soils are initially low in organic matter so the uppermost soil horizons can desiccate rapidly. Even in wet climates, drought may be a major cause of seedling mortality early in primary succession (Cooper 1923; Wood & del Moral 1987; Chapin & Bliss 1989).

The low absolute growth rate of primary successional colonizers is a consequence of small seed size, but not necessarily of a low relative growth rate (RGR). In fact, small seed size generally leads to a high RGR (Fenner 1983; Gross 1984; Chapin *et al.* 1989). When grown under favourable laboratory conditions, early successional floodplain species grow more rapidly than do late successional species (Fig. 3). These patterns were also observed in secondary successional species in Britain (Fig. 4; Grime & Hunt 1975). In Britain, colonizers of primary successional skeletal habitats have a lower RGR than do colonizers of more fertile secondary successional disturbed sites (Fig. 4; Grime & Hunt 1975). Thus, within the colonizer guild, low soil fertility that is generally associated with primary succession appears to select for species with a low RGR (Parsons 1968; Grime 1977; Chapin 1980).

In Alaskan floodplain succession, early successional species (non-N fixers) were more sensitive to variation in nutrient supply than were late successional species (Fig. 3). Consequently, growth rate of the colonizers is particularly slow in the field on infertile early successional sites. However, if a seed of a colonizer lands in a favourable microsite, the seedling can grow rapidly and quickly produce

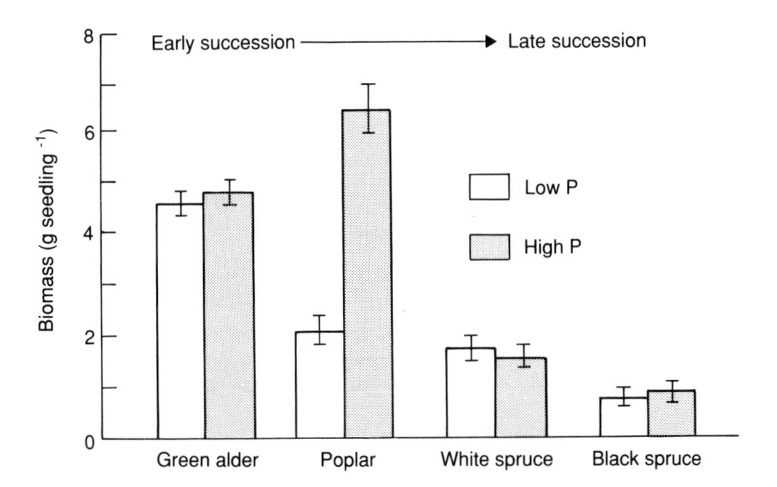

FIG. 3. Seedling biomass of Alaskan primary successional tree species grown under controlled conditions for 9 months under low or optimal phosphate (P) supply. Data are mean ± SE, $n = 5$. Redrawn from Chapin *et al.* (1983).

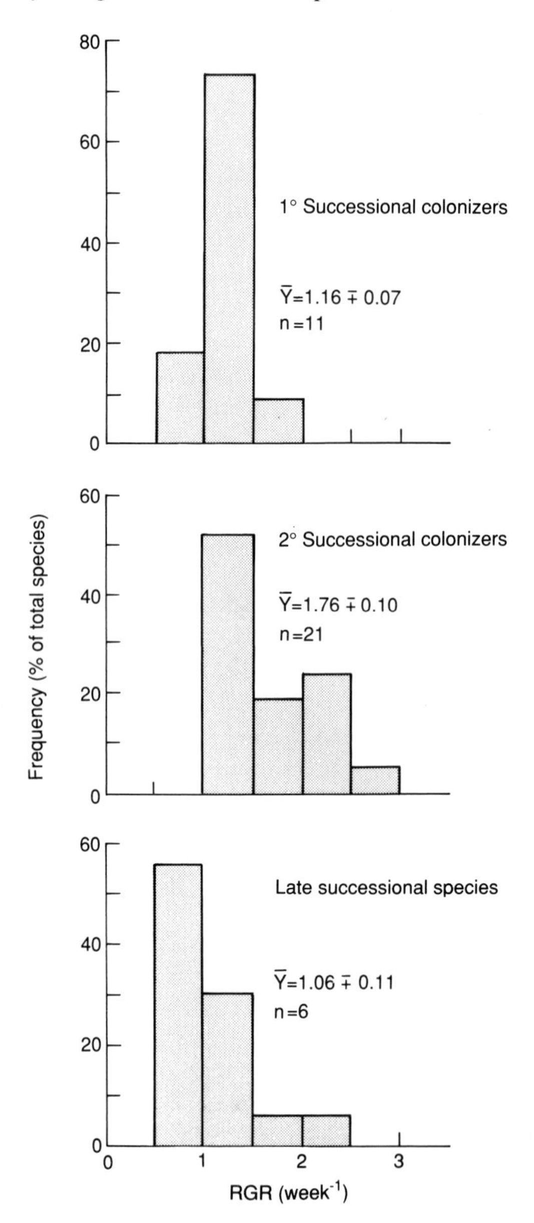

FIG. 4. Frequency distribution of RGR for British species that are colonizers of primary successional (skeletal), secondary successional (disturbed) or late successional (woodland) habitats. Calculated from Grime & Hunt (1975) after classifying species according to Grime *et al.* (1981).

a root system that renders the seedling less susceptible to desiccation. I suggest that microsite conditions are particularly critical to establishment of primary successional colonizers because of their small seed size and sensitivity of growth to nutrient supply. On Arctic gravel disturbances, virtually all establishment occurred at the edges of small stones (Bishop & Chapin 1989b), where there was both sufficient moisture and adequate light. A similar pattern was observed in early succession following glacial retreat (F. S. Chapin, unpubl. data). It is in these extreme sites early in primary succession that facilitative effects of nurse plants are most likely to be observed (Lawrence *et al.* 1967; Wood & del Moral 1987).

NITROGEN FIXATION

One of the most dramatic differences between primary and secondary succession is the prominent role played by symbiotic N fixation in the early stages of primary succession. N fixers play a crucial role in the accumulation of N in the ecosystem and are associated with a reduction in the degree of N limitation of plant growth through succession (Crocker & Major 1955; Palaniappan *et al.* 1979; Vitousek & Walker 1987). In Alaskan floodplain ecosystems, alder fixes half the N in 20 years that eventually accumulates during 250 years of succession (see Fig. 1; Van Cleve *et al.* 1971). On Hawaiian volcanic sites an exotic N-fixing tree *Myrica faya* has completely changed the pattern of primary succession by quadrupling the amount of N entering this N-limited ecosystem (Vitousek *et al.* 1987).

Relatively little is known about the requirements for establishment of N fixers in primary succession. Their relatively large seed size limits arrival rate. Among legumes, scarification is generally a prerequisite for germination (Grime *et al.* 1981). Extremely low N availability inhibits nodule formation in alder (Benecke 1970) and legumes (Sprent & Thomas 1984) and could limit initial establishment of N fixers in early primary succession. In fact, Grubb (1986) suggests that N fixers have large seeds, because only in this way can they ensure sufficient 'starter nitrogen' for initial establishment. Once nodule formation is initiated, N fixers provide sufficient N to support further nodulation and growth and N fertilization reduces the extent of nodule formation. Some small-seeded N fixers (e.g. *Dryas drummondii*) are not consistently nodulated (Fitter & Parsons 1987).

Other factors that are necessary for successful establishment of N fixers include adequate light and P. Light, which is abundant early in primary succession, is necessary because of the high energy demands of N fixation. P, which is required in substantial amounts to produce phosphorylated intermediates that are involved in the energetic transfers of N fixation, is generally most available in early succession (Walker & Syers 1976).

In early secondary succession N fixers are uncommon because the large energetic demands of N fixation (and consequent slow growth rates) place them at a disadvantage when competing with species that get their N from soil (Vitousek & Howarth 1991). In late succession, N fixers tend to decrease in abundance because of reduced light availability. However, in the tropics leguminous trees are important throughout succession.

PLANT NUTRITION

For those species that do not rely on symbiotic N fixation, the acquisition and retention of nutrients are major factors determining survival and growth in early primary succession. Those individuals that acquire inadequate nutrients remain small and are vulnerable to desiccation.

Plants growing in early primary succession typically have low tissue concentrations of both N and P, in contrast to high tissue nutrient concentrations in colonizers of more fertile secondary successional soils (Fig. 5). The low tissue N concentrations in early succession presumably reflect N limitation of growth. The low tissue P concentration could reflect either simple N limitation (which is often characterized by low tissue P concentrations) or low availability of both N and P.

In the Alaskan floodplain, early- and mid-successional species have a higher potential to absorb phosphate (Fig. 6; Walker & Chapin 1986), nitrate and ammonium (F. S. Chapin, unpubl. data) than does the late successional spruce. This could reflect the high potential growth rate and therefore the high nutrient demands of colonizing species (Chapin 1980). Individuals growing in early primary succession might be expected to have higher potential to absorb N than would the same genotype in late succession, because low nutrient status of plants enhances their potential to absorb growth-limiting nutrients (Table 2; Hoagland & Broyer 1936; Harrison & Helliwell 1979).

Absence of mycorrhizae might be an important factor limiting colonization in certain primary successional sites. In mine spoils there is often a low density of spores of mycorrhizal fungi, and plants have a low degree of mycorrhizal infection and consequently a slow growth rate. In postglacial succession, uncolonized sites have a pH of 8·0 (Crocker & Major 1955) which may inhibit ectomycorrhizal associations. Perhaps the slow growth of woody species in early primary succession reflects the lack of adequate mycorrhizal association due to high pH as much as inadequate N supply. It is noteworthy that in postglacial succession, spruce grows as well in the presence of willow (not a N fixer) as in the presence of N-fixing alder (Cooper 1923). Casual observation of roots of Sitka spruce in early succession at Glacier Bay failed to show substantial amounts of ectomycorrhizal association (F. S. Chapin, unpubl. data).

TABLE 2. Effect of N and P stress on N and P uptake by barley. Calculated from Lee (1982) and Lee & Rudge (1987)

Stress	Ion absorbed	Uptake rate by stressed plant (% of control)
N	Ammonium	209
	Nitrate	206
	Phosphate	56
P	Phosphate	400
	Nitrate	35

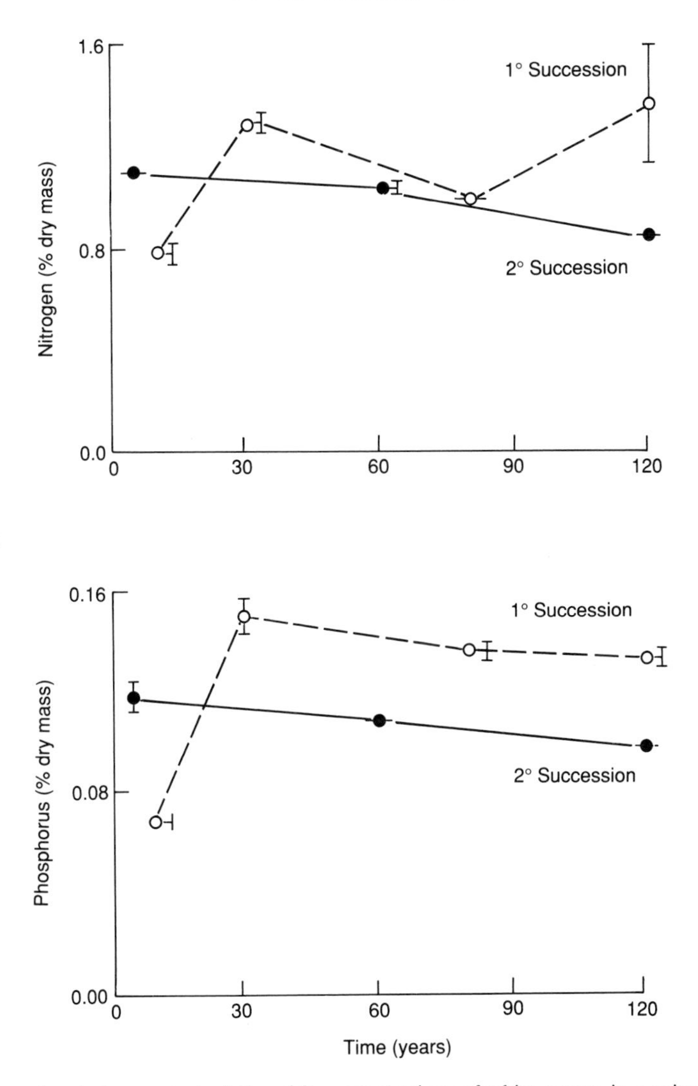

FIG. 5. Successional changes in leaf N and P concentrations of white spruce in a primary successional floodplain (Walker & Chapin 1986) and of black spruce in a post-fire secondary successional ecosystem (F. S. Chapin, unpubl. data) in interior Alaska. Data are mean ± SE, $n = 5$.

PHOTOSYNTHESIS

In secondary succession there are clear differences in photosynthetic characteristics between colonizing and late successional species (see Fig. 7; Bazzaz 1979). Early successional species have the highest potential for photosynthesis under high light because they have thick leaves and a large amount of photosynthetic enzymes per unit leaf area. However, the high protein content associated with this high photosynthetic potential leads to high rates of respiration and consequently a low

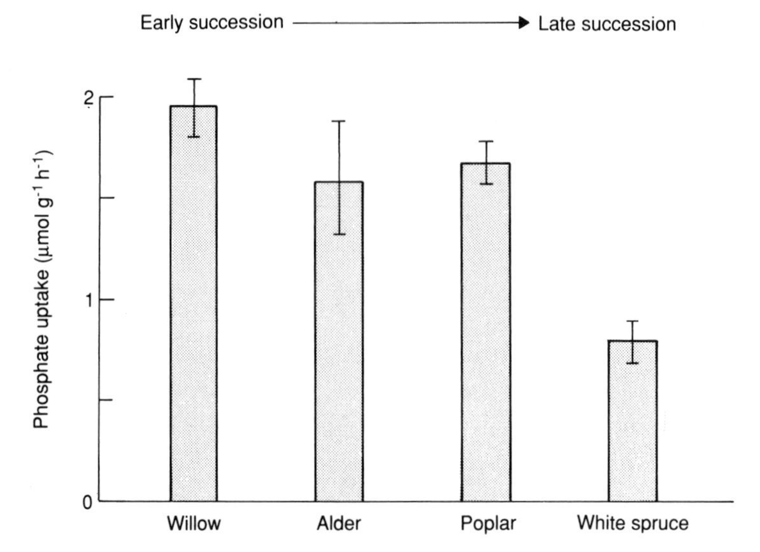

FIG. 6. Rate of phosphate uptake by excised roots of tree seedlings from an Alaskan primary successional floodplain sequence grown in a glasshouse (Walker & Chapin 1986). Data are mean ± SE, $n = 5$.

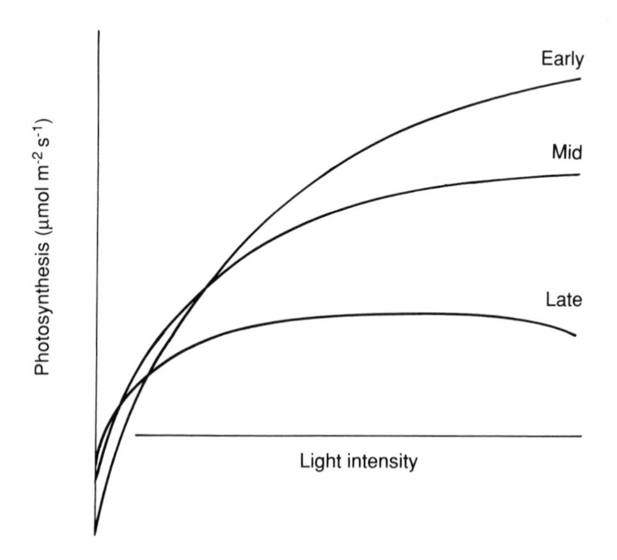

FIG. 7. Idealized light saturation curves for early, mid- and late successional plants. Redrawn from Bazzaz (1979).

rate of net photosynthesis at low light intensity (Fig. 7). By contrast, late successional species with their thinner leaves and lower content of photosynthetic enzymes have a lower photosynthetic rate at high light but maintain a substantial rate of photosynthesis at low light (Fig. 7).

These patterns in secondary succession lead to the following predictions: (a) in primary succession colonizing species will have a higher photosynthetic potential than late successional species, in part reflecting their higher growth rate, a pattern observed in Alaskan floodplain succession (Fig. 8; Walker & Chapin 1986); (b) within a species, plants growing in an early primary successional environment could tend to have a high photosynthetic potential (reflecting the thick leaves that develop in a high-light environment) or a low photosynthetic potential (reflecting low leaf N). The relative importance of these two effects would depend on the extent of change in light v. N availability through succession. In an Alaskan floodplain, where initial soil N availability is relatively high, early successional plants have a higher light-saturated photosynthetic rate than the same species growing in the understorey of a low-light mid-successional forest (Walker & Chapin 1986). However, on early successional moraines at Glacier Bay, Alaska, spruce growing in early succession has a very low leaf N concentration and correspondingly low rates of photosynthesis compared with plants growing in mid- and late succession (F. S. Chapin & C. L. Fastie, unpubl. data). Here, presumably the effect of increasing N through succession is a more important determinant of photosynthetic potential than is the changing light environment.

WATER RELATIONS

Drought may be the major cause of seedling mortality during the first year after germination in early primary succession (Chapin & Bliss 1989), as in many dry

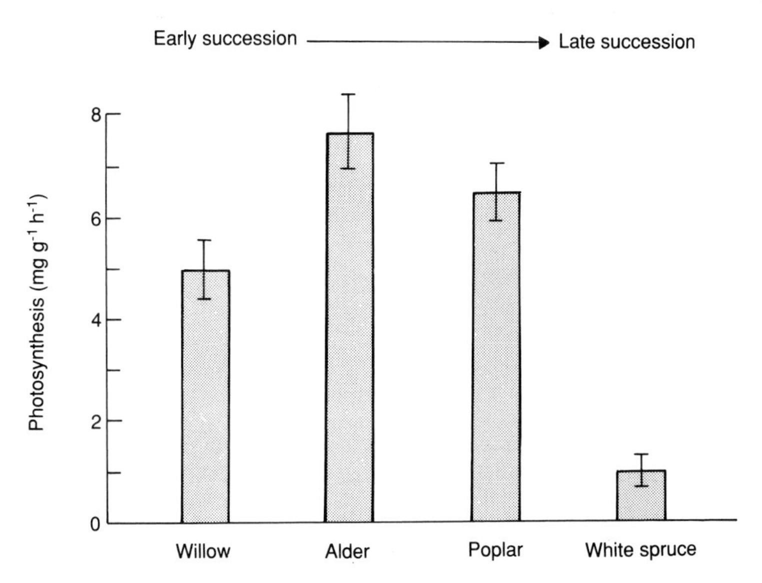

FIG. 8. Photosynthetic rate of tree seedlings from an Alaskan primary successional floodplain sequence after growth for 2 years in an early successional silt site (Walker & Chapin 1986). Data are mean ± SE, $n = 12$.

environments (Jordan & Nobel 1979; Frazer & Davis 1988). In light of the potential importance of drought-induced mortality, it has received surprisingly little attention in studies of primary succession. This would appear to be an important area for future research.

HERBIVORY

Herbivores are often a major cause of plant mortality during primary succession, once seedlings survive the initial establishment phase. Late successional species have an inherently slow growth rate (see Figs 3 & 4), and their growth is reduced even more by the low-light intensity of the late successional environment. Thus, seedlings of late successional species are small and remain vulnerable to mammalian herbivores for a long time. Consequently, they are strongly selected for effective chemical defence against herbivores. By contrast, early successional species have a higher potential growth rate and apparently allocate more energy to growth than to chemical defence, so they are preferred by generalist mammalian herbivores (Fig. 9; Bryant *et al.* 1983). Herbivory by mammals can be a major cause of disappearance of colonizing species during succession (Bryant & Chapin 1986).

COMPETITION

Initial species composition in primary succession depends primarily on factors determining arrival and initial establishment, as discussed above. However, subsequent changes in species composition reflect competitive balance (Tilman 1985) as well as patterns of differential arrival, longevity and growth rate (Egler 1954; Walker *et al.* 1986). Tilman (1985) has suggested that early successional species are competitively superior under conditions of high light and low availability of some soil resources (e.g. N), and late successional species are competitively superior under conditions of low light and high availability of soil resources. The well-known shade tolerance of late successional species is consistent with this resource-ratio hypothesis. However, the patterns of species composition in early succession could reflect either differential arrival and establishment or the outcome of competition. Future studies are required to determine whether competition plays a major role in governing the species composition in early primary succession.

CONCLUSIONS

The two main obstacles that confront a colonizer of primary succession are: (a) dispersal to a site that lacks a buried seed pool; and (b) establishment in a site that is infertile and often dry. To the extent that large seed size restricts effective dispersal but provides the initial capital for seedling establishment, there are trade-offs between traits which promote dispersal and those which promote establishment. Both the general patterns of physiological traits of primary successional colonizers and exceptions to these patterns can be understood in the context of this trade-off (Table 3). Primary successional colonizers generally have small,

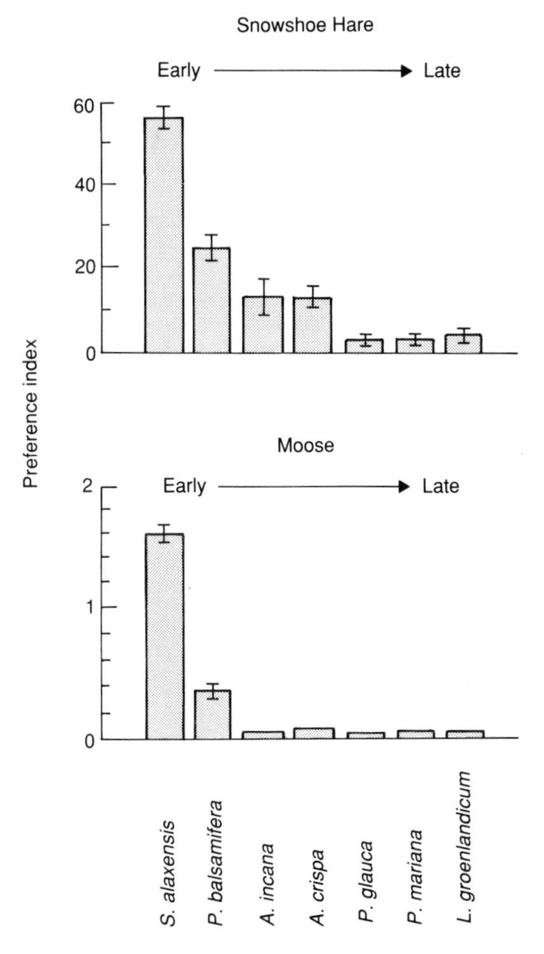

FIG. 9. Preference by two species of herbivores for plant species from an Alaskan primary successional floodplain sequence. Data are mean ± SE. Redrawn from Bryant & Chapin (1986).

wind-dispersed seeds which germinate immediately on arrival, although sites with extremely unstable substrate are often colonized by large-seeded species. As most primary successional colonizers have small seed reserves, their absolute growth rate is slow, and only those seedlings which encounter favourable microsites are likely to survive. Most primary successional colonizers have a relative growth rate that is higher than that of late successional species (reflecting smaller seed size of the colonizers) but lower than that of secondary successional colonizers (reflecting adaptation to dry, infertile soils within the colonizer guild). Major exceptions to this pattern are colonizers of extremely inhospitable sites (e.g. polar climates or toxic mine spoils) where colonizers have particularly low relative growth rates. Photosynthetic potential and nutrient uptake potential (which provide the resources for growth) and palatability to herbivores (which is high in rapidly growing species) are also higher in early than in late successional species but lower in

TABLE 3. Summary of traits of species that colonize primary succession

Trait	Major exceptions
Small seeds	N-fixing species Highly disturbed habitats (dunes, mudflats)
Lack of seed dormancy	Legumes (scarification)
Low absolute growth rate*	Large-seeded species (e.g. legumes)
High relative growth rate*	Toxic soils or extreme climate
High potential to absorb nutrients	Extreme environments
High photosynthetic potential*	Extreme environments
Predominantly trees and shrubs	
Intermediate longevity†	
High palatability to herbivores*	Extreme environments

* Higher than late successional species but lower than colonizers of secondary succession.
† Shorter than late successional species but longer than colonizers of secondary succession.

colonizers of primary than of secondary succession. Many colonizers of primary succession are short-lived trees and shrubs whose longevity is greater than that of herbaceous colonizers of secondary succession but less than the longevity of the late successional species which displace them. These traits of primary successional colonizers comprise an ecological strategy that has proven effective in a diverse array of primary successional ecosystems.

ACKNOWLEDGMENTS

Research leading to these generalizations was funded by the University of Alaska and by National Science Foundation grants DPP-8602568, BSR-8415281, BSR-8702629, and BSR-8817897.

REFERENCES

Baker, H.G. (1972). Seed weight in relation to environmental conditions in California. *Ecology*, **53**, 997–1010.

Bazzaz, F.A. (1979). The physiological ecology of plant succession. *Annual Review of Ecology and Systematics*, **10**, 351–71.

Bazzaz, F.A. & Pickett, S.T.A. (1980). Physiological ecology of succession: A comparative review. *Annual Review of Ecology and Systematics*, **11**, 287–310.

Benecke, U. (1970). Nitrogen fixation by *Alnus viridis* (chaix) DC. *Plant and Soil*, **33**, 30–48.

Bishop, S.C. & Chapin III, F.S. (1989a). Patterns of natural revegetation on abandoned gravel pads in arctic Alaska. *Journal of Applied Ecology*, **26**, 1970–81.

Bishop, S.C. & Chapin III, F.S. (1989b). Establishment of *Salix alaxensis* on a gravel pad in arctic Alaska. *Journal of Applied Ecology*, **26**, 575–83.

Bliss, L.C. & Cantlon, J.E. (1957). Succession on river alluvium in northern Alaska. *American Midland Naturalist*, **58**, 452–69.

Bryant, J.P. & Chapin III, F.S. (1986). Browsing–woody plant interactions during boreal forest plant succession. *Forest Ecosystems in the Alaskan Taiga. A Synthesis of Structure and Function* (Ed. by

K. Van Cleve, F.S. Chapin III, P.W. Flanagan, L.A. Viereck & C.T. Dyrness), pp. 213–25. Springer-Verlag, New York.

Bryant, J.P., Chapin III, F.S. & Klein, D.R. (1983). Carbon/nutrient balance of boreal plants in relation to vertebrate herbivory. *Oikos*, **40**, 357–68.

Chapin D.M. & Bliss, L.C. (1989). Seedling growth, physiology, and survivorship in a subalpine, volcanic environment. *Ecology*, **70**, 1325–34.

Chapin III, F.S. (1980). The mineral nutrition of wild plants. *Annual Review of Ecology and Systematics*, **11**, 233–60.

Chapin III, F.S., Groves, R.H. & Evans, L.T. (1989). Physiological determinants of growth rate in response to phosphorus supply in wild and cultivated *Hordeum* species. *Oecologia*, **79**, 96–105.

Chapin III, F.S., Tyron, P.R. & Van Cleve, K. (1983). Influence of phosphorus supply on growth and biomass allocation of Alaskan taiga tree seedlings. *Canadian Journal of Forest Research*, **13**, 1092–8.

Cooper, W.S. (1923). The recent ecological history of Glacier Bay, Alaska: II. The present vegetation cycle. *Ecology*, **4**, 223–46.

Crocker, R.L. & Major, J. (1955). Soil development in relation to vegetation and surface age at Glacier Bay, Alaska. *Journal of Ecology*, **43**, 427–48.

Davy, A.J. & Figueroa, M.E. (1993). The colonization of strandlines. *Primary Succession on Land* (Ed. by J. Miles & D.W.H. Walton), pp. 113–31. Blackwell Scientific Publications, Oxford.

Egler, F.E. (1954). Vegetation science concepts. I. Initial floristic composition, a factor in old field vegetation development. *Vegetatio*, **4**, 412–7.

Fenner, M. (1983). Relationships between seed weight, ash content and seedling growth in twenty-four species of Compositae. *New Phytologist*, **95**, 697–706.

Fitter, A.H. & Parsons, W.F.J. (1987). Changes in phosphorus and nitrogen availability on recessional moraines of the Athabasca Glacier, Alberta. *Canadian Journal of Botany*, **65**, 210–13.

Frazer, M.M. & Davis, S.D. (1988). Differential survival of chaparral seedlings during the first summer drought after wildfire. *Oecologia*, **76**, 215–21.

Grime, J.P. (1977). Evidence for the existence of three primary strategies in plants and its relevance to ecological and evolutionary theory. *American Naturalist*, **111**, 1169–94.

Grime, J.P. & Hunt, R. (1975). Relative growth rate: Its range and adaptive significance in a local flora. *Journal of Ecology*, **63**, 393–422.

Grime, J.P., Mason, G., Curtis, A.V., Rodman, J., Band, S.R., Mowforth, M.A.G., Neal, A.M. & Shaw, S. (1981). A comparative study of germination characteristics in a local flora. *Journal of Ecology*, **69**, 1017–59.

Gross, K.L. (1984). Effects of seed size and growth form on seedling establishment of six monocarpic perennial plants. *Journal of Ecology*, **72**, 369–87.

Grubb, P.J. (1986). The ecology of establishment. *Ecology and Landscape Design* (Ed. by A.D. Bradshaw, D.A. Loove & A.E. Thorpe), pp. 83–97. Blackwell Scientific Publications, Oxford.

Harrison, A.F. & Helliwell, D.R. (1979). A bioassay for comparing phosphorus availability in soils. *Journal of Applied Ecology*, **16**, 497–505.

Hoagland, D.R. & Broyer, T.C. (1936). General nature of the process of salt accumulation by roots with description of experimental methods. *Plant Physiology*, **11**, 471–507.

Huston, M. & Smith, T. (1987). Plant succession: life history and competition. *American Naturalist*, **130**, 168–98.

Jordan, P.W. & Nobel, P.S. (1979). Infrequent establishment of seedlings of *Agave deserti* (Agavaceae) in the northwestern Sonoran Desert. *American Journal of Botany*, **66**, 1079–84.

Krasny, M.E., Vogt, K.A. & Zasada, J.C. (1988). Establishment of four Salicaceae species on river bars in interior Alaska. *Holarctic Ecology*, **11**, 210–9.

Lawrence, D.B., Schoenike, R.E., Quispel, A. & Bond, G. (1967). The role of *Dryas drummondii* in vegetation development following ice recession at Glacier Bay, Alaska, with special reference to its nitrogen fixation by root nodules. *Journal of Ecology*, **55**, 793–813.

Lee, R.B. (1982). Selectivity and kinetics of ion uptake by barley plants following nutrient deficiency. *Annals of Botany*, **50**, 429–49.

Lee, R.B. & Rudge, K.A. (1987). Effects of nitrogen deficiency on the absorption of nitrate and ammonium by barley plants. *Annals of Botany*, **57**, 471–86.

Miles, J. (1979). *Vegetation Dynamics*. Chapman and Hall, London.

Palaniappan, V.M., Marrs, R.H. & Bradshaw, A.D. (1979). The effect of *Lupinus arboreus* on the nitrogen status of china clay wastes. *Journal of Applied Ecology*, **16**, 825–31.

Parsons, R.F. (1968). The significance of growth-rate comparisons for plant ecology. *American Naturalist*, **102**, 595–7.

Pianka, E.R. (1978). *Evolutionary Ecology*, 2nd edn. Harper and Row, New York.

Roberts, R.D., Marrs, R.H., Skeffington, R.A. & Bradshaw, A.D. (1981). Ecosystem development on naturally colonized china clay wastes. I. Vegetation changes and overall accumulation of organic matter and nutrients. *Journal of Ecology*, **69**, 153–61.

Salisbury, E.J. (1942). *The Reproductive Capacity of Plants; Studies in Quantitative Biology*. Bell, London.

Sprent, J.I. & Thomas, R.J. (1984). Nitrogen nutrition of seedling grain legumes: some taxonomic, morphological and physiological constraints: opinion. *Plant, Cell and Environment*, **7**, 637–45.

Switzer, G.L., Shelton, M.G. & Nelson, L.E. (1979). Successional development of the forest floor and soil surface on upland sites of the East Gulf Coastal Plain. *Ecology*, **60**, 1162–71.

Thompson, K. (1987). Seeds and seed banks. *New Phytologist*, **106** (Suppl.), 23–34.

Thompson, K. & Grime, J.P. (1983). A comparative study of germination responses to diurnally fluctuating temperatures. *Journal of Applied Ecology*, **20**, 14–56.

Tilman, D. (1985). The resource-ratio hypothesis of plant succession. *American Naturalist*, **125**, 827–52.

Tilman, D. (1987). Secondary succession and the pattern of plant dominance along experimental nitrogen gradients. *Ecology*, **57**, 189–214.

Van Cleve, K. & Dyrness, C.T. (1985). The effect of the Rosie Creek fire on soil fertility. *Early Results of the Rosie Creek Fire Research Project 1984*. (Ed. by G.P. Juday & C.T. Dyrness), pp. 7–11. University of Alaska Agriculture and Forestry Experiment Station Miscellaneous Publication 85–2, Fairbanks, Alaska.

Van Cleve, K., Dyrness, C.T. & Viereck, L. (1980). Nutrient cycling in interior Alaska flood plains and its relationship to regeneration and subsequent forest development. *Forest Regeneration at High Latitudes* (Ed. by M. Murray & R.M. Van Veldhuizen), pp. 11–18. USDA Forest Service, Portland, Oregon.

Van Cleve, K., Viereck, L.A. & Schlentner, R.L. (1971). Accumulation of nitrogen in alder (*Alnus*) ecosystems near Fairbanks, Alaska. *Arctic and Alpine Research*, **3**, 101–14.

Viereck, L.A. (1967). Plant succession and soil development on gravel outwash of the Muldrow Glacier, Alaska. *Ecological Monographs*, **36**, 181–99.

Viereck, L.A. (1973). Wildfire in the taiga of Alaska. *Quaternary Research*, **3**, 465–95.

Vitousek, P.M. & Howarth, R.W. (1991). Nitrogen limitation on land and in the sea: How can it occur? *Biogeochemistry*, **13**, 87–115.

Vitousek, P.M. & Walker, L.R. (1987). Colonization, succession and resource availability: ecosystem-level interactions. *Colonization, Succession and Stability* (Ed. by A.J. Gray, M.J. Crawley & P.J. Edwards), pp. 207–23. Blackwell Scientific Publications, Oxford.

Vitousek, P.M., Walker, L.R., Whiteaker, L.D., Mueller-Dombois, D. & Matson, P.A. (1987). Biological invasion by *Myrica faya* alters ecosystem development in Hawaii. *Science*, 238, 802–4.

Walker, L.R. (1989). Soil nitrogen changes during primary succession on an Alaskan floodplain. *Arctic and Alpine Research*, **21**, 341–9.

Walker, L.R. & Chapin III, F.S. 1986). Physiological controls over seedling growth in primary succession on an Alaskan floodplain. *Ecology*, **67**, 1508–23.

Walker, L.R., Zasada, J.C. & Chapin III, F.S. (1986). The role of life history processes in primary succession on an Alaskan floodplain. *Ecology*, **67**, 1243–53.

Walker, T.W. & Syers, J.K. (1976). The fate of phosphorus during pedogenesis. *Geoderma*, **15**, 1–19.

Wood, D.M. & del Moral, R. (1987). Mechanisms of early primary succession in subalpine habitats on Mount St. Helens. *Ecology*, **68**, 780–90.

The vascular plant pioneers of primary successions: Persistence and phenotypic plasticity

A. J. GRAY

Institute of Terrestrial Ecology (Natural Environment Research Council), Furzebrook Research Station, Wareham, Dorset BH20 5AS, UK

SUMMARY

1 The pioneer plant species of primary successions do not display the sets of correlated life-history traits said to characterize early secondary successional or colonizing species. This may be partly a problem of semantics and definition, but is principally a result of the time-scales of comparison and of major differences between primary and secondary successions, particularly in the initial availability of resources.

2 Many (and in some case studies, most) of the pioneers of primary succession are long-lived, iteroparous perennials which are frequently slow-growing and some of which persist into the later stages of succession. Herein lies one of the most intriguing genetical and evolutionary problems concerning pioneers: the persistence of some populations (and perhaps individuals) from open, resource-poor habitats through to crowded, more mesic habitats.

3 The changes which occur during this process may include genetic differentiation resulting from selection for increased competitive ability, biotype, and/or genotype depletion and plastic adjustment of the phenotype. Habitat-correlated differences in phenotypic plasticity in the pioneer sand-dune species *Ammophila arenaria* suggest that genetic differentiation may also include the selection of individuals differing in the phenotypic flexibility of specific traits. Comparison of three different sets of successional (and age-related) populations of *A. arenaria* partly, but not unequivocally, support this hypothesis.

INTRODUCTION

Understanding of the processes of plant community assembly and development has been advanced in recent years by a revitalized interest in colonizing and invading species. Boosted by an international scientific programme on the ecology of biological invasions (Mooney & Drake 1986; Groves & Burdon 1986; Macdonald *et al.* 1986; Kornberg & Williamson 1987; Drake *et al.* 1989), this review of an old topic (Elton 1958; Baker & Stebbins 1965) has raised some new questions and generated the mandatory terminological variety. Thus, before attempting to characterize the genetic and life-history attributes of primary successional species, the following section will cover definitions.

DEFINITIONS AND GENERALIZATIONS

In the sense that they have to establish, all plants are colonizers (Bazzaz 1986, in discussion of Grime 1986). However, a group of species that habitually enters transient or disturbed habitats has been recognized and dubbed 'colonizing species' (Baker & Stebbins 1965). This group of species has been the subject of considerable research and many generalizations. Some of its members have become pest species or 'weeds', particularly when introduced accidentally or deliberately to new continents (e.g. Mediterranean annuals in North America). Therefore, some colonizing species are also 'invading species', entering territories in which they have never before occurred (Mack 1985), although invaders may be defined as a wider group which includes species that can establish in undisturbed situations (Gray 1986, and in discussion of Grime 1986).

The species that establish in the early stages of succession are frequently regarded as a type of colonizing species, for example Bazzaz (1986, p. 102) refers to 'early successional plants (which are one class of colonizing species)'. At the same time, comparisons of their physiological, life-history and genetic characteristics have produced contrasting profiles of early and late successional species (Bazzaz 1979; Bazzaz & Pickett 1980; also references below). Indeed, Huston & Smith (1987, p. 169) describe the 'set of characteristics used to distinguish early- from late-successional species' as 'one of the oldest and most widely accepted generalisations in plant ecology'. Taken together, these two approaches — the characterization of colonizing species and the comparison of early and late successional species — have yielded a set of expectations about the life-history and breeding system attributes of plants found in the early stages of succession. These are typified by statements such as:

> Early successional, weedy species are characterized as short-lived, often annual plants with wind or animal-attached dispersal, seed dormancy, hermaphroditic flowers that are selfed or pollinated by wind or various small insects, and vegetative reproduction. These traits permit colonizing species to exploit open, early successional habitats.
>
> [Loveless & Hamrick (1984, p. 77)]

There are many similar generalizations in the literature.

Generalizations about the genetic attributes of early successional and colonizing species have been rather more equivocal. By genetic attributes I mean here both the set of characteristics which collectively determine the form in which, and the frequency with which, genetic information is passed from generation to generation (the genetic system) and also the result of the interaction of those characteristics and others on the distribution of heritable variation within and among populations (genetic structure). The genetic attributes most frequently ascribed to colonizing species (*sensu lato*) are a uniparental mode of reproduction (self-fertilization, argamospermy or clonal propagation), fixed heterozygosity arising from polyploidy, higher levels of phenotypic plasticity and greater between-population genetic differentiation (presumably stemming from a combination of the mating system, faster population turnover and more founding events). Empirical observations have not always agreed with these expectations, which are partly based on theoretical predictions, and more recent reviews have struck a more cautious note, stressing the variety of genetic attributes shared by colonizers (Brown & Marshall 1981; Jain 1983; Rice & Jain 1985; Barrett & Richardson 1986; Bazzaz 1986; Gray 1986; Brown & Burdon 1987). Indeed, Brown & Marshall (1981, p. 361) conclude that the great interest for geneticists in colonizing species is 'not so much because they form a homogeneous group, but because they display a wide range of evolutionary pathways'.

Similarly, Grubb (1986, 1987) has recently pointed out that there is great diversity of form and life-history among colonizing species (which he defines as 'the first organisms to become established in a succession'). This is particularly true of plants that colonize primary successions, which, following During (1979), Grubb distinguishes as a subset of colonizers called 'pioneers'. In the second paper he traces the linking of ideas about early and late successional plants with those of r- and K-selection (Grubb 1987), an association embedded in many of the generalizations discussed earlier. I will not repeat here Grubb's argument for rejecting this misleading and erroneous association, but in the next section analyse in more detail the spectra of life-histories among species which establish in the early stages of primary successions.

Finally, however, it is necessary to provide a working definition of 'primary succession'. Following Clements (1904), who first recognized two types of succession, I will regard primary succession as *vegetation succession on a land surface which has not formerly, or in the recent past, been vegetated and which initially contains no higher plant propagules*. Although avoiding the problem of microbial colonization, this definition enables the inclusion, along with successions on sand-dunes, salt-marshes, shingle and deglaciated surfaces, of those on volcanic ash, and some quarries and industrial waste tips (and the exclusion of artificially bared surfaces such as road embankments).

LIFE-HISTORY VARIATION IN PIONEER SPECIES OF PRIMARY SUCCESSIONS

Fig. 1 provides an analysis of the distribution of life-history types of plant species which establish in the early stages of two natural and two man-made surfaces satisfying the definition of primary succession given above. Two general points can be made. First, there is a relatively wide spectrum of types, and second, the most common species in these biotopes are iteroparous perennials.

The salt-marsh species in the British flora (Fig. 1a) display an interesting distribution of life-histories in that those of the main, regularly flooded marsh are predominantly (90%) iteroparous perennials, while 70% of those in the upper marsh and paramaritime fringe are annuals or short-lived (Gray 1991). Apart from individuals of *Salicornia* spp. and *Suaeda maritima*, the pioneer plants on salt-marshes may be extremely long-lived (Gimingham 1964; Jefferies 1972; Gray

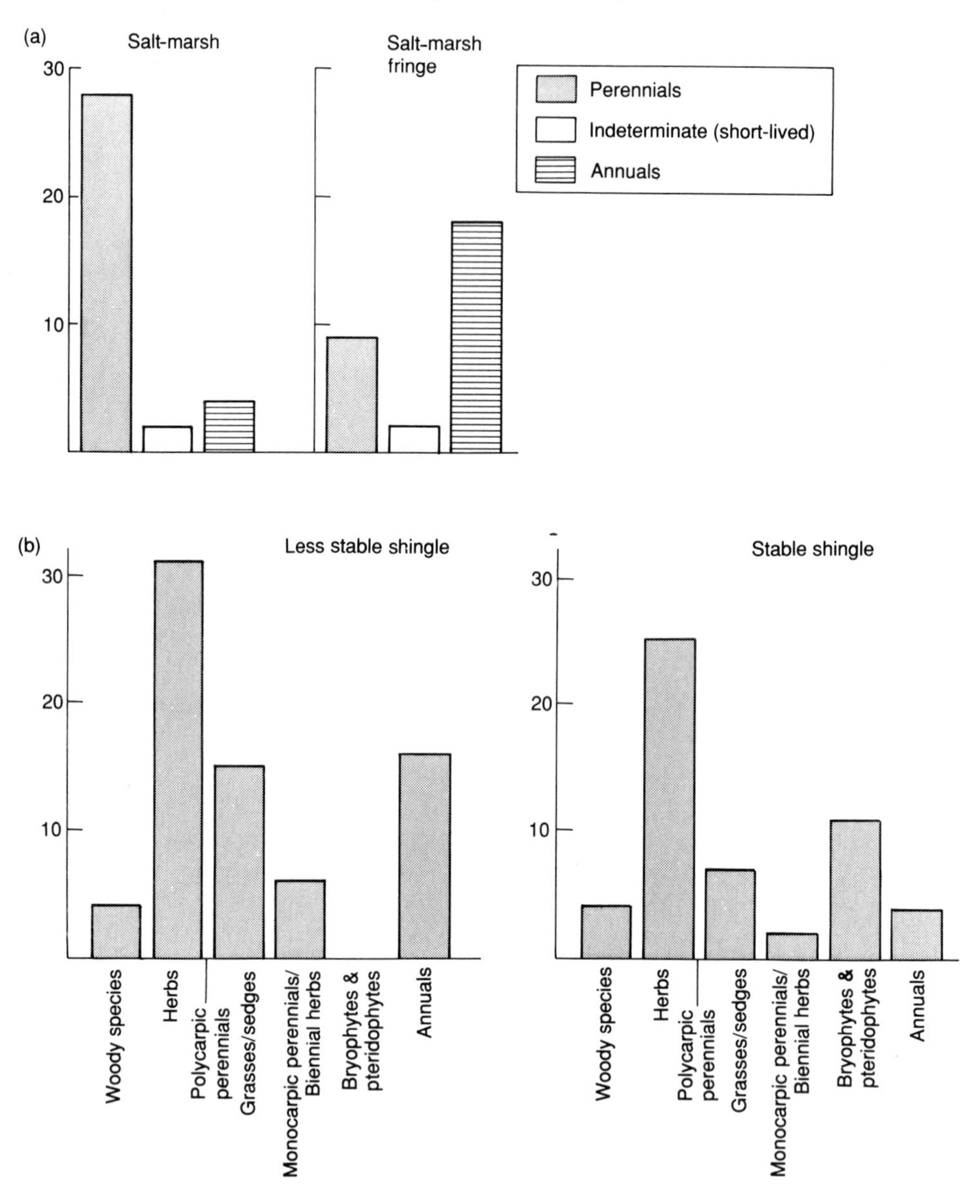

FIG. 1. Life-history characteristics of vascular plant pioneers: (a) the salt-marsh species of the British flora from the regularly flooded marsh (left) and the upper marsh fringe (right) (after Gray 1991). (b) The shingle flora of Britain on less stable (left) and more stable (right) shingle (data from Scott 1963). (*Opposite*) (c) The species naturally colonizing bare chalk and scree in an English quarry (data from Locket 1945). (d) The species naturally colonizing china clay waste tips in western England (data from Roberts *et al.* 1981). The *y* axis on all graphs gives the actual numbers.

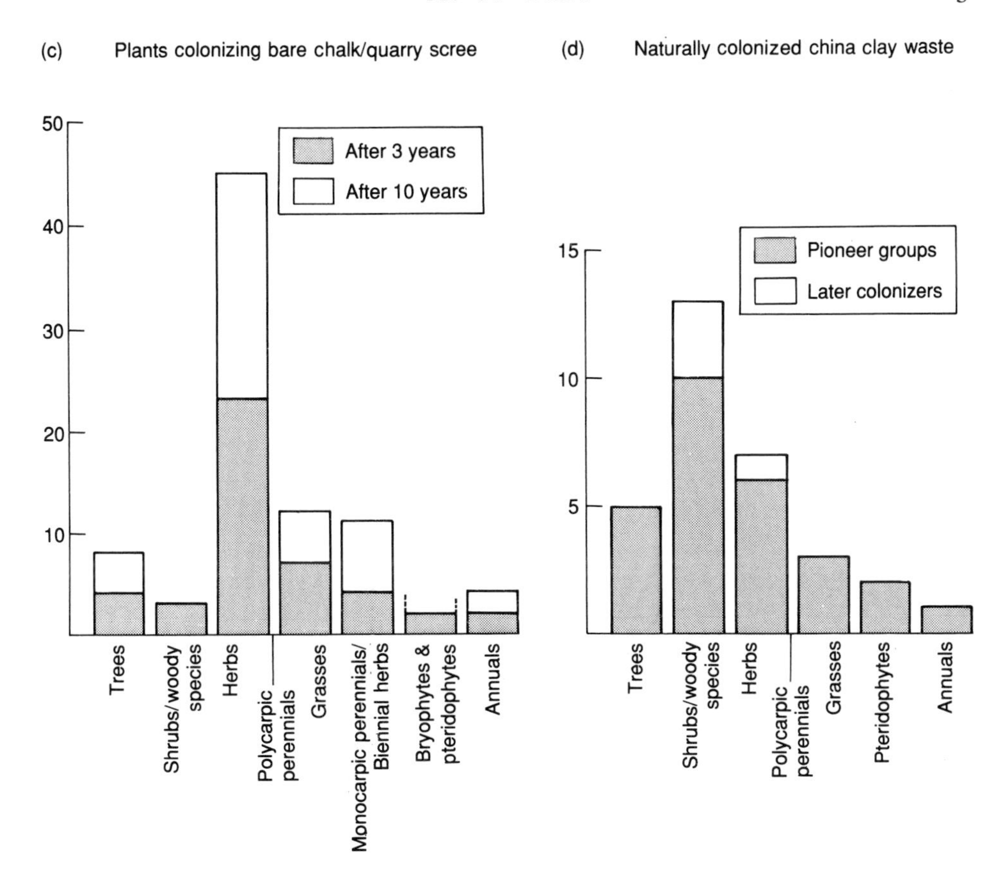

1985). They include clonal grasses (e.g. *Spartina anglica, Puccinellia maritima*) and tap-rooted or woody herbs (e.g. *Limonium vulgare, Halimione portulacoides*). Interestingly, several species of the maritime fringe of salt-marshes, including annuals (e.g. *Atriplex* species) and perennials (*Rumex crispus*), appear as colonists of relatively unstable shingle beaches (Fig. 1b). Scott (1963) stressed the role of beach stability and stone size in regulating shingle floras, and many of the species in his lists of relatively unstable shingle reflect the nature of the matrix (saline mud or sand) in which the shingle is embedded. The importance of fine particles in shingle species' establishment has been demonstrated experimentally by Fuller (1987), and very coarse and/or mobile shingle with a very small fine component supports only a few 'specialist' shingle species, almost invariably long-lived perennials (e.g. *Crambe maritima, Suaeda vera*). Apart from driftline plants, the relatively large number of annuals on shingle beaches are opportunist species more commonly found in other open biotopes, including dry, frequently rocky sites (e.g. *Aira praecox, Catapodium marinum*), saline muds (e.g. *Salicornia* spp., *Suaeda maritima*) and arable wasteland (e.g. *Senecio vulgaris, Anagallis arvensis*).

Scott (1963) notes the important role of encrusting lichens as colonists of relatively stable shingle and lists several species (not shown in Fig. 1).

More than 80% of the species colonizing bare chalk in a quarry in southern central England (Locket 1945) were relatively long-lived perennials such as trees and shrubs or polycarpic grasses and herbs (the latter comprising the biggest group with 54%; Fig. 1c). Only four annuals are listed among the 86 species recorded during the first 10 years. Annual plants are also rare among those species colonizing abandoned waste tips produced by the china clay industry in western England (Fig. 1d). Roberts *et al.* (1981) list only one (*Lamium purpureum*) from the tips of largely sand-sized quartz particles in Cornwall. They observed an age-related sequence with trees appearing in a latter group of colonizers following a stage in which shrubs (e.g. *Lupinus arboreus, Calluna vulgaris, Ulex europaeus* and *Sarothamnus scoparius*) in particular, but also perennial herbs, were especially common. At least one of these pioneer species, *Ulex europaeus*, persists through the later successional stages into woodland, and was found on the oldest site (116 years).

Undoubtedly the presence of shrubs, including dwarf shrubs such as *Calluna vulgaris* and *Erica cinerea*, as waste-tip colonizers reflects their abundance in the surrounding countryside, a point made by Grime (1986) in his analysis of the flora of spoil habitats in the Sheffield region. All but two of the 20 most common vascular plant species on spoil were perennial, and over half were grasses widespread in the region as a whole. However, it was possible to distinguish a group of specialist colonizers, generally scattered through the region and typified by ephemerals such as *Chaenorhinum minus, Minuarta verna* and *Senecio viscosus*. Most of these colonizers depend on efficient distribution by wind or by human activity.

There are parallels between spoil heap colonization and that on volcanic islands or lava flows. The pioneer plants on both represent the most efficiently dispersed (large production of generally small, often wind-dispersed, propagules) of the most widely available nearby species which can establish in and tolerate the extreme edaphic (and climatic) conditions. Thus, 11 of the 24 species of higher plants found on Rakata, the largest of the Krakatoa islands, 3 years after the active volcano of 1883, were ferns (Treub, in Whittaker *et al.* 1989). On Anak Krakatoa, a new volcanic island establishing between 1927 and 1930, ferns were not a part of the pioneer flora, although colonization has been interrupted repeatedly by volcanic eruptions and has been dominated by sea-dispersed species. In the sub-Alpine habitats on Mount St Helens studied by Wood & del Moral (1987), the post-volcanic substrates were very patchily colonized, largely because of poor dispersal of tolerant plants and poor tolerance of site conditions by well-dispersed plants. The early successional plants on the debris of the avalanche on Mount St Helens were largely wind-dispersed (Dale 1989).

The importance of substrata in determining which pioneer plants colonize deglaciated areas was emphasized by Grubb (1986, 1987), who gives several examples. Primary successions on deglaciated areas in Alaska are described by

Cooper (1931) and Crocker & Major (1955), and again emphasize the importance of efficient dispersal and the predominance of perennial species among the pioneers, which include mosses, a horsetail and a fern, but most notably several species of willows. *Equisetum* and *Salix* species are also the pioneers on silt bars of the glacially fed Tanana River in Alaska (Walker *et al.* 1986).

Finally, a contrast is provided by the salt-tolerant annual species of *Salicornia*, and *Atriplex* which colonized the sandy flats of the newly enclosed Lauwerzee polder in northern Holland (Joenje 1979) and the wind-dispersed perennials, such as *Cirsium arvense* and *Tussilago farfara*, which invaded the mudflats of the newly reclaimed Ijselmeer polders to the south-west (Bakker 1960). Succession to a stage dominated by perennials was delayed in the Lauwerzee polder by wildfowl grazing (Joenje 1985).

WHY DO PIONEERS OF PRIMARY SUCCESSIONS DIFFER FROM COLONIZING SPECIES?

The divergence with respect to life-history among pioneer species illustrated in the previous section will probably be reflected in a variety of genetic attributes. Other aspects of the genetic system, for example pollination modes and mating systems, are likely to be equally variable. For example, of the 43 early colonist species of bare chalk listed by Locket (1945), and shown in Fig. 1c, 29 are insect-pollinated and 14 wind-pollinated, 11 are known to be self-incompatible and a further six thought to be mainly outbreeding, while there are nine known self-compatible species with a further five having at least some apomictic forms. Although the breeding systems of many species are imperfectly known, it would be labouring the point to make similar analyses of other groups of pioneers.

Instead, in this section I will consider briefly why vascular plant pioneers of primary successions do not display the sets of correlated life-history traits believed to characterize early successional or colonizing species, before, in the final section, discussing some genetic and evolutionary questions which a study of pioneers may help to answer.

There are at least three reasons why pioneers öf primary successions and colonizing species differ. One may be semantic. For example, both Bazzaz (1986) and Grubb (1987) use different (and perfectly acceptable) working definitions of 'colonizing species' from that used here (and in Gray 1986). Thus, different subsets of species may be being compared by different reviewers.

A related, and more important, aspect is the time-scale of comparison. Where groups of similar taxa are being compared, it is valid and reasonable to classify some as early successional or colonizing in relation to others, as for example in comparing North American conifer species (Hamrick 1983 and references therein; Govindaraju 1984). Such comparisons provide valuable insights into the selective forces shaping patterns of differentiation in natural populations. Similarly, on Alaskan floodplains the life-history traits of the early successional willow and alder may be compared with those of poplar and spruce, which are later successional trees (Walker *et al.* 1986). Interestingly, differences in these traits (notably in growth rates and longevity) between the four species were sufficient to explain the observed patterns of successional replacement. Whether such differences provide the major cause of the phenomenon of succession or not (Drury & Nesbit 1973; Fagerstrom & Agren 1979; Huston & Smith 1987; Walker & Chapin 1987), the life-history traits of early and late species in any undisturbed successional replacement may show a similar contrast. Thus, 'early successional' species from different communities may be very dissimilar.

Third, there are manifest differences between primary and secondary successions (Gray *et al.* 1987) and most concepts and generalizations about colonizing and early successional species appear to have been derived from studies of secondary successions (notably in old fields and forests). The main thread running through the above descriptions of primary successions has been the extremely low resource levels, notably of nutrients and water, in the predominantly abiotic early stages, and the importance of substratum type, stability and particle size in controlling the type of pioneer. By contrast, most secondary successions occur on relatively productive soils, often containing banked seeds. This contrast, and the type of species replacement each produces, has been described in terms of his C–S–R strategy model by Grime (1987), who classifies many of the pioneers of primary successions as stress-tolerators.

PERSISTENCE, GENETIC DIFFERENTIATION AND PHENOTYPIC PLASTICITY

What, then, are the most interesting features of pioneer plants from a genetical and evolutionary viewpoint? I have discussed some of these in detail elsewhere (Gray 1985, 1987) and propose only a brief summary here.

Pioneer species present an intriguing problem; principally because of their persistence. In contrast to colonizing species, which have been characterized as ephemeral, escapist or fugitive, the pioneers of primary succession persist through, and may help to shape, successional changes going on around them. Whether they persist by mainly genetic differentiation or by phenotypic plasticity (which are not alternatives) is likely to depend on a number of interacting factors, the most important of which are the relationship between individual and population longevity and the corresponding rates of successional and population changes. Genetic tracking of successional change is most likely to be displayed by annual or short-lived perennials, which invade the early stages and persist for periods exceeding average individual longevity (Gray 1987). The conditions under which genetic differentiation in response to successional change can be demonstrated to have occurred are rather limited, particularly in communities along successional gradients. However, some studies have provided evidence for increased competitive ability of plants in successionally older populations, e.g. in *Puccinellia maritima* (Gray 1987), *Poa annua* (Law *et al.* 1977) and *Ambrosia trifida* (Hartnett *et al.* 1987). In other studies, a feature of change during succession has been the biotype, and in some cases genotype, depletion which occurs in populations of pioneers invading (or being sown into) formerly open habitats (Gray *et al.* 1979; McNeilly & Roose 1984; Aarssen & Turkington 1985; Gray 1987).

It was suggested in an earlier review (Gray 1987) that succession may be a potent force retaining plasticity for many traits in plants. Pioneer species whose populations contain individuals which persist into the later stages of primary succession are expected to display particularly high levels of phenotypic plasticity. This is generally because the transition from resource-poor, open environments to more mesic, crowded environments occurs during the lifetime of the individual. Unfortunately, the hypothesis that pioneer species are more phenotypically plastic than late successional species is difficult to test. First, as Bradshaw's (1965) seminal account pointed out, plasticity is not a general property of the plant but is trait-specific and is under the control of genes which are not necessarily linked to those controlling variation in the trait itself. Indeed subsequent experimental work has demonstrated both the genetic control of trait plasticity (Khan *et al.* 1976; Jain 1978; Nicholls & McNeilly 1979; Schiener & Lyman 1989) and the uncoupling, and potentially separate evolution, of genes controlling variation in plasticity and in the trait itself (Schlichting 1986; Schlichting & Levin 1986).

Nevertheless, those clonal pioneer species in which plasticity has been measured demonstrate a remarkable ability to respond if supplied experimentally with resources. Far from being slow-growing stress tolerators, grasses such as *Danthonia spicata, Puccinellia maritima* and *Ammophila arenaria* show very high plasticity for traits measuring various aspects of individual yield. In *D. spicata* grown in six environments differing in light levels and water supply, between 40 and 96% of the total variation in vegetative traits such as the number, size and biomass of culms and leaves was due to plastic variation (Schiener & Goodnight 1984). Plants for this experiment were collected from five populations in sites differing in successional age from 0 to 69 years. Although genetic differentiation is a major feature of adjacent populations of *Puccinellia maritima*, there is also considerable plasticity for characters such as plant height and flowering tiller production in some populations (Gray 1985). In one experiment, *Ammophila arenaria* displayed particularly high levels of plasticity, average variation between environments (two dune and two experimental garden sites) ranging from twofold for plant height to 18-fold for rhizome weight and over 30-fold for total yield (Gray 1985).

The pattern of phenotypic plasticity in *Ammophila arenaria* is particularly interesting because it appears to be habitat-correlated, plants from foredune environments being generally more responsive or plastic than those from mature dunes. The results of an experiment to further test this possibility are summarized in Fig. 2. In this, four individuals (genets) were collected from four dune ridges of different age at three sites where the approximate age of each ridge was known. They were grown in standard compost for 30 months, during which they were divided and cloned-on twice before four clones (ramets) from each individual were subjected to different nutrient treatments (A. J. Gray & H. E. Ambrosen, unpubl. data). As the results in Fig. 2 indicate, individuals from early successional

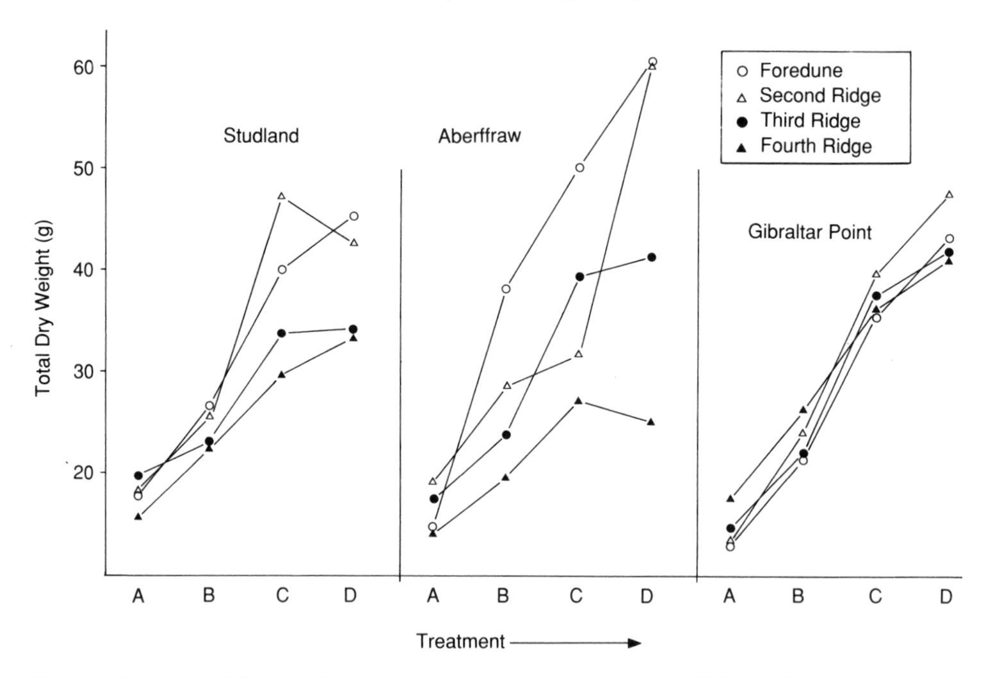

FIG. 2. Response of *Ammophila arenaria* clones to nutrient treatment. Points with the same symbol on each graph indicate the mean performance of clones of the same four individuals. There were significant treatment effects between ridges in plants from Studland ($P < 0.05$) and Aberffraw ($P < 0.05$) but not from Gibraltar Point. Nutrient treatments were: A, dune sand with no nutrient addition; B, half dune sand and half John Innes No. 3 compost (JI3) with no nutrient addition; C, JI3 + 5 g per pot of a slow-release fertilizer (Enmag) added after 5 months; D, JI3 + 15 g Enmag after 5 months and 5 g Enmag after 8 months. Plants harvested after 12 months' growth.

stages were more responsive to nutrient addition than those from late successional stages in two out of the three dune systems. Although there was within-ridge variation in an individual (genet) response as measured by joint regression analysis (Finlay & Wilkinson 1963) the overall pattern suggests that selection for genes conferring high yield plasticity may occur in the pioneer stages of at least some dune successions. However, it is not possible to discount at least two alternative explanations for this habitat-correlated plasticity. These are the effects of ageing, in which the results for *Ammophila arenaria* may parallel those for *Spartina anglica* across its successional stages (Thompson 1989; Thompson *et al.* 1993), and the effects of pathogen accumulation, shown by van der Putten *et al.* (1989) and van der Putten (1993) to be important in reducing *Ammophila arenaria* vigour in Dutch dune systems. The conditions of the experiment described above suggest that harmful soil organisms are unlikely to have been an important factor but do not rule out a build-up of systemic pathogens such as viruses.

Whatever the major cause of habitat-correlated plasticity, it is easy to envisage the importance in pioneer individuals of an ability to utilize intermittent and unpredictable pulses of nutrients. Selection for this ability may be a general feature of the evolution of long-lived pioneer species. Conversely, genes conferring plasticity in growth rate and assimilation pattern will increase tolerance of low nutrient conditions and thus facilitate both invasion of the early stages of primary successions and persistence as conditions ameliorate.

ACKNOWLEDGMENTS

I gratefully acknowledge the help of Liz Warman who produced the figures and Ros Weller who typed the manuscript.

REFERENCES

Aarssen, L.W. & Turkington, R. (1985). Within-species diversity in natural populations of *Holcus lanatus, Lolium perenne* and *Trifolium repens* from four different aged pastures. *Journal of Ecology*, **73**, 869–86.

Baker, H.G. & Stebbins, G.L. (Eds) (1965). *The Genetics of Colonizing Species*. Academic Press, New York.

Bakker, D. (1960). A comparative life-history study of *Cirsium arvense* (L.) Scop. and *Tussilago farfara* L., the most troublesome weeds in the newly reclaimed polders of the former Zuiderzee. *The Biology of Weeds* (Ed. by J.L. Harper), pp. 205–22. Oxford University Press, Oxford.

Barrett, S.C.H. & Richardson, B.J. (1986). Genetic attributes of colonizing species. *The Ecology of Biological Invasions: An Australian Perspective* (Ed. by R.H. Groves & J.J. Burdon), pp. 21–33. Australian Academy of Science, Canberra.

Bazzaz, F.A. (1979). The physiological ecology of plant succession. *Annual Review of Ecology and Systematics*, **10**, 351–71.

Bazzaz, F.A. (1986). Life history of colonizing plants: some demographic, genetic, and physiological features. *Ecology of Biological Invasions of North America and Hawaii* (Ed. by H.A. Mooney & J.A. Drake), pp. 96–110.

Bazzaz, F.A. & Pickett, S.T.A. (1980). The physiological ecology of tropical succession: a comparative review. *Annual Review of Ecology and Systematics*, **11**, 287–310.

Bradshaw, A.D. (1965). Evolutionary significance of phenotypic plasticity in plants. *Advances in Genetics*, **13**, 115–55.

Brown, A.H.D. & Burdon, J.J. (1987). Mating systems and colonizing success in plants. *Colonization, Succession and Stability* (Ed. by A.J. Gray, M.J. Crawley & P.J. Edwards), pp. 113–31. Blackwell Scientific Publications, Oxford.

Brown, A.H.D. & Marshall, D.R. (1981). Evolutionary changes accompanying colonization in plants. *Evolution Today* (Ed. By G.G.E. Scudder & J.L. Reveal), pp. 351–63. Proceedings of the Second International Congress of Systematic and Evolutionary Biology. Carnegie-Mellon University, Pittsburg.

Clements, F.E. (1904). *The Development and Structure of Vegetation*. Botanical Survey of Nebraska, 7. The Botanical Seminar, Lincoln, Nebraska.

Cooper, W.S. (1931). A third expedition to Glacier Bay, Alaska. *Ecology*, **12**, 61–95.

Crocker, R.L. & Major, J. (1955). Soil development in relation to vegetation and surface age at Glacier Bay, Alaska. *Journal of Ecology*, **43**, 427–48.

Dale, V.H. (1989). Wind dispersed seeds and plant recovery on the Mount St Helens debris avalanche. *Canadian Journal of Botany*, **67**, 1434–41.

Drake, J.A., Mooney, H.A., di Castri, F., Groves, R.H., Kruger, F.J., Rejmanek, M. & Williamson, M. (Eds) (1989). *Biological Invasions. A Global Perspective*. Wiley, Chichester.

Drury, W.H. & Nesbit, I.C.T. (1973). Succession. *Journal of the Arnold Arboretum*, **54**, 331–68.

During, H.J. (1979). Life strategies of bryophytes: a preliminary review. *Lindbergia*, **5**, 2–18.

Elton, C.S. (1958). *The Ecology of Invasions by Animals and Plants*. Methuen, London.

Fagerstrom, T. & Agren, G. (1979). Theory for coexistence of species differing in regeneration properties. *Oikos*, **33**, 1–10.

Finlay, K.W. & Wilkinson, G.N. (1963). The analysis of adaptation in a plant breeding programme. *Australian Journal of Agricultural Research*, **14**, 742–54.

Fuller, R.M. (1987). Vegetation establishment on shingle beaches. *Journal of Ecology*, **75**, 1077–89.

Gimingham, C.H. (1964). Maritime and sub-maritime communities. *The Vegetation of Scotland* (Ed. By J.H. Burnett), pp. 67–142. Oliver & Boyd, Edinburgh.

Govindaraju, D.R. (1984). Mode of colonization and patterns of life history in some North American conifers. *Oikos*, **431**, 271–6.

Gray, A.J. (1985). Adaptation in perennial coastal plants — with particular reference to heritable variation in *Puccinellia maritima* and *Ammophila arenaria*. *Vegetatio*, **61**, 179–88.

Gray, A.J. (1986). Do invading species have definable genetic characteristics? *Philosophical Transactions of the Royal Society of London, B*, **314**, 655–74.

Gray, A.J. (1987). Genetic change during succession in plants. *Colonization, Succession and Stability* (Ed. by A.J. Gray, M.J. Crawley & P.J. Edwards), pp. 273–93. Blackwell Scientific Publications, Oxford.

Gray, A.J. (1991). Management of coastal communities. *Scientific Management of Temperate Communities for Conservation* (Ed. by I.F. Spellerberg, F.B. Goldsmith & M.G. Morris), pp. 227–43. Blackwell Scientific Publications, Oxford.

Gray, A.J., Crawley, M.J. & Edwards, P.J. (Eds) (1987). *Colonization, Succession and Stability*. Blackwell Scientific Publications, Oxford.

Gray, A.J., Parsell, R.J. & Scott, R. (1979). The genetic structure of plant populations in relation to the development of salt marshes. *Ecological Processes in Coastal Environments* (Ed. by R.L. Jefferies & A.J. Davy), pp. 43–64. Blackwell Scientific Publications, Oxford.

Grime, J.P. (1986). The circumstances and characteristics of spoil colonization within a local flora. *Philosophical Transactions of the Royal Society of London, B*, **314**, 637–54.

Grime, J.P. (1987). Dominant and subordinate components of plant communities: implications for succession, stability and diversity. *Colonization, Succession and Stability* (Ed. by A.J. Gray, M.J. Crawley & P.J. Edwards), pp. 413–28. Blackwell Scientific Publications, Oxford.

Groves, R.H. & Burdon, J.J. (Eds) (1986). *The Ecology of Biological Invasions: An Australian Perspective*. Australian Academy of Science, Canberra.

Grubb, P.J. (1986). The ecology of establishment. *Ecology and Landscape Design* (Ed. by A.D. Bradshaw, D.A. Goode & E. Thorp), pp. 83–98. Blackwell Scientific Publications, Oxford.

Grubb, P.J. (1987). Some generalizing ideas about colonization and succession in green plants and fungi. *Colonization, Succession and Stability* (Ed. by A.J. Gray, M.J. Crawley & P.J. Edwards), pp. 81–102. Blackwell Scientific Publications, Oxford.

Hamrick, J.L. (1983). The distribution of genetic variation within and among natural plant populations. *Genetics and Conservation* (Ed. by C.M. Schonewald-Cox, S.M. Chambers, B. MacBride & L. Thomas), pp. 335–48. Benjamin/Cummings, Menlo Park, California.

Hartnett, D.C., Hartnett, B.B. & Bazzaz, F.A. (1987). Persistence of *Ambrosia trifida* populations in old fields and responses to successional changes. *American Journal of Botany*, **74**, 1239–48.

Huston, M. & Smith, T. (1987). Plant succession: life history and competition. *American Naturalist*, **130**, 168–98.

Jain, S.K. (1978). Inheritance of phenotypic plasticity in soft chess (*Bromus mollis* L.) Gramineae. *Experientia*, **34**, 835–6.

Jain, S.K. (1983). Genetic characteristics of populations. *Ecological Studies: Analysis and Synthesis* (Ed. by H.A. Mooney & M. Gordon), pp. 240–58. Springer-Verlag, Berlin.

Jefferies, R.L. (1972). Aspects of salt-marsh ecology with particular reference to inorganic plant nutrition. *The Estuarine Environment* (Ed. By R.S.K. Barnes & J. Green), pp. 41–85. Applied Science Publishers, London.

Joenje, W. (1979). Plant succession and nature conservation of newly embanked flats in the Lauwerszeepolder. *Ecological Processes in Coastal Environments* (Ed. by R.L. Jefferies & A.J. Davy), pp. 617–34. Blackwell Scientific Publications, Oxford.

Joenje, W. (1985). The significance of waterfowl grazing in the primary vegetation succession on embanked sandflats. *Vegetatio*, **62**, 399–406.

Khan, M.A., Antonovics, J. & Bradshaw, A.D. (1976). Adaptation to heterogeneous environments. III The inheritance of response to spacing in flax and linseed *Linum usitatissimum*. *Australian Journal of Agricultural Research*, **27**, 649–59.

Kornberg, H. & Williamson, M.H. (Eds) (1987). *Quantitative Aspects of the Ecology of Biological Invasions*. The Royal Society, London.

Law, R., Bradshaw, A.D. & Putwain, P.D. (1977). Life history variation in *Poa annua*. *Evolution*, **31**, 233–46.

Locket, G.H. (1945). Observations on the colonization of bare chalk. *Journal of Ecology*, **33**, 205–9.

Loveless, M.D. & Hamrick, J.L. (1984). Ecological determinants of genetic structure in plant populations. *Annual Review of Ecology and Systematics*, **15**, 65–95.

Macdonald, I.A.W., Kruger, F.J. & Ferrar, A.A. (Eds) (1986). *Ecology and Management of Biological Invasions in South Africa*. Oxford University Press, Cape Town.

Mack, R.N. (1985). Invading plants: their potential contribution to population biology. *Studies in Plant Demography: A Festschrift for John L. Harper* (Ed. by J. White), pp. 127–42. Academic Press, London.

McNeilly, T. & Roose, M.L. (1984). The distribution of perennial rye grass genotypes in swards. *New Phytologist*, **98**, 503–13.

Mooney, H.A. & Drake, J.A. (Eds) (1986). *Ecological Biological Invasions of North America and Hawaii*. Springer-Verlag, Berlin.

Nicholls, M.K. & McNeilly, T. (1979). Sensitivity of rooting and tolerance to copper in *Agrostis tenuis* Sibth. *New Phytologist*, **83**, 653–64.

Rice, K. & Jain, S. (1985). Plant population genetics and evolution in disturbed habitats. *The Ecology of Natural Disturbance and Patch Dynamics* (Ed. by S.T.A. Pickett & P.S. White), pp. 287–303. Academic Press, Orlando, FL.

Roberts, R.D., Marrs, R.H., Skeffington, R.A. & Bradshaw, A.D. (1981). Ecosystem development on naturally-colonized china clay wastes. I. Vegetation changes and overall accumulation of organic matter and nutrients. *Journal of Ecology*, **69**, 153–61.

Schiener, S.M. & Goodnight, C.J. (1984). The comparison of phenotypic plasticity and genetic variation in populations of the grass *Danthonia spicata*. *Evolution*, **38**, 845–55.

Schiener, S.M. & Lyman, R.F. (1989). The genetics of phenotypic plasticity. 1. Heritability. *Journal of Evolutionary Biology*, **2**, 95–107.

Schlichting, C.D. (1986). The evolution of phenotypic plasticity in plants. *Annual Review of Ecology and Systematics*, **17**, 667–93.

Schlichting, C.D. & Levin, D.A. (1986). Phenotypic plasticity: an evolving plant character. *Biological Journal of the Linnean Society*, **29**, 37–47.

Scott, G.A.M. (1963). The ecology of shingle beach plants. *Journal of Ecology*, **51**, 517–27.

Thompson, J.D. (1989). *Population variation in* Spartina anglica *C.E. Hubbard*. PhD thesis, University of Liverpool.

Thompson J.D., McNeilly, T. & Gray, A.J. (1993). The demography of clonal plants in relation to successional habitat change: The case of *Spartina anglica*. *Primary Succession on Land* (Ed. by J. Miles & D.W.H. Walton), pp. 193–207. Blackwell Scientific Publications, Oxford.

van der Putten, W.H., van der Werf-Klein Breteler, J.T. & van Dijk, C. (1989). Colonization of the root zone of *Ammophila arenaria* by harmful soil organisms. *Plant and Soil*, **120**, 213–23.

van der Putten, W.H. (1993). Soil organisms in coastal foredunes involved in degeneration of *Ammophila arenaria*. *Primary Succession on Land* (Ed. by J. Miles & D.W.H. Walton), pp. 273–81. Blackwell Scientific Publications, Oxford.

Walker, L.R. & Chapin III, F.S. (1987). Interactions among processes controlling successional change. *Oikos*, **50**, 131–5.

Walker, L.R., Zasada, J.C. & Chapin III, F.S. (1986). The role of life history processes in primary succession on an Alaskan floodplain. *Ecology*, **67**, 1243–53.

Whittaker, R.J., Bush, M.B. & Richards, K. (1989). Plant recolonization and vegetation succession on the Krakatau Islands, Indonesia. *Ecological Monographs*, **59**, 59–123.

Wood, D.M. & del Moral, R. (1987). Mechanisms of early primary succession in subalpine habitats on Mount St Helens. *Ecology*, **68**, 780–90.

The demography of clonal plants in relation to successional habitat change: The case of *Spartina anglica*

J. D. THOMPSON,*‡ T. MCNEILLY* AND A. J. GRAY†

* *Department of Environmental and Evolutionary Biology, University of Liverpool, PO Box 147, Liverpool L69 3BX; and* † *Institute of Terrestrial Ecology (Natural Environment Research Council), Furzebrook Research Station, Wareham, Dorset BH20 5AS, UK*

SUMMARY

1 The zonation of plant communities across estuarine salt-marshes represents an ideal opportunity to study intraspecific variation in perennial plant populations which occur in the different stages of a primary succession. A species which has a wide ecological amplitude over estuarine salt-marsh zones in the British Isles is *Spartina anglica* C. E. Hubbard. This species occurs as a pioneer colonist of bare intertidal flats, forms dense swards in many middle marsh sites and is a component of the mixed species assemblage which characterizes upper level, mature marsh sites.

2 This chapter describes the demographic variation which has been observed in studies of successionally adjacent pioneer, sward and mature populations of *S. anglica*. Then, to outline the possible causes of such variation recent experimental work is described, in which the survival and growth of clones is examined using:

(a) 10 populations sampled from the full latitudinal and successional distribution of this species in the British Isles grown in a common garden; and

(b) reciprocal transplants between three successionally adjacent populations.

3 In the field, the sward population showed a lower survivorship of tiller cohorts relative to the pioneer and mature populations. Tillers in the pioneer population increased in number over the study period, and had greater propensity to flower as annuals, than tillers in the mature population, which were more likely to flower in their second year. Under common garden conditions and in the reciprocal transplant experiment, clone survival and clonal growth were lower in the successionally mature populations, which also showed a longer time to flowering for tillers produced in the first year of the study.

4 These results suggest an inverse correlation between plant vigour and successional status, indicative of somatic variation related to clonal age, rather than to the genetically based differentiation of successionally adjacent populations. The implications of these trends for long-term succession and the 'die-back' of mature populations of *S. anglica* are discussed.

‡ Present address: Centre d'Ecologie Fonctionelle et Evolutive, Centre Louis Emberger, CNRS, Route de Mende, BP 5051, 34033 Montpellier Cedex, France.

INTRODUCTION

Genetically based population variation requires environmental factors to show marked spatial variation but also to be temporally predictable (Hamrick & Allard 1972; Bradshaw 1984). In such conditions differentiation will occur where selection pressures are strong enough to overcome the effects of gene flow. Estuarine salt-marshes are spatially heterogeneous, but temporally predictable for a range of physical environmental factors, ultimately determined by the elevation of the marsh surface and tidal inundation (Davy & Smith 1988). This range of conditions frequently produces a mosaic of plant communities over an individual marsh surface and, for individual species which occur over the range of different environments, represents an ideal opportunity to study intraspecific population variation in relation to successional habitat change.

For several salt-marsh species, genetic differentiation has been correlated with habitat differences, e.g. *Plantago maritima* (Gregor 1938), *Aster tripolium* (Gray 1974), *Salicornia* spp. (Davy & Smith 1988), *Armeria maritima* (Preston 1981), *Puccinellia maritima* (Gray & Scott 1980) and *Spartina patens* (Silander 1985a). There is also increasing evidence that variation in response to such successional change may be related to differences in phenotypic plasticity (Gray 1985, 1993). Nevertheless, 'there is a surprising lack of empirical evidence that successional change is a major factor maintaining genetic diversity in species populations' (Gray 1987). The difficulties of establishing such a cause–effect relationship between the selection pressures imposed by successional habitat change and patterns of population differentiation include: (a) the confounding nature of spatial and temporal relations; (b) the inability to disentangle any selective effects of succession from the genetic consequences of changes in abundance; and (c) the inherent difficulty which arises from the fact that current successional changes may not be the cause of any observed differentiation. Furthermore, for clonal plant species, the persistence of clones in the face of successional development would suggest that phenotypic plasticity will be a key component of any variation between and within populations, and that succession may have a strong directional component which acts to maintain these differences.

A species which has had a significant impact on the primary succession of plant communities in estuarine salt-marshes throughout the British Isles, due to both widespread introduction for reclamation and coastal defence purposes and natural dispersal, is *Spartina anglica* C. E. Hubbard (Charman 1990). *Spartina anglica* is an allopolyploid, rhizomatous grass that was produced during the middle of the nineteenth century by: (a) the hybridization of *S. maritima* and *S. alterniflora* in Southampton Water; and (b) subsequent chromosome doubling in the sterile hybrid, *S. x townsendii*, which restored fertility and produced an amphidiploid with $2n = 120-124$ (Marchant 1968; Raybould 1989).

As a result of vigorous rhizome growth and tiller production following establishment at a site, *S. anglica* has a wide ecological amplitude across individual marshes, frequently occurring in three successional stages: (a) as pioneer populations colonizing bare, low-lying intertidal flats; (b) rapid clonal growth of established tussocks leads to their coalescence into the dense, monospecific sward populations which characterize many *Spartina* dominated marshes; and (c) mature populations of scattered tussocks are frequent in the mixed species assemblage of higher elevation salt-marsh. The latter may represent independent colonization of pans and channel edges or the remnants of a declining sward population. Hence a successional sequence from pioneer to sward to mature populations can be observed on individual marshes. In that the low-lying tidal flats colonized by pioneer populations of *S. anglica* have had little or no vegetation cover during the 12 000 years since the last glaciation, this provides a clear example of primary succession and provides an excellent illustration of a pioneer species which colonizes and persists in the face of successional development.

The ecology of clonal populations, which rely on vegetative propagation as their means of persistence and spread, can be markedly influenced by variation in tiller dynamics (e.g. see Noble *et al.* 1979; Watkinson *et al.* 1979; Lovett-Doust 1981; Fetcher & Shaver 1983). In this respect, the population ecology of perennial species such as *S. anglica*, which occupy adjacent seral stages, can be considered, in relation to successional change, as two interrelated sets of mechanisms. The first determines the pattern of initial colonization and the manner in which populations of *S. anglica* come to dominate a particular site. The second involves those factors responsible for the decline and replacement of the dominant populations. In established populations of *S. anglica* (Taylor & Burrows 1968; Hill 1984) as in many perennial species (Harper 1977), seedling establishment is a rare event. Consequently, the development of the dense monospecific sward populations which dominated many estuarine salt-marshes in the British Isles is frequently a result of vegetative clonal growth following episodic colonization (e.g. see Gray & Pearson 1984). The localized success and fitness of a genotype will thus be dependent on its ability to spread and persist by vegetative propagation.

This chapter investigates the nature of demographic variation among populations of *S. anglica* to examine the hypothesis that successional habitat change may influence the dynamics and structure of natural populations. Two issues are addressed: (a) the influence of successional change on the contrasting patterns of demography which natural populations can exhibit; and (b) the relative importance of genetic variation and phenotypic plasticity as components of this variation. We first review previous studies and our recent work, which has documented variation in the population dynamics of natural populations of *S. anglica*. We then describe, and discuss the results of, experimental investigations done to ascertain the causes of variation among natural populations.

THE CONTRASTING DYNAMICS OF NATURAL TILLER POPULATIONS

In the British Isles, successionally adjacent populations of *S. anglica* often show significant variation in the dynamics of tiller populations. On the Dee Estuary,

Taylor & Burrows (1968) and Hill (1984) reported that shoot production occurs over a longer period, and to a greater extent in any one year, on the lower marsh than on the upper marsh. Hill (1984) also found that tillers in the upper marsh tended to overwinter and flower as biennials whereas in the low marsh a greater proportion of tillers flowered as annuals (and thus did not overwinter). Percentage flowering at a particular time was also greater in the low marsh sites. Likewise, on the Ribble estuary Mullins & Marks (1987) found that flowering was earlier in pioneer populations than in mature populations, with the result that the percentage seed set was significantly less in mature populations where inflorescence production occured too late to provide enough time for successful seed set. This corroborated previous findings at a different site on the Ribble Estuary (Marks & Truscott 1985) that, although a mature population produced significantly more spikelets than an adjacent pioneer population, seed set was proportionately greater in the pioneer population.

To examine further the significance and consistency of these patterns of variation in tiller dynamics, a 3-year demographic investigation was done in the adjacent pioneer sward and mature zone populations of *S. anglica* on the Dee Estuary at Parkgate (see Thompson 1989). Tillers were grouped into cohorts based on time of emergence, and the variation between the populations in survival, emergence and phenological status of tillers of each cohort was examined. Tiller survival (see Pyke & Thompson 1986) was found to be significantly different (PROC LIFETEST (SAS 1985), $P<0{\cdot}05$) among populations for three tiller cohorts and in relation to a population × quadrat interaction for a fourth cohort. This was predominantly due to a lower survival of tillers in the sward population during the winter months in comparison with tillers emerging at the same time in the pioneer and, to a lesser extent, mature populations. It was clear that the probability of death for individual *S. anglica* tillers was concentrated during the winter months, and not when new tillers were emerging, such as found for other clonal plants (Noble *et al.* 1979; Watkinson *et al.* 1979). There was thus no evidence that birth and death processes in *S. anglica* tiller populations are linked, mortality being more dependent on the season and the successional status of the populations.

In the mature zone quadrats, tiller production only occurred during May to August, whereas it continued until December in the sward and pioneer zone quadrats during 1987 and in the pioneer zone in 1988. In the sward and mature zone populations, tiller mortality was not balanced by the birth of new tillers, hence there was a decline in tiller numbers, whereas in the pioneer population, births exceeded deaths and there was a gradual increase in tiller numbers at the end of each calendar year. Furthermore, although not significantly distinct, the percentage of tillers which flowered in their first year, and which flowered over the whole study period, was highest in the pioneer zone quadrats. In August 1988 the tiller population in the sward also included a high proportion of very young tillers, whereas in the pioneer and mature zone quadrats there were more older tillers, particularly in the mature population where second-year tillers were most numerous. This reflects a greater tendency to overwinter and flower as biennials in the mature population, a result consistent with previous findings (Hill 1984). Consequently, although flowering did not begin earlier in the growing season (i.e. from May to December in this species) in the seaward population (as found in other studies), it did appear to occur earlier in the life-history of individual tillers. This trend is similar to that reported by Reinartz (1984), who showed that early successional populations of *Verbascum thapsus* in a garden tended to flower as annuals, whereas those from late successional populations mostly flowered after more than 1 year's growth.

The low survival of tillers in the sward population, the imbalance between births and deaths of tillers, and the lower percentage flowering in the sward and mature populations, and the later flowering in the mature population, are symptoms of a decline in vigour of clonal material in the successionally advanced populations. These differences between the populations corroborate those previously reported on the Dee Estuary (Hill 1984), suggesting that they are a consistent feature of demographic variation in this species. But are there any differences between the populations in the dynamics of whole clones rather than individual tillers and could such differences have a genetic basis? The inability to distinguish individual clones in natural populations due to extensive clonal growth by rhizome and tiller propagation made it necessary to investigate this issue in experimental trials using clones collected from the natural populations grown as spaced plants.

CLONE DYNAMICS IN EXPERIMENTAL POPULATIONS

A garden experiment

As part of a detailed investigation into the causes of population variation in *S. anglica*, clones were sampled from 10 populations representing the full latitudinal and intertidal distribution of the species in the British Isles (see Thompson *et al.* 1991a for details). These populations comprised three adjacent populations (pioneer, sward and mature) from each of the Dee and Ribble estuaries in north-west England, a mature population from Brands Bay in Poole Harbour, two expanding populations in Wales from the Menai Straits and the Mawddach estuary and one expanding population from the Cromarty Firth in Scotland. Clones were washed free of all original substrate and were propagated by repeated separation and replanting of new tillers in sand-filled pots for 8 months. After this time, uniformly sized clonal fragments, comprising single tillers with one developing rhizome bud, were separated from each clone and planted individually in sand-filled pots. Two fully randomized replicate blocks, each containing one replicate of each of 30 clones from each population (other than the Cromarty Firth population for which only seven clones were sampled due to its small size), were laid out in a polythene tunnel house at the University of Liverpool Botanical Gardens. Clone and tiller survival, and the emergence of new tillers on each clone were monitored bimonthly, and inflorescence emergence recorded at 6–8 day intervals from April 1986 until December 1988.

There were significant differences in the pattern of clone survival between populations (PROC LIFETEST (SAS 1985), log-rank $\chi^2 = 74{\cdot}47$, $P < 0{\cdot}001$) such that the four successionally mature populations, i.e. the Dee Sward, Dee Mature, Ribble Mature and Brands Bay populations, had lower survival rates than the remaining populations (Fig. 1a). Repeated measures ANOVA of mean tiller number per plant after 8, 16, and 20 months growth indicated that there were significant differences ($P < 0{\cdot}05$) due to populations, clones, blocks, time of survey and a population × time interaction. The effect due to time was simply a

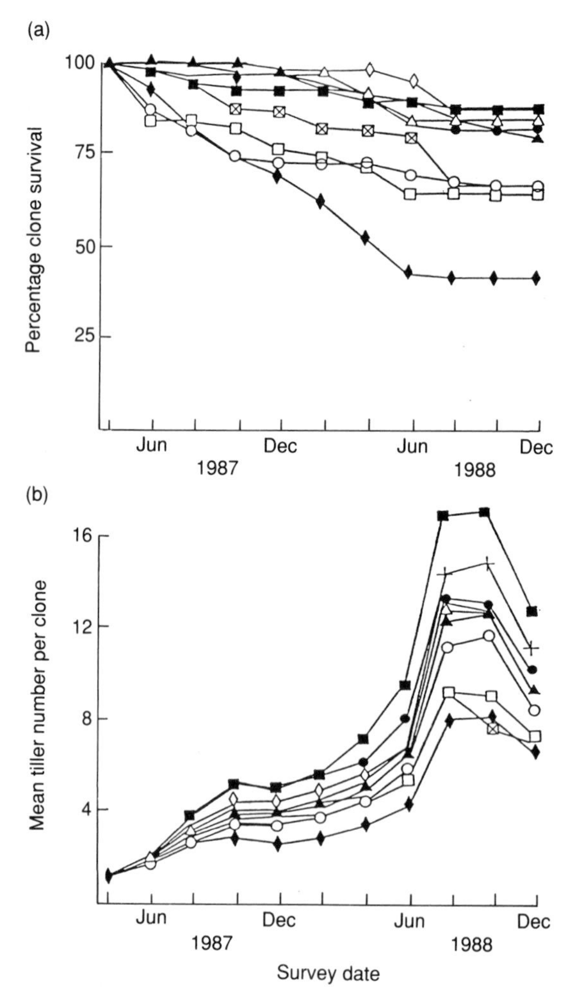

FIG. 1. The clonal dynamics of 10 populations of *S. anglica* grown in a common garden: (a) percentage clone survival; (b) mean tiller number per plant. Redrawn from a composite figure in Thompson *et al.* (1991a). Each symbol represents an individual population and is the same for both graphs: (□) Brands Bay, (◇) Mawddach, (△) Afon Menai, (●) Dee Pioneer, (◆) Dee Sward, (⊠) Dee Mature, (▲) Ribble Pioneer, (■) Ribble Sward, (○) Ribble Mature, (+) Cromarty Firth.

result of clonal growth. The population × time interaction was due to differences between populations in the magnitude of the changes in tiller number rather than the populations showing changes in their rank order (Fig. 1b). Interestingly, populations with the lowest mean tiller number per plant were also those with a lowest expectation of life, as shown by the four lowest lines in Fig. 1a and b. The four mature populations also had a significantly longer time to flower emergence in tillers produced during the first year of the experiment (Thompson *et al.* 1991a), causing a greater proportion to have a biennial life-history compared with the successionally younger populations in which time to flower emergence was earlier in tillers produced in the first year.

A reciprocal transplant experiment

During the same period that ramets were transplanted to the garden (see above) similarly treated and uniformly-sized clonal fragments containing a single tiller with a developing rhizome bud were taken from 10 clones of the pioneer, sward and mature populations sampled from the Dee Estuary. Three clonal replicates were then reciprocally transplanted into each of the three original collection sites (see Thompson *et al.* 1991b for further details). In the sward and mature sites, transplants were introduced into both cleared plots and undisturbed vegetation (i.e. two treatments). Tiller survivorship and new tiller emergence were recorded periodically over two growing seasons from May 1987 until September 1988.

Significantly lower ($P<0.05$) survival of all populations occurred in the pioneer and cleared sward sites during the initial establishment phase and in the pioneer sites during the winter, such that less than 30% of the clones transplanted to the pioneer site survived to the end of the experiment, compared with 65–70% in the sward and mature sites. The mean survival of the pioneer population was significantly greater than the other populations when averaged across all transplant sites (Fig. 2), although this difference was consistent across sites it was only significant in the mature zone transplant site (Thompson *et al.* 1991b).

Mean tiller number per plant was significantly influenced ($P<0.05$) by sites × clones (within populations) and sites × treatments interactions, by the main effects of populations, clones (within populations), and treatments, and by an interaction of these effects with time. The effect of populations was due to a significantly higher tiller number per plant in the pioneer population in all the site-treatment combinations, except the pioneer site, the harshest environment for plant growth (see Fig. 3). These differences between the populations increased with time, hence the populations × time interaction. The significant sites × clones (within populations) interaction indicated that clones of *S. anglica* respond differently when grown in different sites, i.e. have different plasticity in tiller number over these environments. In a simplified ANOVA model with the effects due to sites and treatments reclassified as a single effect ('environments'), clones within each of the pioneer and sward populations, but not the mature population, were found to differ significantly in their response to environmental variation ($P<0.01$). The

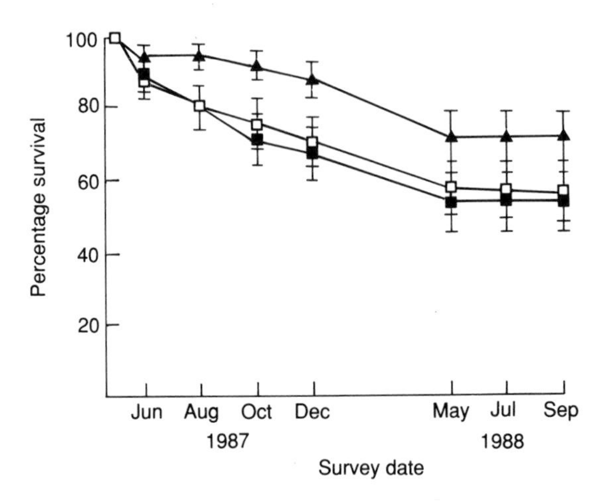

FIG. 2. Percentage survival (with 95% confidence limits) of replicated clones of three populations of *S. anglica* reciprocally transplanted into the pioneer site and into cleared and intact plots in the sward and mature sites on the Dee Estuary at Parkgate. Pioneer (▲), sward (■) and mature (□) populations.

norms of reaction for each clone in each environment (see Thompson *et al.* 1991b) showed that clones in the pioneer population had a much improved performance in the cleared plots in the mature site. There were noticeable differences between clones such that the norms of reaction lines diverged and frequently crossed, hence the significant clones × environment interaction. There was much less variation in response to environmental change in the clones of the sward and mature populations whose norms of reaction were flatter and essentially similar.

Clearly, then, the two experimental trials showed essentially similar results, manifested as a poor survival and clonal growth, slower development of tillers and lack of differences in plasticity between clones in the successionally older populations. When taken together these results suggest that a decline in vigour may occur during the process of successional habitat change. This may cause a reduced ability of clones to establish, survive, expand and flower quickly following vegetative fragmentation in natural conditions. Such variation may have a profound effect on the successional development of *Spartina*-dominated salt-marsh, and it is thus necessary to consider the possible causes of such variation.

WHAT IS THE BASIS OF THE VARIATION?

Prima facie, the consistent inverse correlation between vigour and successional status of the populations of *S. anglica* described here suggests genetic differentiation between the populations. However, if successional change has provided the requisite selection to produce differentiation between these populations, we would have predicted that clones transplanted into their original collection site would out-perform clones from different sites, the so-called 'home-site advantage', i.e. selection against 'alien' clones, relative to 'home' clones in a particular transplant

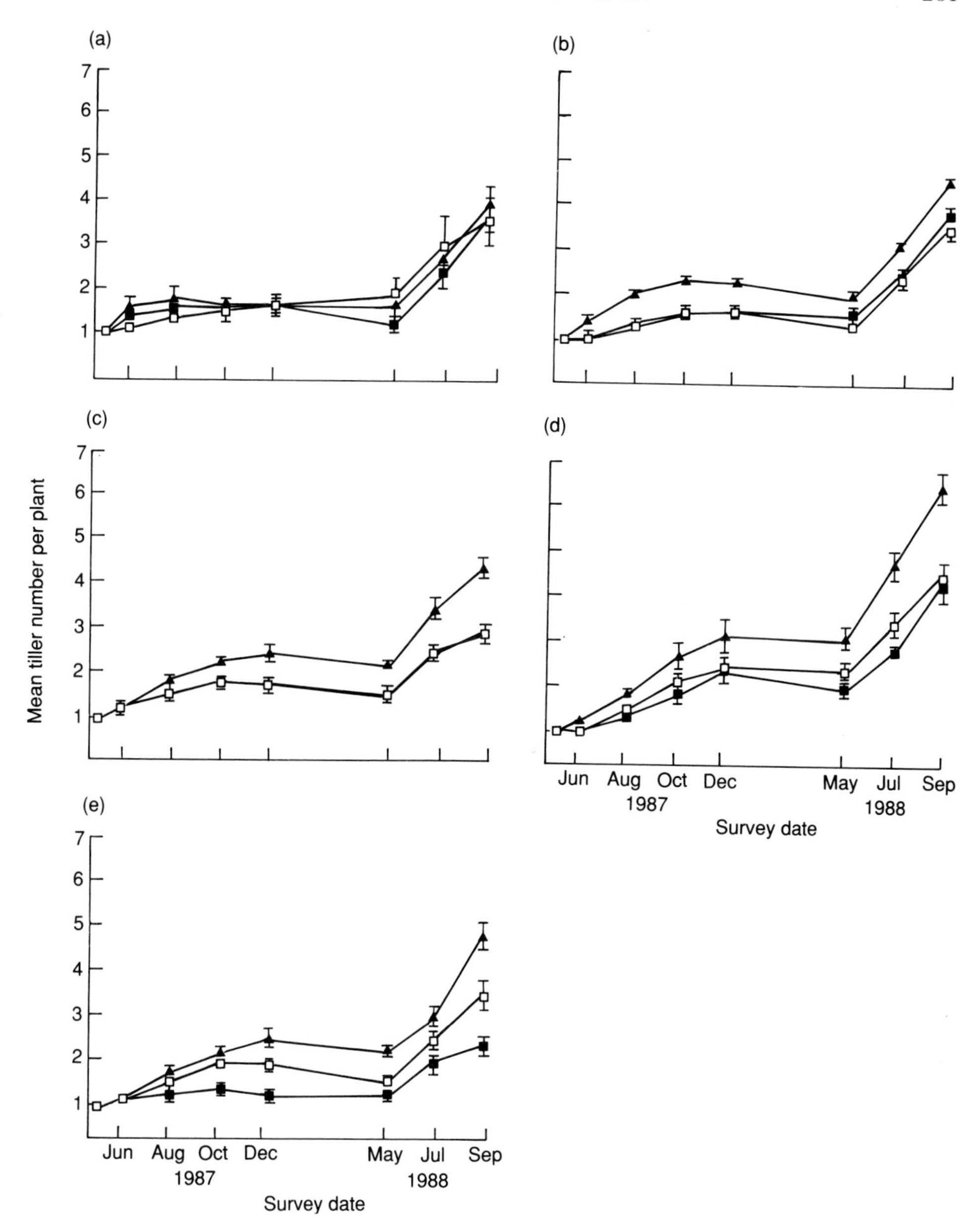

FIG. 3. Mean tiller number per plant (with standard errors) for clones of three populations of *S. anglica* reciprocally transplanted into the pioneer site and into cleared and intact plots in the sward and mature sites on the Dee Estuary at Parkgate. (a) Pioneer site; (b) sward site — cleared plots; (c) sward site — intact vegetation; (d) mature site — cleared plots; (e) mature site — intact vegetation. Pioneer (▲), sward (■) and mature (□) populations. Redrawn from a composite figure in Thompson *et al.* (1991b).

environment. For the successional populations of *S. anglica* that we studied this hypothesis can be rejected. What then is the cause of such variation?

A decline in tiller productivity was observed in relation to the presence of *Puccinellia maritima* in the mature site, which suggests that *P. maritima* may affect *S. anglica* abundance in the mature marsh (as found by Scholten & Rozema 1990). This suggests that the mature zone population of *S. anglica* is a successional remnant of sward degeneration and invasion by *Puccinellia maritima*. It follows that, in the absence of seedling input into the sward and mature populations, the increase in successional status is equivalent to an increase in clonal age. We are therefore led to ask if the decline in vigour could be due to a somatic, age-related change in clone vitality, rather than genetically based differences? Breese *et al.* (1965) found that it was possible to select for high and low rates of tiller production from within a single clone of *Lolium perenne* if the clone was seedling derived, but not if the clone had been vegetatively propagated over a number of years. With increasing age somatic variation had become 'fixed'. Following this reasoning, as the successional development of *S. anglica* marsh occurs clones may thus undergo a loss of flexibility, caused by such a 'fixation', and hence suffer a decline in vigour and plasticity.

It is particularly interesting to consider the decline in vigour of *S. anglica* populations in relation to that observed for the two species of *Ammophila* (marram grass) which inhabit coastal sand-dunes in the United States and British Isles. *Ammophila* populations exhibit a decline in flowering and vigour over time and in dunes where sand deposition has finished, a phenomenon known as 'the *Ammophila* problem' (Marshall 1965). Many explanations (see Eldred & Maun 1982) have been put forward to account for the debility, all related to the cessation of sand deposition, e.g. a lack of nutrients (Marshall 1965), an age-related decline in the efficiency of the roots and loss of cortex (Marshall 1965; Hope-Simpson & Jefferies 1966; Wallen 1990), competition (Watkinson *et al.* 1979), genetic differentiation (Laing 1967) and harmful biotic soil factors (van der Putten *et al.* 1988). There has been much debate about the relative importance of these factors, for example Hope-Simpson & Jefferies (1966) reported debility in the absence of competitors and van der Putten *et al.* (1988) and Hope-Simpson & Jefferies (1966) showed that plants can be rejuvenated on transfer to garden soil, ruling out genetic differences. This also argues against an age-related effect, but perhaps the age effect is initiated by an environmental change (Wallen 1990).

The situation is very similar to the phenomenon of 'die-back', evidenced by soft rotting of the rhizome, which has been frequently observed in successionally mature populations of *S. anglica* on the south coast of England (Goodman *et al.* 1959; Haynes & Coulson 1982) and in populations of *Spartina alterniflora* in the United States (Mendelssohn 1980; Mendelssohn & McKee 1988). It is thought to result from anaerobiosis and toxicity, rather than infection by saprophytic fungi (Goodman 1959, 1960; Goodman *et al.* 1959; Goodman & Williams 1961), but the absence of definitive evidence for an environmental cause means that the age-related decline in vigour reported here may influence the degeneration. Interestingly, *Spartina* and *Ammophila* share another difference between successional populations. The lower plasticity for tiller number in the sward and mature populations of *S. anglica* when reciprocally transplanted is similar to the trend reported for biomass plasticity in successional populations of *Ammophila arenaria* (Gray 1985, 1993).

However, it has been argued that perennial plants do not age; only the shoots and leaves age and die, and these are replaced (Watkinson & White 1985; van der Putten *et al.* 1988). But in species such as *S. anglica*, the rhizomatous base from which shoots establish persists for long periods and hence the underground source of meristem production does physically age. The longevity of *S. anglica* rhizomes (Hubbard 1965; Ranwell 1972), may increase the possibility of reaching an age at which vigour may decline. Furthermore, the ability to gain access to resources over a wide area, which prolific clonal growth allows, may lead to senescence in later life (Grime 1987). In addition, older rhizome segments may have very low metabolic rates and the build-up of rhizomatous tissue may thus become a metabolic drain on resources (Kershaw 1960).

Similar patterns of phasic development and decline in natural populations of clonal plants have been reported in *Pteridium aquilinum* (Watt 1971) and *Carex arenaria* (Noble *et al.* 1979) which can be related to successional status, and thus possibly influenced by ageing of clonal material. Environmental differences, due to waterlogging and anaerobiosis, will obviously be important in the field population studied here, but the persistence of differences in a common garden suggests that other factors, such as age, are involved. But what of the alternatives? Other factors which may be involved include the following.

First, a key component of the success of colonizing perennials is the rapid multiplication of individuals with an appropriate genotype (Baker 1974) and the development of a large, interconnected, clonal network which may endow an ability to buffer environmental change and persist across a range of successional conditions and serve to co-ordinate growth in the face of environmental stress (Hutchings 1979; Pitelka & Ashmun 1985). Indeed, in *Spartina alterniflora* the storage of translocates in the rhizome during winter, and subsequent translocation in the spring, has been shown to be essential to facilitate regrowth of the sward (Lytle & Hull 1982). In species such as *S. anglica*, which inhabit a range of successional habitats, the proportion of the plant which is underground increases as succession proceeds. The dependency which this entails might impair the ability of fragments to regenerate following fragmentation, and thus cause the poor vigour of young tillers in the mature populations, whose development is conditioned to drawing reserves from a large, integrated rhizome base. However, this alternative is not supported by the finding that the Ribble sward population, which had a very dense rhizome base in the field, was among the most vigorous populations in the common garden trial. Furthermore, the conditions of the experiments, in which cloning and replanting of material was done during the 8 months before transplantation to the experimental sites, suggests that neither this possible effect, nor the possible occurrence of harmful soil organisms, caused the differences in vigour.

A second possibility is that the decline in vigour could be due to viral or endophyte accumulation. This could be expected due to clone longevity and the fact that plant viruses are frequently non-lethal, but systemic (Nooden 1980; Silander 1985b). Although symptomless in natural conditions, infection can affect performance if the plant is stressed, such as would be the case following transplantation. It is worth noting that a virus causing leaf mottling in *Spartina* spp. has been reported (Jones 1980).

Finally, it is possible that there may be genetic differences between the populations which are selectively neutral, i.e. due to founder effects and genetic drift. However, the consistency of the trend for poor vigour in mature populations from geographically distinct estuaries suggests that this possibility is unlikely to be the ultimate basis of the variation. Indeed, the lack of electrophoretic enzyme variation (Raybould 1989) and morphological variation in a common garden (Thompson *et al.* 1991a) compared with the same clones in natural conditions (Thompson 1990), and the high proportion of bivalent pairing at meiosis (Marchant 1968) suggests that there is a paucity of genetic variation in this species and that very little variation can be generated by recombination.

CONCLUSIONS

Although the results of the studies described here indicate consistent differences in the clonal dynamics of populations of *S. anglica*, they provide no evidence that natural selection may have acted to produce genetically different populations adapted to the different stages of succession in adjacent salt-marsh zones. The pattern of differences between the populations suggests that clones from populations of different successional age show an age-related decline in vigour. This has important ramifications for understanding the processes influencing salt-marsh succession. The ability to establish and expand by vigorous clonal growth clearly facilitates the rapid spread of developing populations. However, the loss of vigour as succession proceeds may be a factor influencing the frequently observed 'die-back' of *Spartina* swards.

Whatever the cause, the patterns of variation introduce a new perspective to the history and spread of *Spartina anglica*. When *S. anglica* first began to spread along the south coast at the turn of the century (which it continues to do at a remarkable rate in several estuaries in the north-west of England and Wales) it was recognized that 'the grass is evidently wonderfully adapted to life on mudflats' (Stapf 1908). Due to its innate tolerance of the physical conditions imposed by frequent periods of tidal immersion, which have kept bare a habitat too harsh for other species, *S. anglica* is particularly suited to the colonization of low-lying intertidal mudflats. Eighty years after Stapf's comments, the pattern of mortality as the plants attain 'old age' appears to be just as crucial in determining the demographic behaviour of successional populations, the 'die-back' of mature populations, and the successional transition to mixed salt-marsh communities.

ACKNOWLEDGMENTS

Our thanks to Flemming Ulf-Hansen for his help with site preparation and advice on statistical analysis, to José Escarré, Professor A. D. Bradshaw and an anonymous reviewer for their comments on the manuscript, to Georges Michaloud for access to a Macintosh microcomputer, and to the Royal Society for the Protection of Birds for permission to locate experimental sites on the Dee Estuary. This work was funded by a Natural Environment Research Council (CASE) Studentship.

REFERENCES

Baker, H.G. (1974). The evolution of weeds. *Annual Review of Ecology and Systematics*, **5**, 1–24.

Breese, E.L., Hayward, M.D. & Thomas, A.C. (1965). Somatic selection in perennial ryegrass. *Heredity*, **20**, 367–79.

Bradshaw, A.D. (1984). Ecological significance of genetic variation between populations. *Perspectives on Plant Population Ecology* (Ed. by R. Dirzo & J. Sarukhan), pp. 213–28. Sinauer, Sunderland, MA.

Charman, K. (1990). The current status of *Spartina anglica* in Britain. Spartina anglica — *A Research Review* (Ed. by A.J. Gray & P.E.M. Benham), pp. 11–14. HMSO, London.

Davy, A.J. & Smith, H. (1988). Life-history variation and environment. *Plant Population Ecology* (Ed. by A.J. Davy, M.J. Hutchings & A.R. Watkinson), pp. 1–22, Symposia of the British Ecological Society, 28. Blackwell Scientific Publications, Oxford.

Eldred, R.A. & Maun, M.A. (1982). A multivariate approach to the problem of decline in vigour of *Ammophila*. *Canadian Journal of Botany*, **60**, 1371–80.

Fetcher, N. & Shaver, G.R. (1983). Life histories in tillers of *Eriophorum vaginatum* in relation to tundra disturbance. *Journal of Ecology*, **71**, 131–48.

Goodman, P.J. (1959). The possible role of pathogenic fungi in 'die back' of *Spartina townsendii* agg. *Transactions of The British Mycological Society*, **42**, 409–15.

Goodman, P.J. (1960). Investigations into 'die-back' in *Spartina townsendii* agg. II. The morphological structure and composition of the Lymington sward. *Journal of Ecology*, **48**, 711–24.

Goodman, P.J. & Williams, W.T. (1961). Investigations into 'die-back' in *Spartina townsendii* agg. III. Physiological correlates of 'die-back'. *Journal of Ecology*, **49**, 391–8.

Goodman, P.J., Braybrooks, E.M. & Lambert, J.M. (1959). Investigations into 'die-back' in *Spartina townsendii* agg. I. The present status of *Spartina townsendii* in Britain. *Journal of Ecology*, **47**, 651–77.

Gray, A.J. (1974). The genecology of saltmarsh plants. *Hydrological Bulletin*, **8**, 152–65.

Gray, A.J. (1985). Adaptation in perennial coastal plants — with particular reference to heritable variation in *Puccinellia maritima* and *Ammophila arenaria*. *Vegetatio*, **61**, 179–88.

Gray, A.J. (1987). Genetic change during succession in plants. *Colonization, Succession and Stability* (Ed. by A.J. Gray, M.J. Crawley & P.J. Edwards), pp. 273–98, Symposium of The British Ecological Society, 26. Blackwell Scientific Publications, Oxford.

Gray, A.J. (1993). The vascular plant pioneers of primary succession: Persistence and phenotypic plasticity. *Primary Succession on Land* (Ed. by J. Miles & D.W.H. Walton), pp. 179–91. Blackwell Scientific Publications, Oxford.

Gray, A.J. & Pearson, J.M. (1984). *Spartina* marshes in Poole Harbour, Dorset, with particular reference to Holes Bay. Spartina anglica *in Great Britain* (Ed. by J.P. Doody), pp. 11–14. Focus on Nature Conservation No. 5. Nature Conservancy Council, Peterborough.

Gray, A.J. & Scott, R. (1980). A genecological study of *Puccinellia maritima* Huds. (Parl.). I. Variation estimated from single plant samples from British populations. *New Phytologist*, **85**, 89–107.

Gregor, J.W. (1938). Experimental taxonomy. II. Initial population differentiation of *Plantago maritima* L. of Britain. *New Phytologist*, **37**, 15–49.

Grime, J.P. (1987). Dominant and subordinate components of plant communities: implications for succession, atability and diversity. *Colonization, Succession and Stability*. (Ed. by A.J. Gray, M.J. Crawley & P.J. Edwards), pp. 413–28. Symposium of the British Ecological Society, 26. Blackwell Scientific Publications, Oxford.

Hamrick, J.L. & Allard, R.W. (1972). Microgeographical variation in allozyme frequencies in *Avena barbata*. *Proceedings of the Natural Academy of Sciences* USA, **69**, 2100–4.

Harper, J.L. (1977). *Population Biology of Plants*. Academic Press, London.

Haynes, F.N. & Coulson, M.G. (1982). The decline of *Spartina* in Langstone Harbour, Hampshire. *Proceedings of the Hampshire Field Club Archeological Society*, **38**, 5–18.

Hill, M.I. (1984). Population studies on the Dee Estuary. Spartina anglica *in Great Britain* (Ed. by J.P. Doody), pp. 50–2. Focus on Nature Conservation No. 5. Nature Conservancy Council, Peterborough.

Hope-Simpson, J.F. & Jefferies, R.L. (1966). Observations relating to vigour and debility in marram grass (*Ammophila arenaria* (L.) Link.). *Journal of Ecology*, **54**, 271–4.

Hubbard, J.C.E. (1965). *Spartina* marshes in southern England. VI. Pattern of invasion in Poole Harbour. *Journal of Ecology*, **53**, 799–813.

Hutchings, M.J. (1979). Weight-density relationships in ramet populations of clonnal perennial herbs, with special reference to the −3/2 power law. *Journal of Ecology*, **67**, 21–33.

Jones, P. (1980). Leaf mottling of *Spartina* species caused by a newly recognised virus, spartina mottle virus. *Annals of Applied Biology*, **94**, 77–81.

Kershaw, K.A. (1960). Cyclic and pattern phenomena as exhibited by *Alchemilla alpina*. *Journal of Ecology*, **48**, 443–53.

Laing, C.C. (1967). The ecology of *Ammophila breviligulata*. II. Genetic change as a factor in population decline on stable dunes. *American Midland Naturalist*, **77**, 495–500.

Lovett-Doust, J. (1981). Intraclonal variation and competition in *Ranunculus repens*. *New Phytologist*, **89**, 495–502.

Lytle, R.W. Jr & Hull, R.J. (1982). Photoassimilate distribution in *Spartina alterniflora* Loisel. II. Autumn and winter storage and spring regrowth. *Agronomy Journal*, **72**, 938–42.

Marchant, C.J. (1968). Evolution in *Spartina* (Graminae). II. Chromosomes, basic relationships and the problem of *S.* x *townsendii* agg. *Journal of the Linnean Society (Botany)*, **60**, 381–409.

Marks, T.C. & Truscott, A.J. (1985). Variation in seed production and germination of *Spartina anglica* within a zoned salt marsh. *Journal of Ecology*, **73**, 695–705.

Marshall, J.K. (1965). *Coynephorus canescens* (L.) P. Beauv. as a model for the *Ammophila* problem. *Journal of Ecology*, **53**, 447–63.

Mendelssohn, I.A. (1980). The influence of soil drainage on the growth of salt marsh cord grass *Spartina alterniflora* in North Carolina. *Estuarine and Coastal Marine Science*, **11**, 27–40.

Mendelssohn, I.A. & McKee, K. (1988). *Spartina alterniflora* die back in Louisiana: time course investigation of soil waterlogging effects. *Journal of Ecology*, **76**, 509–21.

Mullins, P.H. & Marks, T.C. (1987). Flowering phenology and seed production of *Spartina anglica*. *Journal of Ecology*, **74**, 1037–48.

Noble, J.C., Bell, A.D. & Harper, J.L. (1979). The population biology of plants with clonal growth. I. The morphology and structural demography of *Carex arenaria*. *Journal of Ecology*. **67**, 983–1008.

Nooden, L.D. (1980). Senescence in the whole plant. *Senescence in Plants* (Ed. by K.V. Thiman), pp. 219–58. CRC Press, Boca Raton, FL.

Pitelka, L.F. & Ashmun, J.W. (1985). The physiology and ecology of connections between ramets in clonal plants. *Population Biology and Evolution of Clonal Organisms* (Ed. by J.B.C. Jackson, L.W. Bliss & R.E. Cook), pp. 399–435. Yale University Press, New Haven, CT.

Preston, M. (1981). *Survival, growth and reproduction in* Armeria maritima. PhD thesis, University of Oxford.

Pyke, D.A. & Thompson, J.N. (1986). Statistical analysis of survival and removal rate experiments. *Ecology*, **67**, 240–5.

Ranwell, D.S. (1972). *Ecology of Salt Marshes and Sand Dunes*. Chapman & Hall, London.

Raybould, A.R. (1989). *The population genetics of* Spartina anglica *C.E. Hubbard*. PhD thesis, University of Birmingham.

Reinartz, J.A. (1984). Life history variation of common mullein *(Verbascum thapsus)*. II. Plant size, biomass partitioning and morphology. *Journal of Ecology*, **72**, 913–25.

SAS Institute Incorporated (1985). *SAS User's Guide: Statistics Version* 5. Cary, NC.

Scholten, M. & Rozema, J. (1990). The competitive ability of *Spartina anglica* on Dutch salt marshes. Spartina anglica *— A Research Review* (Ed. by A.J. Gray & P.E.M. Benham), pp. 39–47. HMSO, London.

Silander, J.A. Jr (1985a). The genetic basis of the ecological amplitude of *Spartina patens*. II. Variance and correlation analysis. *Evolution*, **39**, 1034–52.

Silander, J.A. Jr (1985b). Microevolution in clonal plants. *Population Biology and Evolution of Clonal Organisms* (Ed. by J.B.C. Jackson, L.W. Bliss & R.E. Cook), pp. 107–52. Yale University Press, New Haven, CT.

Stapf, O. (1908). *Spartina townsendii. Gardeners Chronicle*, **1099**, 33–5.

Taylor, M.C. & Burrows, E.M. (1968). Studies on the biology of *Spartina* in the Dee Estuary, Cheshire. *Journal of Ecology*, **56**, 795–809.

Thompson, J.D. (1989). *Population variation in* Spartina anglica *C.E. Hubbard*. PhD thesis, University of Liverpool.

Thompson, J.D. (1990). Morphological variation among natural populations of *Spartina anglica*. Spartina anglica *— A Research Review* (Ed. by A.J. Gray & P.E.M. Benham), pp. 26–33. HMSO, London.

Thompson, J.D., McNeilly, T. & Gray, A.J. (1991a). Population variation in *Spartina anglica*. I. Evidence from a common garden experiment. *New Phytologist*, **117**, 115–28.

Thompson, J.D., McNeilly, T. & Gray, A.J. (1991b). Population variation in *Spartina anglica*. II. Reciprocal transplants among successionally adjacent populations. *New Phytologist*, **117**, 129–39.

van der Putten, W.H., van Dijk, C. & Troelstra, S.R. (1988). Biotic soil factors affecting the growth and development of *Ammophila arenaria. Oecologia*. **76**, 313–20.

Wallen, B. (1990). Changes in structure and function of *Ammophila* during primary succession. *Oikos*, **34**, 227–38.

Watkinson, A.R. & White, J. (1985). Some life-history consequences of modular construction in plants. *Philosophical Transactions of the Royal Society of London Series B*, **313**, 31–51.

Watkinson, A.R., Huiskes, A.H.L. & Noble, J.C. (1979). The demography of sand dune species with contrasting life cycles. *Ecological Processes in Coastal Environments* (Ed. by R.L. Jefferies & A.J. Davy), pp. 95–112, Symposium of The British Ecological Society, 19. Blackwell Scientific Publications, Oxford.

Watt, A.S. (1971). Contributions to the ecology of bracken (*Pteridium aquilinum*). VIII. The marginal and the hinterland plant, a study in senescence. *New Phytologist*, **70**, 967–86.

The role of nitrogen fixation in primary succession on land

J. I. SPRENT
Department of Biological Sciences, University of Dundee, Dundee DD1 4HN, UK

SUMMARY

1 The possible role of nitrogen-fixing organisms in primary succession is evaluated, principally for sites on sandy shores (intertidal and above high tide) and volcanic lava.
2 Cyanobacteria are common on most sites. How much nitrogen they fix and how much of this is made available to subsequent stages in succession is unclear. At its maximum level, cyanobacterial fixation could support growth of some angiosperms.
3 Legume and non-legume nodulated plants are comparatively rare in the early stages of primary succession. They may, however, establish successfully when seed is introduced. The presence of an endophyte rarely appears to be a limiting factor in establishment.
4 The possible consequences of nodulated plant colonization during early succession are briefly discussed.

INTRODUCTION

Nitrogenase, the enzyme that converts N_2 gas into ammonium, is big, slow and expensive to operate (Sprent & Raven 1985). It follows that nitrogen-fixing organisms will only have a significant role to play when nitrogen is the major factor limiting colonization and there is a plentiful supply of energy (either direct, e.g. light, or indirect, e.g. reduced carbon). These conditions do not always occur in the early stages of succession and even when they do, nitrogen-fixing organisms do not necessarily occur.

This chapter considers where, when, why and which nitrogen-fixing organisms are involved in primary succession. It will do so by discussing a selection of examples covering a range of nitrogen-fixing organisms and some major natural pioneer habitats. For more detail of the biology of the organisms involved, see Sprent & Sprent (1990). An exhaustive survey of the occurrence of nitrogen-fixing organisms in different primary successions is given by Walker (1993).

SANDY SHORES

By their very nature, sandy shores have an interface with the intertidal zones.

Recent work has analysed the laminated microbial mats which occur in temperate and tropical beaches. They consist of a series of layers, the exact order of which may vary according to physicochemical properties (oxygen and light penetration, etc.) but which usually have layers of green algae/cyanobacteria, purple sulphur bacteria and sulphate-reducing bacteria. Occasionally the green algal/cyanobacterial component is absent (Herbert *et al.* 1988). Fig. 1 illustrates the main features of a classical microbial mat, found in many sheltered beaches, such as the island of Mellum in the bay of Bremerhaven in the North Sea (Stal 1985). These, theoretically at least, are self-perpetuating.

The upper layer contains cyanobacteria, the exact species composition of which varies: these fix both carbon dioxide and nitrogen. The lower cells die, releasing organic carbon, which is metabolized by aerobic organisms, eventually generating anaerobic conditions. Sulphate, which is plentiful, can then be used by sulphate-reducing species, generating sulphide, which in turn forms a substrate for the anoxic photosynthetic purple sulphur bacteria. In an accreting system the organic nitrogen of such mats could form a rich substrate for land-colonizing plants. Unlike many other situations involving nitrogen-fixing cyanobacteria, no heterocystous forms were found by Stal (1985) at Mellum, possibly because of the high levels of sulphide (Malin & Pearson 1988).

SAND ABOVE THE TIDEMARK

In terms of primary colonization, coastal sand is arguably most important. However, current concern over 'desertification' and its reversal has enhanced consideration of nitrogen fixation in inland sandy areas (which may or may not be saline as well as arid). Consideration of both inland and coastal sandy areas allows some separation of factors, which may influence the occurrence and type of nitrogen-

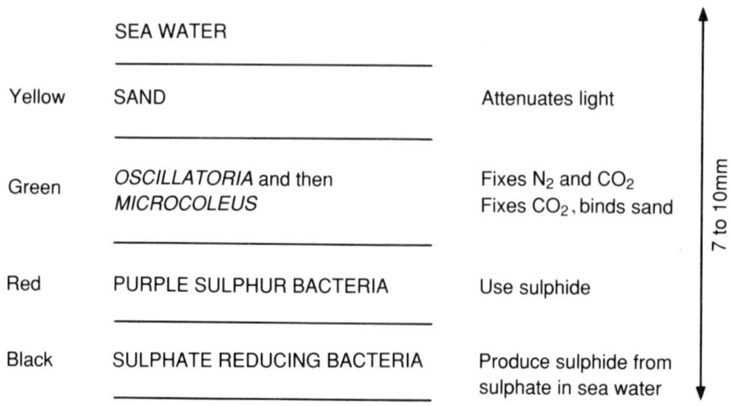

	SEA WATER	
Yellow	SAND	Attenuates light
Green	*OSCILLATORIA* and then *MICROCOLEUS*	Fixes N_2 and CO_2 Fixes CO_2, binds sand
Red	PURPLE SULPHUR BACTERIA	Use sulphide
Black	SULPHATE REDUCING BACTERIA	Produce sulphide from sulphate in sea water

7 to 10mm

FIG. 1. Some features of the microbial mat on Mellum island (Stal 1985). For variations see Herbert *et al.* (1988).

fixing organisms found. The major factors are texture, water, mineral content (including nitrogen), salinity, mobility and their interactions.

The effects of soil texture and cation exchange capacity on the occurrence of the nitrogen-fixing cyanobacterium *Anabaena variabilis* were considered by Skujins *et al.* (1987). Texture was varied by combining acid-washed and neutralized quartz sand with 20 or 50% of either bentonite (cation exchange capacity 90 meq 100 g^{-1}, particle size 0·01–1 μm) or kaolinite (CEC 8 meq 100 g^{-1}, size 0·1–5 μm). All mixtures were saturated with Ca^{2+}, washed with excess nitrogen-free nutrient, and buffered at pH 7·5 with 0·01 mol m^{-3} phosphate buffer. Data are given in Table 1. Although growth and nitrogenase activity were detectable in pure sand, rates were very low. Highest carbon and nitrogen yields were obtained when the surface area of the substrate was highest (particle size smallest) and cation exchange capacity highest. Thus although all treatments had adequate water and minerals other than nitrogen, there were large differences in yield. These were thought to relate to surface factors, in particular the interface between particles and the surface polymers of the cyanobacteria; these in turn could affect processes such as ion uptake.

Water content also has a major effect on cyanobacterial activity. Some species such as *Nostoc flagelliforme* and *N. commune* from China (Scherer *et al.* 1984) are very desiccation tolerant and after 2 years of drought can rapidly resume respiration (30 min after rewatering), photosynthesis (6–8 hours) and nitrogen fixation (120–150 hours). Such plants clearly fill an ecological niche. Do they have a role in succession? This depends upon how much nitrogen they fix and release in a form which can be used by other organisms and whether there is enough water to support these other organisms. A succession proceeding towards vascular plants, with the possible exception of certain ephermerals, requires more water than is present in true deserts without a deep water-table accessible to phreatophytes.

Salinity represents no real problem for cyanobacteria, many of which are marine. Where there is much sand movement cyanobacteria may be unable to colonize either on their own or in symbiosis with fungi, forming lichens, which may also be very desiccation tolerant. Lichens may be found in deserts, in dune

TABLE 1. Growth (measured as carbon and nitrogen content) of *Anabaena variabilis* on sand–clay mixtures. Carbon and nitrogen content were determined after 12 days, nitrogenase activity (acetylene reduction) after 10 days. The carbon and nitrogen contents of 100% sand were at the limits of the assay and are thus given in parentheses. After Skujins *et al.* (1987)

Substrate			mg g^{-1} substrate		C_2H_4/nmol $hour^{-1}$ 10 cm^{-2} surface
S	B	K	C	N	
100	0	0	(0·18)	(0·07)	110
80	20	0	1·01	0·26	530
50	50	0	2·68	0·65	680
80	0	20	0·53	0·13	260
50	0	50	0·87	0·21	210

S = sand; B = bentonite; K = kaolinite.

slacks and on coastal rocks. Although various micro-organisms have good particle-binding properties, fixation of sand-dunes tends to be linked with angiosperms such as marram grass; nitrogen-fixing micro-organisms may be associated with roots of such plants, but true symbiotic nitrogen-fixing organisms such as legumes tend to be absent. Grubb (1986) suggests that this may be because they cannot produce 'an exceedingly light and outstandingly dispersible seed' which also contains sufficient 'starter nitrogen'. He notes occasional exceptions including *Lupinus densiflorus* on the coastal range of California. The particular properties of the genus *Lupinus* in this general context are now considered.

One of the success stories in the use of *Lupinus* has been with *L. arboreus* on coastal sand-dunes in the North Island of New Zealand. Here the sand is about 0·008% nitrogen-deficient by any standards. Other nutrients are generally not deficient (see Gadgil 1983, and references therein). Although foredune stabilization is generally achieved using marram grass (given some nitrogen fertilizer) prior to sowing lupins, the latter can successfully bind dunes. Lupins establish readily from seed, nodulate copiously and fix enough nitrogen to support growth of *Pinus radiata* which is planted after the lupins have been killed. Sufficient rhizobia are present in the sand, inoculation being without effect — perhaps it should be noted that *Lupinus* nodulates with a wide variety of rhizobia. This situation would appear to vindicate Grubb's suggestion. The only apparent restriction on successful colonization by lupins in this system is the presence of seed. Is there anything special about lupins? They generally produce long tap roots, favouring both anchorage and mineral uptake from deep layers. They do not normally form mycorrhizal associations (Tester *et al.* 1987); this would be considered by many to be a disadvantage on nutrient-poor soils. However, they readily form proteoid roots which may be an alternative way of obtaining nutrients (see review by Lamont 1982). Proteoid roots appear to be formed mainly by plants (especially the Proteaceae), which have a relatively low demand for nutrients. They have the advantage of not having to supply an endophyte with carbon and being produced locally on parts of root systems where nutrients are available (e.g. litter fragments). They would thus appear to use the host plant's carbon resources effectively, although a cost–benefit analysis of the type carried out for nodulated or mycorrhizal plants does not appear to have been made.

The role of root-associated nitrogen-fixing micro-organisms in the colonization of sand-dunes has been considered many times, but really critical data are rather scant. Non-photosynthetic species require a carbon source — usually a plant. In his biological flora of *Ammophila arenaria* Huiskes (1979) gives references to claims for nitrogen fixation in association with its roots. Some of these should be regarded with caution, e.g. those involving *Azotobacter*, which is not salt tolerant. The case for the saltmarsh plant *Spartina alterniflora* is rather better made (Whiting *et al.* 1986), the authors pointing out the tight coupling between host carbon dioxide fixation and nitrogenase activity. In such systems it appears that the nitrogen-fixing organisms use soluble carbon compounds synthesized recently by the host. Is it also possible that significant nitrogen fixation could be associated with dead plant material? Recent interest in the proper management of organic waste such as straw has produced evidence that cellulolytic organisms (fungi or bacteria) can live in mixed culture with nitrogen-fixing organisms (Veal & Lynch 1984). Whether such associations occur naturally in pioneer habitats is not known. A further possibility has been raised by the recent evidence of cellulolytic anaerobic nitrogen-fixing bacteria in fresh water and forest soils (Leschine *et al.* 1988). In view of the comparatively small amounts of nitrogen required in early successional stages (see later) any of these organisms could have a significant role in succession.

A rather different system has been reported for inland sand-dunes and sand plains in northern Canada (Nelson *et al.* 1986). Here *Hudsonia tomentosa* (Cistaceae) is a major colonizing species. It is found in sands with poor water retention but relatively good light penetration. It is semi-evergreen and buried leaves and fruits may aid water retention. Associated with this plant (within 1 cm of plant parts), but not with other vascular plants in the same area, were several nitrogen-fixing cyanobacteria including *Nostoc* spp. Unlike the cyanobacteria in desert areas, which form crusts, those associated with *Hudsonia* were found in discrete grains or aggregates of grains sometimes forming green bands down to 22 cm. Whether these would have sufficient light to photosynthesize is very doubtful. Nitrogen-fixation rates were thought to be more water limited than light limited in the field. Although their nitrogen-fixing activity was low, they were considered to be significant in the autecology of *Hudsonia*. Why these cyanobacteria do not colonize areas near other vascular plants in this region is not clear.

Casuarina spp., especially *C. equisetifolia* can have a significant role to play in colonizing base-rich sand-dunes. Like *Lupinus, Casuarina* can form proteoid roots; it may also be mycorrhizal. Discussion is deferred to the next section.

VOLCANIC LAVA

There are relatively well-documented accounts of colonization of sites following volcanic eruption, the most comprehensive being for Krakatoa. The results have been reviewed by Whittaker *et al.* (1989). There are a number of elements in common between the early colonization of Krakatoa and that of Surtsey Island (off Iceland) and La Soufrière in Guadeloupe (Fritz-Sheridan 1987). The first colonizers on all three were mainly (or exclusively) cyanobacteria, including nitrogen-fixing forms, probably brought in by wind. The next stages, however, varied both within and between sites. We shall compare the tropical sites of Guadeloupe and Krakatoa. La Soufrière is a mountain 1456 m high and an eruption in 1976–77 sterilized the south-east flank of the dome. Near the summit, which is shrouded in cloud for most of the year, the soil surface temperature is generally in the range 15–18°C. On the rare clear days it rises to 26–42°C. Whereas on the other volcanoes studied several different species of cyanobacteria were found, the La Soufrière site was mainly colonized by the heterocystous genus *Scytonema*. This nitrogen-fixing organism is very gelatinous with good water-holding properties, although it dries out to a thin papery sheet on clear days. Its gelatinous nature when hydrated also enables it to resist rain scouring. *Scytonema* not only fixes nitrogen, but also provides a substratum for propagules of the next colonizer, the nitrogen-fixing lichen *Stereocaulon virgatum*. This sequence of two nitrogen-fixing forms is unusual. The lichen develops to a height of 10 cm, reflects heat and enables *Scytonema* to grow at the base of its canopy. The major reason for the success of this system appears to be the persistent cloud cover, maintaining hydration.

On Krakatoa, colonization patterns varied between coastal and inland parts. Inland the successions appear to have begun with gelatinous mats of several cyanobacteria, including nitrogen-fixing forms. These were quickly followed at all levels (the highest point being 735 m) by ferns, rather than lichens. Subsequent development towards forest or scrub varied with site, altitude and occurrence of further eruptions (Whittaker *et al.* 1989). At one site, where the inland forest was dominated by *Timonius compressicaulis*, the thorny creeper *Elaeagnus latifolia* was common on the ground and in the canopy. *Elaeagnus* is an actinorhizal genus (like *Casuarina*) and it would be interesting to know if this species was fixing nitrogen and why the site (presumably) had areas with lower levels of combined nitrogen than others. At coastal sites there is a type of vegetation known as pes-caprae, consisting of low-growing herbs and creepers, including *Ipomoea pes-caprae*. However, also included are some legume species, such as *Canavalia maritima* and *Vigna marina*. Coastal erosion apparently restricts the expanse of pes-caprae. In areas where accretion occurs, *Casuarina equisetifolia* may take on a successional role and persist to form mature woodland.

The excellent review by Whittaker *et al.* (1989) for Krakatoa enables us to examine the suggestion by Grubb (1986) that a limitation on legumes as colonizers may be because their necessarily nitrogen-rich seeds are not light enough for rapid dispersal. Fig. 2 shows how the various species of plant (including cyanobacteria) recorded on all islands, except Anak Krakatoa, are dispersed. Note the preponderance of wind and sea dispersal in the early stages, with birds and bats becoming more important as the populations of these animals build up. The authors suggest that cyanobacteria and ferns arrived by wind and that when they had modified the nature of the soil and microclimate sufficiently, wind-dispersed angiosperms became incorporated. Sea-dispersed plants also arrived early on. Of the 103 species in this category recorded between 1883 and 1893, one-third are legumes. Of the 34 legumes in the species list, 30 are known to be potentially nodulated, one cannot (*Cassia siamea*), two are doubtfully nodulated and for one there is no information (Faria *et al.* 1989). The reason that many of these do not establish well, at least in the early stages, is thought to be the eroding state of the shore. Thus it appears that legume colonization is possible if a seed source sufficiently close for sea dispersal is available. The volcanic islands of Krakatoa are within 40 km of the main Indonesian islands, whereas it is over 2000 km from the east coast of Australia to the sands in New Zealand described above. The natural arrival of *Casuarina equisetifolia*, which has winged seeds, is by wind. It is native to Indonesia. Its role in stabilizing coastal areas in Krakatoa is similar to

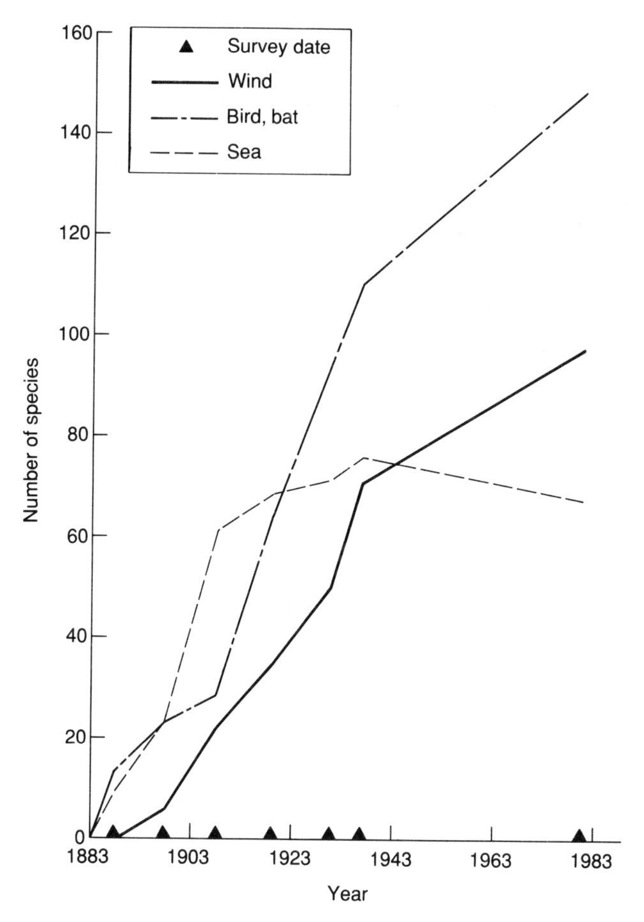

FIG. 2. How colonizing plant species arrived at Krakatoa following volcanic eruption. From Whittaker *et al.* (1989).

that in other areas where it is native or has been exploited for the improvement of arid and saline areas in various parts of the tropics and subtropics (National Research Council 1984).

In the Hawaiian Islands there is a considerable legume flora. However, these native species do not appear to be able to colonize the young volcanic substrates on the island of Hawaii. This could be because of seed dispersal problems, although this possibility has not been explored. These volcanic sites were studied by Vitousek *et al.* (1987) who found that of all the major and minor plant nutrients only addition of nitrogen led to increased growth of the native tree *Metrosideros polymorpha*. The site clearly has major nitrogen-deficiency problems, as do most sites of recent volcanic activity. In this case, however, there is considerable invasion by *Myrica faya*, an exotic plant originally brought to the Hawaiian Islands by Portuguese immigrants. In the absence of *Myrica*, most nitrogen input is from rainfall (5 kg N ha^{-1} $year^{-1}$). In similar areas with *Myrica* an additional input of 18 kg N ha^{-1} $year^{-1}$ was estimated by Vitousek *et al.* (1987). Further, the *Myrica* sites contained a considerable amount of biologically available soil nitrogen, whereas non-*Myrica* sites had little. *Myrica* potentially leads to substantial modification of the succession; the whole system is reviewed in Vitousek & Walker (1989). The seed of *Myrica* is dispersed by a variety of birds including exotic species. Although the plants were heavily nodulated, the source of the *Frankia* inoculum is unknown. It would be tempting to suggest, from this *Myrica* study and the work on tree lupins described earlier, that the arrival of the host rather than the endophyte is the limiting factor in establishing legume and non-legume nodulated plants in pioneer situations which are nitrogen-deficient. However, this may not always be the case, since for example, European soils lack the rhizobia necessary to nodulate soybeans (*Glycine max*).

ROCKY AREAS

These occur on coasts, mountains (including volcanoes) and elsewhere. Nitrogen-fixing cyanobacteria occur on, in and under rocks, as long as the substrate is low in nitrogen, sufficient in other minerals, wet at least some of the time and usually not acid. Table 2 summarizes some of the habitats and their limitations. For further details see Hoffmann (1989) and references therein. The cyanobacteria frequently are associated with microalgae and often with lichens. Depending on the environment they can trap mineral and other particles which can lead to the formation of pockets of 'soil' which can be further colonized. How much nitrogen is fixed and what proportion enters the soil pockets with precipitation and dry deposition are largely a matter of conjecture at this stage.

GENERAL CONSIDERATIONS

Almost all primary successions have nitrogen-fixing cyanobacteria among their first colonizers. Fixing both carbon dioxide and nitrogen is a clear advantage if nitrogen is the major limiting nutrient. The need for light may be a disadvantage and it should not be assumed that because visibly green cyanobacterial cells are present they are actively photosynthesizing (symbiotic forms may be dark green

TABLE 2. Rock habitats colonized by cyanobacteria

Type of habitat	Description	Comments on cyanobacteria
Epilith	Exposed surface	Most tolerate temperature extremes and desiccation
Hypolith	Under translucent pebbles down to about 4 cm	May be active down to 0·005% of incident light
Endolith	Colonize or make cavities	Sites very variable in irradiance. Some species may bore into limestone

inside coralloid roots of cycads 1 m below ground). Stal (1985) and Herbert *et al.* (1988) found a rapid attenuation of light penetrating sand, reaching zero at 8–10 mm depth. The total amount of nitrogen fixed by such organisms is likely to be small and may or may not be made available to later successional stages; it may, for example, be leached out following a drying–wetting cycle or lost by denitrification. Thus, the presence of cyanobacteria is not guaranteed to lead to a build-up of combined nitrogen in an ecosystem. Studies by Stal (1985) and others agree that an upper limit of 15 kg N ha^{-1} $year^{-1}$ may be fixed by cyanobacteria. They have a low carbon : nitrogen ratio (about 5–15) and are therefore likely to be mineralized rapidly following death (probably more rapidly than symbiotically fixing structures such as nodules). If we assume a carbon : nitrogen ratio of 5 and a carbon content of 45% of dry weight, then this level of nitrogen fixation would yield 16·6 g dry weight cyanobacterial cells m^{-2}. For colonizing similar to marram grass, the carbon : nitrogen ratio is likely to be closer to 45 : 1 (i.e. overall plant nitrogen concentration is about 1%). Thus, if cyanobacterial nitrogen were all made available for later plant colonizers, it would support about 150 g dry matter m^{-2}. If supplemented by nitrogen from other sources (rain, sea-spray) cyanobacterial fixation could suffice for subsequent angiosperm colonization to a level which is frequently observed.

It appears that much pioneering vegetation remains nitrogen-limited and yet nodulated legumes and actinorrhizal plants are often not among primary colonizers. Evidence cited above for *Lupinus arboreus* and *Myrica faya* supports Grubb's (1986) suggestion that this may be because of lack of seed. However, it seems somewhat surprising that such an obvious niche for nitrogen fixation has not been more widely exploited. We shall end with some speculation as to the consequences such exploitation could produce. In both the examples cited here, as well as in those where nitrogen-fixing plants have been used to hasten recolonization of mine spoils and other man-made wastes (e.g. Dancer *et al.* 1977a, b), accumulation of nitrogen in the ecosystem has been greatly accelerated. Is this necessarily a good thing?

If the supply of combined nitrogen is the rate-limiting step can other resources (phosphorus, potassium, trace elements) be better husbanded? Removal of nitrogen limitation could lead to more rapid cycling of all nutrients with possible losses to the ecosystem in wetter areas. The occasional occurrence (e.g. of *Casuarina equisetifolia*) of nitrogen-fixing plants in primary succession is usually associated with some type of environmental stress such as drought. Amounts of nitrogen fixed may then be limited by water supply. Where nitrogen-fixing plants now occur naturally they are commonly either a small component of a mixture of species or, if dominant or co-dominant, in arid or semi-arid areas, or as part of regeneration after fire. In all cases the total amount of nitrogen fixed per hectare is low (considerably less than 100 kg ha^{-1} $year^{-1}$). For example, Sheehy (1989) calculated that the net nitrogen input from white clover to a grazed pasture in England needs to be about 41 kg ha^{-1} $year^{-1}$ and this could be met by clover occupying about 10% of the surface area of the pasture. Similarly, in stands of

Acacia spp. in Africa and Australia the total nitrogen fixed (in as much as this can be measured!) is likely to be well below 50 kg ha^{-1} year^{-1} (Sutherland & Sprent 1993). High rates of fixation (500 kg ha^{-1} year^{-1}) may occur in pure stands of lucerne or grain legumes where most of the nitrogen is removed with the crop. This, ecologically, is a wholly artificial situation.

High rates of fixation in natural systems — such as those following the aggressive colonization by *Leucaena leucocephala* in parts of Hawaii (L. Handley-Raven, unpublished data) could result in excess mineralization and enhanced nitrate is groundwater. In a balanced system nitrogen lost from a legume should be taken up by other plants. This may require a more varied flora than normal in primary colonization. Unfortunately, although there is recent evidence of nitrogen loss from growing (as opposed to damaged or senescent) legumes (Wacquant *et al.* 1989), its quantification, like that of nitrogen fixation in the field, is virtually non-existent. Until we have realistic figures for such processes, the potential of nitrogen-fixing legumes and actinorhizal species in succession remains largely a matter of speculation. Good land management requires that these problems are properly addressed.

REFERENCES

Dancer, W.S., Handley, J.F. & Bradshaw, A.D. (1977a). Nitrogen accumulation in kaolin mining wastes in Cornwall. I. Natural communities. *Plant and Soil*, **48**, 153–67.

Dancer, W.S., Handley, J.F. & Bradshaw, A.D. (1977b). Nitrogen accumulation in kaolin mining wastes in Cornwall. II. Forage legumes. *Plant and Soil*, **48**, 303–14.

Faria, S.M. de, Lewis, G.P., Sprent, J.I. & Sutherland, J.M. (1989). Occurrence of nodulation in the Leguminosae. *New Phytologist*, **111**, 607–19.

Fritz-Sheridan, R.P. (1987). Nitrogen fixation on a tropical volcano, La Soufrière. II. Nitrogen fixation by *Scytonema* sp. and *Stereocaulon virgatum* Ach. during colonization by phreatic material. *Biotropica*, **19**, 297–300.

Gadgil, R.L. (1983). Biological nitrogen fixation in forestry — research and practice in Australia and New Zealand. *Biological Nitrogen Fixation in Forest Ecosystems: Foundations and Applications* (Ed. by J.C. Gordon & C.T. Wheeler), pp. 317–32. Martinus Nijhoff/Dr W. Junk, The Hague.

Grubb, P.J. (1986). The ecology of establishment. *Ecology and Design in Landscape* (Ed. by A.D. Bradshaw & M.J. Chadwick), pp. 83–97. Symposium of the British Ecological Society, 24. Blackwell Scientific Publications, Oxford.

Herbert, R.A., Tughan, C.S., De Wit, R. & van Gemerden, H. (1988). Development of laminated mat microbial ecosystems in temperate marine environments. *Perspectives in Microbial Ecology* (Ed. by F. Megusar & M. Gantar), pp. 105–11. Slovene Society for Microbiology, Ljubljana.

Hoffmann, L. (1989). Algae of terrestrial habitats. *Botanical Review*, **55**, 77–105.

Huiskes, A.H.L. (1979). Biological flora of the British Isles: *Ammophila arenaria* (L.) Link. *Journal of Ecology*, **67**, 363–82.

Lamont, B. (1982). Mechanisms for enhancing nutrient uptake in plants, with particular reference to Mediterranean South Africa and Western Australia. *Botanical Review*, **48**, 597–689.

Leschine, S.B., Holwell, K. & Canale-Parola, E. (1988). Nitrogen fixation by anaerobic cellulolytic bacteria. *Science*, **242**, 1157–9.

Malin, G. & Pearson, H.W. (1988). Aerobic nitrogen fixation in aggregate-forming cultures of the nonheterocystous cyanobacterium *Microcoleus chtonoplastes*. *Journal of General Microbiology*, **134**, 1755–63.

National Research Council (1984). *Casuarinas: Nitrogen-Fixing Trees for Adverse Sites*, pp. 118. National Academy Press, Washington, DC.

Nelson, S.D., Bliss, L.C. & Mayo, J.M. (1986). Nitrogen fixation in relation to *Hudsonia tomentosa*: a pioneer species in sand dunes, northeast Alberta. *Canadian Journal of Botany*, **64**, 2495–501.

Scherer, S., Ernst, S.A., Chen, T-W. & Böger, P. (1984). Rewetting of drought-resistant blue-green algae: Time course of water uptake and reappearance of respiration, photosynthesis, and nitrogen fixation. *Oecologia*, **62**, 418–23.

Sheehy, J.E. (1989). How much dinitrogen fixation is required in grazed grassland? *Annals of Botany*, **64**, 159–61.

Skujins, J., Nikkanen, T. & Henriksson, E. (1987). Research note: responses of cyanobacterial growth and N_2-fixation to textural components of soils. *Arid Soil Research and Rehabilitation*, **1**, 195–8.

Sprent, J.I. & Raven, J.A. (1985). Evolution of nitrogen-fixing symbioses. *Proceedings of the Royal Society of Edinburgh, B*, **85**, 215–37.

Sprent, J.I. & Sprent, P. (1990). *Nitrogen Fixing Organisms. Pure and Applied Aspects*. Chapman & Hall, London.

Stal, L.J. (1985). *Nitrogen-fixing cyanobacteria in a marine microbial mat*. PhD thesis, Rijksuniversiteit of Groningen.

Sutherland, J.M. & Sprent, J.I. (1993). Nitrogen fixation by legume trees. *Symbiosis in Nitrogen Fixing Trees* (Ed. by N.S. Seibba Rau & C. Rodriguez-Barrueco). Oxford and IBM Publishing Co., New Delhi (in press).

Tester, M., Smith, S.E. & Smith, F.A. (1987). The phenomenon of 'nonmycorrhizal' plants. *Canadian Journal of Botany*, **65**, 419–31.

Veal, D.A. & Lynch, J.M. (1984). Associative cellulolysts and dinitrogen fixation by co-cultures of *Trichodermes hanzianum* and *Clostridium butyricum*. *Nature*, **310**, 695–7.

Vitousek, P.M. & Walker, L.R. (1989). Biological invasion by *Myrica faya* in Hawaii: plant demography, nitrogen fixation, and ecosystem effects. *Ecological Monographs*, **59**, 247–65.

Vitousek, P.M., Walker, L.R., Whiteaker, L.D., Mueller-Dombois, D. & Matson, P.A. (1987). Biological invasion by *Myrica faya* alters ecosystem development in Hawaii. *Science*, **238**, 802–4.

Wacquant, J.P., Ouknider, M. & Jacquard, P. (1989). Evidence for a periodic excretion of nitrogen by roots of grass – legume associations. *Plant and Soil*, **116**, 57–68.

Walker, L.R. (1993). Nitrogen fixers and species replacements in primary succession. *Primary Succession on Land* (Ed. by J. Miles & D.W.H. Walton), pp. 249–72. Blackwell Scientific Publications, Oxford.

Whiting, G.J., Gandy, E.L. & Yoch, D.C. (1986). Tight coupling of root-associated nitrogen fixation and plant photosynthesis in the salt marsh grass *Spartina alterniflora* and carbon dioxide enhancement of nitrogenase activity. *Applied and Environmental Microbiology*, **52**, 108–13.

Whittaker, R.J., Bush, M.B. & Richards, K. (1989). Plant recolonization and vegetation succession on the Krakatau Islands, Indonesia. *Ecological Monographs*, **59**, 59–123.

Primary succession on man-made wastes: The importance of resource acquisition

R. H. MARRS* AND A. D. BRADSHAW†

** Ness Botanic Gardens, University of Liverpool Environmental and Horticultural Research Station, Ness, Neston, South Wirral L64 4AY; and † Department of Environmental and Evolutionary Biology, University of Liverpool, PO Box 147, Liverpool L69 3BX, UK*

SUMMARY

1 This chapter reviews the principles of primary succession on man-made wastes, concentrating on china clay waste as a model system. Three developmental phases have been identified: (a) colonization, (b) intermediate development, and (c) a mature ecosystem.

2 The critical importance of acquisition of resources, particularly of nitrogen in the intermediate development phase, is the key to the development of mature ecosystems. The evidence for the central role of nitrogen in this process is reviewed, with special reference to the mechanisms through which nitrogen can be accumulated and cycled efficiently.

3 A major contention of this chapter is that man-made wastes have a role as an experimental analogue for testing general hypotheses relating to primary succession; a hypothetical example and three case studies are presented as illustrations.

INTRODUCTION

Primary succession is the term used to describe the different sequences of communities that develop on freshly created raw substrates, where there are no existing biota. While it can be argued that all ecosystems are undergoing a never-ending primary succession (Miles 1987), usually the term is restricted to the initial phases from a virgin surface, covering colonization and subsequent development toward a notional equilibrium 'climax' state. Thus, we have a working definition to separate primary from secondary succession, which is used to describe successions that have been initiated from pre-existing communities, in which the soils contain organic matter and carry some biota (including dormant propagules). There are examples that do not fit easily into this simple framework, for example where an ecosystem, including the soil organic matter, has been destroyed by fire leaving only the mineral soil. Nevertheless, in this chapter, we shall confine our attention to the initial development of ecosystems on those freshly created raw mineral substrates created by humans which contain neither organic matter nor any living organism.

It is generally accepted that we know much less about primary successions than secondary ones. There are several reasons for this:

1 On a world-wide basis, there are few areas where primary succession is currently proceeding (Miles 1987).
2 Even where there is an active primary succession occurring, the creation of fresh material may be slow, for example where the succession is brought about by retreat of a glacier (Müller 1977; Haeberli 1985).
3 The initial phases of colonization and development are generally very slow, and may take at least 100 years, that is much beyond the working life of a single investigator.

The last two problems have been circumvented most often by adopting a chronosequence approach. Community change and ecosystem development have been inferred from stands of different successional age for which some independent estimate of age is possible. This approach is a useful one, but is fraught with difficulties (Miles 1979). In this chapter it is argued that the processes and mechanisms involved in primary succession can often be elucidated from studies of man-made wastes, where an inferential approach can be combined with experimental studies. Moreover, additional information can be derived from experiments on the restoration of man-made wastes, because effectively this is the science of accelerating primary successions artificially.

Inspection of the environmental factors limiting plant growth on a range of man-made wastes (Fig. 1) shows that there are many similarities with the raw substrates found in natural primary successions. Wastes derived from organic materials, e.g. landfill sites, are not included since organic matter has effectively been imported from another ecosystem and thus any succession should not be counted as truly primary. The man-made wastes on which primary successions do occur often have adverse physical properties, with problems relating to texture, stability, temperature, water retention and severe nutrient deficiency. Some wastes have additional problems of an unusual nature, including excess salinity, metal toxicity and extreme pH values, which may be infrequent in nature (Bradshaw & Chadwick 1980). In this chapter we examine in detail 'natural' and artificially accelerated primary successional development on one type of man-made waste — china clay sand waste — relating this where possible to similar information from both natural primary successions and land restoration schemes elsewhere, in an attempt to derive general principles. China clay wastes are ideal for such general studies; they comprise gravel-sized particles which are poorly compacted and have a high porosity. Chemically they are inert, have a low pH (4·0–5·0), low cation exchange capacity ($\leqslant$0·2 mmol kg^{-1}), are deficient in most plant nutrients except potassium and have no toxicity problems (Bradshaw *et al.* 1975; Roberts *et al.* 1982; Marrs *et al.* 1983).

China clay wastes are thus an extreme example of a nutrient-poor waste and we are going to emphasize findings from studies on these wastes that seem to have significance for systems developing in primary successions elsewhere. We argue that different man-made wastes afford almost unlimited opportunities for studying ecosystem development during primary successions, and give some illustrations.

MATERIALS	PHYSICAL: Texture/ structure	Stability	Water supply	Surface temperature	CHEMICAL: Nutrients Macro	Nutrients Micro	pH	Toxic materials	Salinity
Colliery spoil	OOO	OOO/o	O/o	o/•••	OOO	o	OOO/o	o	o/••
Strip mining	OOO/o	OOO/o	OO/o	o/•••	OOO/o	o	OOO/o	o	o/••
Fly ash	OO/o	o	o	o	OOO	o	•/•••	••	o/••
Oil shale	OO	OOO/o	OO	o/••	OOO	o	OO/o	o	o/••
Iron ore mining	OOO/o	OO/o	O/o	o	OO	o	o	o	o
Bauxite mining	OO/o	o	o	o	OO	o	o	o	o
Heavy metal wastes	OOO	OOO/o	OO/o	o	OOO	o	OOO/ •	•/•••	o/•••
Gold wastes	OOO	OOO	O	o	OOO	o	OOO	o	o
China clay wastes	OOO	OO	OO	o	OOO	o	O	o	o
Acid rocks	OOO	o	OO	o	OO	o	O	o	o
Calcareous rocks	OOO	o	OO	o	OOO	o	•	o	o
Sand and gravel	O/o	o	o	o	O/o	o	O/o	o	o
Coastal sands	OO/o	OOO/o	O/o	o	OOO	o	o	o	o/•
Land from sea	OO	o	o	o	OO	o	o/•	o	•••
Urban wastes	OOO/o	o	o	o	OO	o	o	o/••	o
Roadsides	OOO/o	OOO	OO/o	O/o	OO	o	O/o	o	o/••

DEFICIENCY		EXCESS	
OOO	Severe	•	Slight
OO	Moderate	••	Moderate
O	Slight	•••	Severe
o	Adequate		

Relative to the establishment of a soil/plant ecosystem appropriate to the material: variations in severity are due to variation in materials and situations.

FIG. 1. The physical and chemical characteristics of different types of derelict land materials (Bradshaw & Chadwick 1980).

NATURAL PRIMARY SUCCESSIONS ON MAN-MADE WASTES

In most primary successions that have been studied, including those on glacial moraines, sand-dunes and china clay wastes, the process can be divided, in broad terms into three main stages (Bradshaw 1983):

1 Initial colonization phase.
2 Intermediate development phase, in which nutrient accumulation is crucial.
3 Late development phase, including further colonization and replacement.

Nitrogen figures prominently in the first two phases (Robertson & Vitousek 1981; Tilman 1985, 1986; Vitousek & Walker 1987), and this fact can be derived from first principles. First, nitrogen is required in larger amounts by plants than any other mineral element. It is an essential element for many aspects of plant metabolism and concentrations in plants tend to be between 1 and 3% (Allen *et al.* 1974). Moreover, although the concentration of potassium is commonly about the same as that of nitrogen, the difference in atomic weight means that the plant tissues contain three to four times as many atoms of nitrogen than potassium and between eight and 10 times the number of atoms of any other element (Epstein 1972). Second, nitrogen is held in most soils only in organic matter and is released mainly by decomposition processes. Since in temperate climates organic matter turnover is slow, there must be a large capital of nitrogen so that a sufficient supply is released to meet the annual requirements of vegetation. As an example, in temperate systems litter decomposition rates are in the order of 0·0625 per year or in other words, k (*sensu* Jenny *et al.* 1949) is 1/16 (Olsen 1963). Many ecosystems have an annual plant requirement of 100 kg N ha^{-1}, based on dry matter production of 5000 kg ha^{-1} and a nitrogen concentration of 2%. To provide this supply with $k = 1/16$ requires a capital of 1600 kg N ha^{-1}. There is very little nitrogen present in the raw substrates of primary succession. For example, in china clay wastes there is as little as 5–20 kg N ha^{-1} within the surface 21 cm. For natural primary successions on glacial moraines initial values can be in the same order of magnitude: Jenny (1980) quoted a value of 28 kg N ha^{-1}. There is, therefore, an inescapable chronic shortage of nitrogen, especially when compared with developed soils which usually contain 2000–10 000 kg N ha^{-1}.

It would be foolish to believe that nitrogen is the only factor of importance in successions on man-made wastes. Depending on the starting material, other elements such as potassium, calcium, magnesium and especially phosphorus may also play a role. Common experience suggests that potassium is not usually limiting, because of its widespread occurrence in clay minerals, and that calcium and magnesium are only limiting in certain materials, although often any positive effects are associated with their effects on pH. By contrast, phosphorus deficiencies are common. Certain clays used in brick-making provide some of the few exceptions; the phosphorus levels on these wastes can be high and have a major impact on the rate of natural colonization of these materials (Dutton & Bradshaw 1982). Normally, on most wastes plant growth is severely limited unless phosphorus is given in combination with nitrogen (Bradshaw *et al.* 1978; Smith & Bradshaw 1979; see Fig. 2).

Initial colonization phase

Initial colonization is stochastic in nature and strongly limited by what is available in the vicinity. This is made clear by the studies of colonization of alkaline waste heaps in North Yorkshire, England (Bradshaw 1983). In temperate regions, the first colonists are generally micro-organisms, green and blue-green algae, followed by lichens, bryophytes and then higher plants. Successful colonists tend to be adapted to the inhospitable nature of the raw mineral substrates, where they experience extremes of heat and drought as well as low mineral status (Grubb 1987).

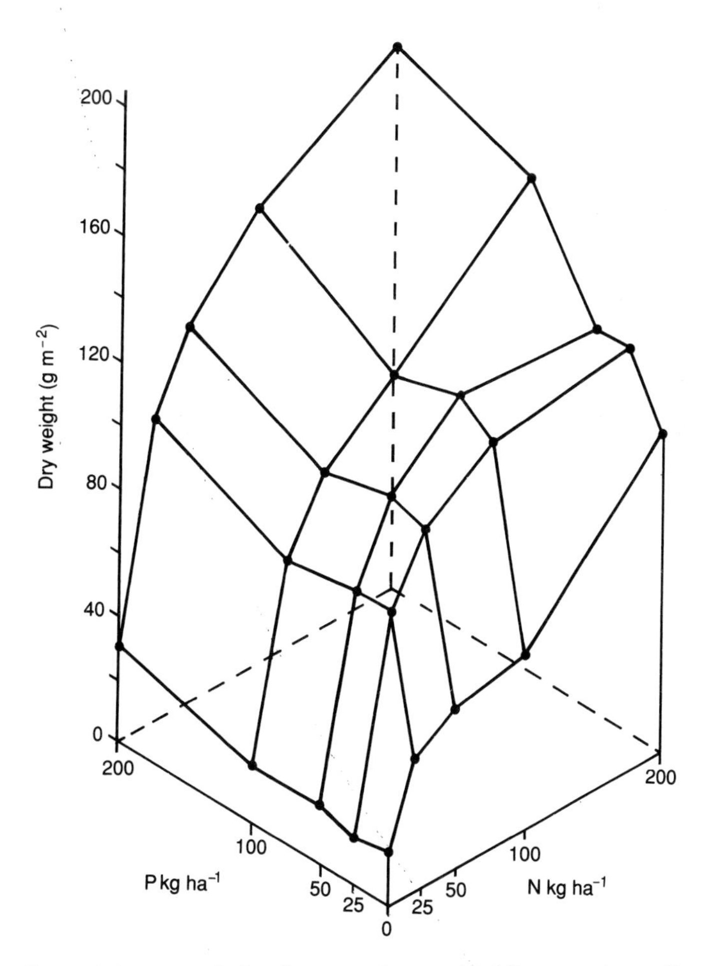

FIG. 2. The effects of nitrogen and phosphorus on the growth of *Festuca ovina* on limestone fines (Bradshaw *et al.* 1978).

Intermediate development phase

If no nitrogen-fixing vascular species invade during the primary succession, nitrogen takes a very long time to accumulate, because inputs must be derived from atmospheric sources. As nitrogen inputs in rainfall in Europe tend to range from $<10\,\mathrm{kg\,N\,ha^{-1}\,year^{-1}}$ in relatively unpolluted areas to $>40\,\mathrm{kg\,N\,ha^{-1}\,year^{-1}}$ in more polluted areas (Brimblecombe & Stedman 1982; Heil & Diemont 1983; During & Willems 1986; Bobbink & Willems 1987; Fowler 1987; Ashenden & Bell 1989), accumulation of the theoretical target capital of $1600\,\mathrm{kg\,N\,ha^{-1}}$ will take 32–160 years if there is a 100% efficiency of capture. As capture-efficiency is likely to be well below 100% in primary successions, especially during the early phases when vegetation is sparse, it might take several hundred years to reach the hypothetical target, especially in unpolluted areas. What else might happen? There appear to be only three possible strategies that will allow plants to colonize and accumulate nutrients.

1 Reduced growth rate (RGR). If a species were to have a relatively low growth rate, its annual nutrient requirement, given the same tissue nutrient concentration, will be correspondingly reduced. This is perfectly possible, since a wide range of growth rates has been recorded and species from mine wastes had exceptionally low mean growth rates in the tests performed by Grime & Hunt (1975). An alternative hypothesis could be that initial colonists could have lower nitrogen concentrations. However, the range of variation in this character appears to be much less than that of RGR (Chapin 1980).

2 Scavenging species. If a species were to have a normal growth rate, but had a widely ramifying root system and maintained itself at low densities, it might be able to scavenge enough nitrogen from a large area and concentrate it into a central shoot system. The growth achieved per unit area exploited would be low, but on an individual plant basis would be normal. Such species have not been formally recognized or their detailed physiological characteristics described, but their possibility has been noted by Grubb (1986). Woody plants with a diffuse root system are likely candidates. The theoretical value of a diffuse root system has been discussed by Robinson (1986) and *Reynoutria japonica* appears to possess the appropriate characters (Hirose 1986). On china clay wastes the shrubs *Salix atrocinerea* and *S. caprea* form low density communities and appear to fit the same role, and the grass *Holcus lanatus* may also fall into the same category.

3 Nitrogen fixation. This is the most obvious method of overcoming nitrogen shortage, with an unlimited supply available in the atmosphere providing the problems of energetics and relationships with nitrogen-fixing micro-organisms can be overcome. Generally, invasion by nitrogen-fixing higher plants is a feature of most primary successions. Studies of successions on natural materials, glacial moraines and sand-dunes plus two on man-made materials — ironstone spoil (Leisman 1957) and china clay waste (Roberts *et al.* 1980) — confirm this view, with a wide range of nitrogen-fixing species involved (Table 1). Both leguminous and non-leguminous species such as *Dryas* spp. and *Alnus* spp. are common, and

TABLE 1. Examples of nitrogen-fixing species which have been found on primary successions

Substrate	Nitrogen-fixing species	Site location
Glacial moraines (Crocker & Major 1955)	*Alnus crispa* *Dryas drummondii*	Glacier Bay, Alaska, USA
Glacial moraines (Viereck 1966)	*Astragalus alpinus* *Astragalus tananaica* *Astragalus nutzotinesis* *Dryas drummondii* *Dryas integrifolia* *Sheperdia canadensis*	Muldrow Glacier, Alaska, USA
Glacial moraines (Friedel 1938a, b; Richard 1968)	*Lotus corniculatus* *Trifolium badium* *Trifolium thalli*	Hintereisferner and Aletsch Glacier, Europe
Glacial moraines (A. D. Bradshaw, unpubl. data)	*Coriaria* spp.	New Zealand
China clay sand waste (Roberts *et al.* 1981)	*Lotus corniculatus* *Lupinus arboreus* *Sarothamnus scoparius* *Ulex eurpaeus* *Ulex gallii*	Cornwall, UK
Ironstone spoils (Leisman, 1957)	*Melilotus alba* *Trifolium repens*	Minnesota, USA

where nitrogen-fixation rates have been estimated, significant contributions have been found (Lawrence *et al.* 1967; Skeffington & Bradshaw 1980).

In theoretical terms the nitrogen accumulation process in primary succession is perhaps the most often quoted example of facilitation (*sensu* Connel & Slatyer 1977), and we can guess empirically the times and nitrogen capitals required for sufficient facilitation by estimating the point at which late successional non-nitrogen-fixing species invade (Table 2). These values show some variability, but are at least in the same order of magnitude as our theoretical target. In our china clay study we found that 1000 kg N ha^{-1} (700 kg N ha^{-1} in the soil) and 1800 kg N ha^{-1} (1200 kg N ha^{-1} in the soil) were the levels at which non-nitrogen-fixing *Salix* scrub and *Betula–Quercus* woodland developed respectively. The critical capital in the sand-dune system is slightly lower than the others at 400 kg N ha^{-1}, but this value has been computed for the shallower depth of 10 cm, and in this particular succession, there were no nitrogen-fixing species of higher plants recorded (Olsen 1958).

Given the obvious advantages of nitrogen fixation, and given the preceding arguments about the crucial role of nitrogen accumulation in primary successions, we might expect a profusion of leguminous species in the early stages. Although legumes are a feature of most primary successions, they do not necessarily predominate. There are several reasons for this. First, even though the energy

TABLE 2. Estimates of target nitrogen contents and the time taken to achieve these targets on four raw substrates

Type of ecosystem	Time to develop non-nitrogen-fixing vegetation (years)	Nitrogen content (kg N ha^{-1})		Source
Glacial moraines	100	Soil (0–30 cm)	1200	Crocker & Major (1955)
		Litter	1000	
Sand-dunes	21	Soil (0–10 cm)	400	Olsen (1958)
Ironstone spoils	100	Soil (0–21 cm)	600	Leisman (1957)
China clay waste	>70	Soil (0–21 cm)	700	Roberts *et al.* (1981)
		Vegetation and litter	300	Marrs *et al.* (1983)
	>120	Soil (0–21 cm)	1200	Roberts *et al.* (1981)
		Vegetation and litter	600	Marrs *et al.* (1983)

costs of nitrogen-fixing have not been elucidated fully, they are certainly considerable. In one of the potential sequences of reactions, the initial step breaking the nitrogen–nitrogen bond requires 209·3 kJ mol^{-1}, and the energy for this process must be derived from photo- or chemosynthesis (Sprent 1987). Where growth is already limited by a range of other environmental factors, there may be little energy available after the requirements of metabolism and growth have been met. Second, fixation of nitrogen requires more than minimal amounts of the plant nutrients, especially phosphorus, which may not be available in the early successional stages. This is certainly true for china clay wastes (Bradshaw *et al.* 1975). Third, it must be remembered that only certain plant groups have evolved the necessary symbiotic relationships with nitrogen-fixing micro-organisms, and clearly both plant and endophyte must colonize for nitrogen fixation to begin. The range of nitrogen-fixing species is therefore limited, but this does not appear to be a sufficient explanation for the paucity of such species in many primary successions. The first and second problems are more likely to be the rate-limiting processes.

The precise nature of phosphorus deficiency on the natural successions found on man-made wastes has not been well-documented. There is, of course, no equivalent to nitrogen fixation. Instead, apart from the slow rate of input from atmospheric sources, usually $<0{\cdot}5$ kg ha^{-1} year^{-1}, the major inputs come from the gradual weathering and transformation to more available forms of whatever phosphorus is available in the mineral substratum. This process is very clear in the long-term New Zealand chronosequence on sand described by Williams & Walker (1969).

Under such circumstances two of the strategies described for coping with limited nitrogen, which do not involve fixation, are appropriate. A reduced growth rate will obviously reduce the demand of phosphorus as well as that of nitrogen. Efficient scavenging will also be effective, but not so effective as for nitrogen, because: (a) the mobility of phosphorus in the soil is much lower than that of nitrate; and (b) the atmospheric inputs are a lower proportion of the annual requirements (3·3 compared with 10% for nitrogen). As a result, a diffuse root system on its own will not help phosphorus supply, although roots which ramify through the soil will help. Scavenging for phosphorus may be enhanced by mycorrhizal associations, which are a common feature of plants growing in nutrient-poor soils. Mycorrhizae are often found in roots growing on man-made wastes during restoration (Daft & Hacskaylo 1976; Allen & Allen 1980; Lambert & Cole 1980; Stahl *et al.* 1988), but we have no information on their abundance on restored china clay wastes. Nevertheless, the supply of phosphorus, like that of nitrogen is likely to be a general problem in most primary successions on man-made wastes, and may be longer lasting.

Late development phase

Very little research has been done on this phase of primary successions, but presumably the processes involved are similar to those under normal secondary successions, except that the species have first to arrive at the site. On our china clay sites, the late successional woodland species were found growing in very immature successions, and even in some instances on almost bare sand. However, these established seedlings did not grow successfully. Again the nitrogen supply appears to be a major factor limiting the growth of trees. This effect was demonstrated by growing *Acer* spp. at varying distances from nitrogen-fixing *Alnus* spp. *Acer* trees growing >3 m from the *Alnus* showed hardly any growth increment, while those growing <1 m away showed substantial height increments (Fig. 3). Similar effects have been found under *Lupinus arboreus* on man-made wastes (Palaniappan *et al.* 1979) and on sand-dunes (Gadgil 1971a, b).

LAND RESTORATION OR 'ARTIFICIALLY ACCELERATED' PRIMARY SUCCESSIONS

The long time period associated with ecosystem development during primary successions is not generally acceptable for land restoration schemes, where, unless the early stages of succession are favoured by wildlife conservation constraints, speed is of the essence. The objective of most restoration schemes is to create visually acceptable, self-sustaining ecosystems as quickly and as cheaply as possible. Although this can often be done at great expense with the addition of topsoil or organic wastes, such as sewage sludge, in the context of artificially accelerating primary successions this importation of organic material and nutrients can be regarded as cheating and thus will be disregarded here.

One of the most obvious ways to accelerate a primary succession is to remove much of the stochasticity from the initial colonization phase by supplying the seed of appropriate species, plus fertilizers. This is normal practice, but the paucity of the nitrogen supply soon exerts its effect and the established vegetation becomes moribund. Usually this moribund vegetation remains alive, but growth is very poor. The vegetation as a whole is unacceptable from an aesthetic viewpoint, and

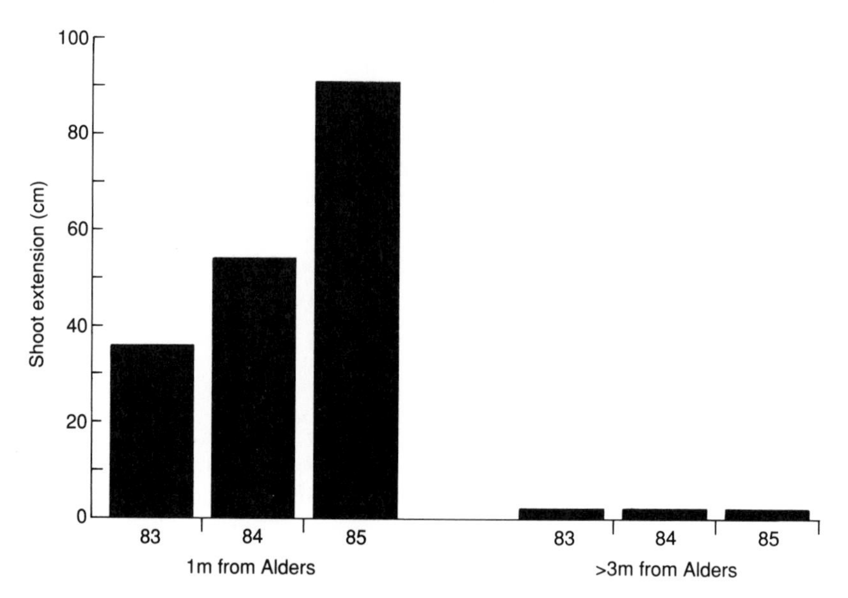

FIG. 3. Shoot extension of 10-year-old *Acer pseudoplatanus* on china clay wastes between 1983 and 1985 in relation to proximity to *Alnus* spp. (unpublished data supplied from A. J. Kendle & A. D. Bradshaw).

may be so poorly established as to provide only a weak, unstable plant cover (Bradshaw & Chadwick 1980). Grassland is often the starting point in many restoration schemes and the acute deficiency of nitrogen is clearly evident when nitrogen is added to these moribund swards (Fig. 4). Good growth and a pleasing visual appearance occurs when nitrogen is added.

A simple balance sheet of nitrogen requirements illustrates the problem. A single starter nitrogen application in most restoration schemes is about 100 kg N ha^{-1}. This level equates with about 1 year's requirements of our hypothetical ecosystem. Since we have predicted that 1600 kg ha^{-1} total soil nitrogen capital is required to maintain an annual 100 kg ha^{-1} supply of nitrogen, we must be about 1500 kg N ha^{-1} short of the required capital. Even the *Salix* shrub ecosystem which required only 700 kg ha^{-1} would have a deficit of 600 kg N ha^{-1}. Thus, it should be no surprise that nitrogen accumulation and cycling is a major problem in the restoration on man-made wastes, and the restoration strategy must include appropriate treatments to: (a) accumulate a sufficient nitrogen capital; and (b) make sure that this capital is cycled efficiently.

Principles of nitrogen accumulation

Nitrogen accumulation can be accelerated in two ways:

1 By fertilizer addition. As we have already demonstrated, applications to man-made wastes will need to be repeated at least annually or the vegetation becomes

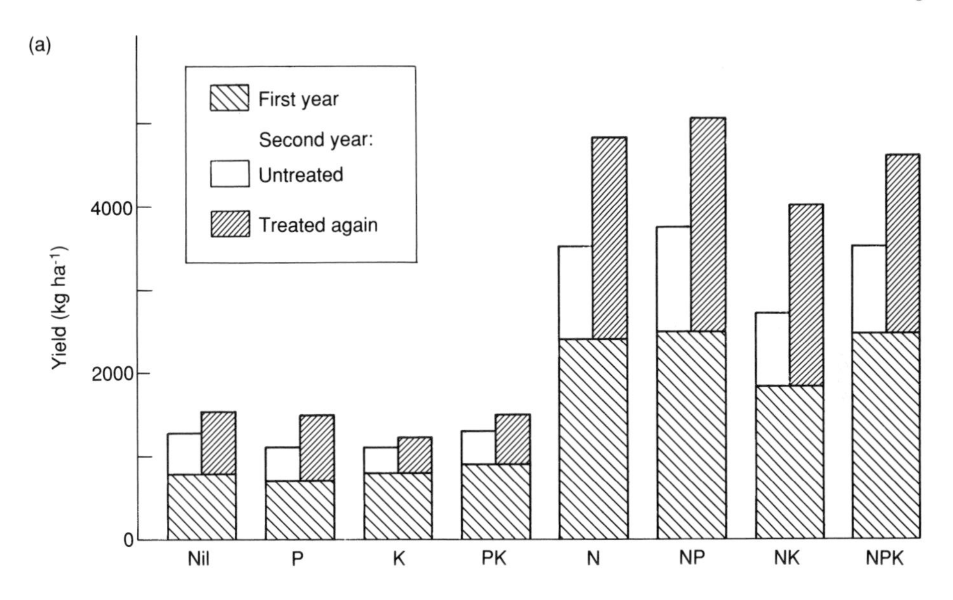

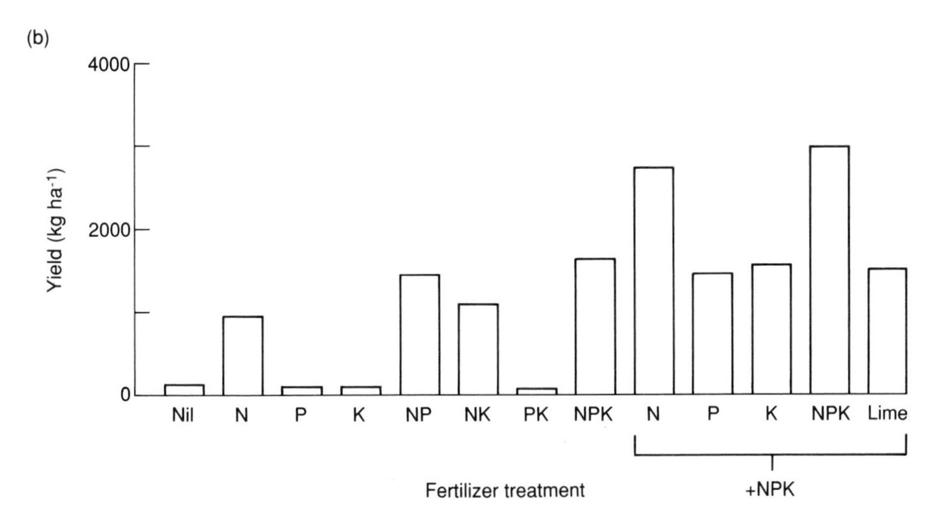

FIG. 4. Regrowth of moribund grass swards following the application of fertilizer on: (a) colliery spoil, with 50 : 22 : 41 kg ha^{-1} of N : P : K, with yields assessed by a single autumn cut (Bloomfield *et al.* 1982); and (b) china clay waste, with 50 : 22 : 42 : 26 : 1000 : 100 kg ha^{-1} of N,P,K,Mg : lime : micronutrients, with yield assessed by a single cut 4 months after treatment (Bradshaw *et al.* 1975).

moribund. Moreover, fertilizers are relatively expensive and there can be substantial losses through leaching and run-off (Marrs & Bradshaw 1980, 1982). During the vegetation establishment phase nitrogen loss on raw china clay wastes can be spectacular, with almost complete removal of soluble nitrogen within a few weeks (Dancer 1975). Similar results have been found on some colliery spoil wastes, where the clay contents are much greater than those of china clay wastes (Elias *et al.* 1982).

2 By biological fixation. Nitrogen fixation can make substantial inputs to restored ecosystems on man-made wastes. Acetylene reduction studies (Skeffington & Bradshaw 1980) measured relatively large inputs from three leguminous species grown on china clay sand waste under field conditions: 27, 49 and 72 kg N $ha^{-1} year^{-1}$ by *Ulex europaeus, Trifolium repens* and *Lupinus arboreus*, respectively. In well-limed and fertilized plots (given phosphorus, potassium and magnesium) net nitrogen accumulation rates of up to 157 kg N $ha^{-1} year^{-1}$ were found (Dancer *et al.* 1977). In other conditions on derelict land, legumes may not achieve these levels, but their continuing contribution over many years, without significant repeated fertilizer costs, can be extremely important (Jefferies *et al.* 1981a, b; Elias *et al.* 1982). Significant contributions can also be obtained from non-leguminous nitrogen-fixing species such as *Alnus* spp. (see Fig. 3).

Clearly, to obtain a rapid accumulation of nitrogen cheaply it is sensible to use a combination of fertilizers and legumes. Where this is done it is recommended that fertilizer nitrogen levels be reduced to below 50 kg N $ha^{-1} year^{-1}$, as inputs greater than this have adverse effects on nitrogen fixation (Skeffington & Bradshaw 1980). Where agricultural forage legumes are used, inputs of lime and phosphorus may still be needed to maintain a large legume component and hence a high nitrogen fixation rate (Handley *et al.* 1978; Roberts *et al.* 1982). As both of these elements may be subject to large leaching losses (Marrs & Bradshaw 1980), additions of these elements may need to be continued for some time.

Estimates of the times required to achieve the theoretical target of 1600 kg N ha^{-1} for china clay wastes are given in Table 3, assuming a 100% capture-efficiency. Measured capture-efficiencies of nitrogen from established swards gave values of 95%, showing that once vegetation is established the ecosystems are extremely nitrogen tight (Marrs & Bradshaw 1980). Where both legumes and fertilizers are used in combination the time required to reach the hypothetical target is reduced from >130 to between 7 and 20 years.

TABLE 3. Times required for developing grassland ecosystems to accumulate the theoretical target of 1600 kg N ha^{-1} under a range of reclamation management scenarios (Marrs & Bradshaw 1982)

Scenario	Measured minimum and maximum inputs (kg N $ha^{-1} year^{-1}$)	Time to reach 1600 kg N ha^{-1} target (years)
Natural inputs only	9–12	133–177
+Legumes and −nitrogen fertilizers	9–144	11–177
−Legumes and +nitrogen fertilizers	81–84	20
+Legumes and +nitrogen fertilizers	81–216	7–20

Optimizing nitrogen efficiency

Once ecosystems have become established, we can gain a first approximation of success by comparing the proportions of accumulated capital within the main ecosystem pools (Table 4). In temperate ecosystems the greatest proportion (>90%) of the capital (>5000 kg N ha^{-1}) is found in the soil, with very small amounts in the vegetation. On naturally colonized china clay wastes the capital is lower, but the greater proportion is still in the soil pool (50–70%). In recently reclaimed sites, where the succession has been accelerated by use of fertilizers and legumes, the capital is small, but the main pool is in the roots with the soil being a subsidiary one.

Generally, the soil tends to have the largest nitrogen store, because the inflow : outflow ratio is greatest and in particular the outflow (nitrogen mineralization) is very slow. This limitation may be of little consequence if the capital size is large, but where the capital is small its effect is dramatic. In normal soils with a mineralization rate of approximately 2% (Reuss & Innis 1977) and a nitrogen capital of between 2000 and 10 000 kg N ha^{-1} a supply from mineralization of 40–200 kg N ha^{-1} $year^{-1}$ could be expected. Clearly, the lower supply rate is inadequate to meet the demand of our theoretical ecosystem (100 kg N ha^{-1}), yet the higher rate would support a much greater productivity. Rates of soil nitrogen mineralization in reclaimed china clay wastes are low, 1–10 μg g^{-1} compared with 28–40 μg g^{-1} for normal soils (Bradshaw *et al.* 1975; Lanning & Williams 1979; Roberts *et al.* 1980). However, these results were derived directly from standard incubation tests (Keeney & Bremner 1967) and take no account of the nitrogen capital. Skeffington & Bradshaw (1981) calculated a turnover index, by expressing their mineralization test data as a percentage of the total nitrogen capital (Fig. 5). Although in some of the younger swards the turnover index was greater than in adjacent pastures, after about 6 years there was little difference between the reclaimed wastes and pastures. These results support the view that it is the low capital which results in reduced nitrogen supplies during primary successions rather than a reduced mineralization rate *per se*. Where a nitrogen capital of 400 kg N ha^{-1} has been accumulated we can expect a supply only of about 8 kg N ha^{-1} $year^{-1}$ from soil mineralization, which is still 92 kg N ha^{-1} $year^{-1}$ short of our theoretical requirement. Similar data have been found on natural primary successions elsewhere (Robertson & Vitousek 1981).

The quality of organic matter also affects turnover, with an increased turnover found on legume-rich compared with legume-poor sites (Skeffington & Bradshaw 1981). High rates of nitrogen mineralization under legumes on man-made wastes have been demonstrated under a range of species and on different waste materials (Palaniappan *et al.* 1979; Jefferies *et al.* 1981a). This is explicable in terms of the carbon : nitrogen ratio of the organic matter produced by legumes and grass species. Where nitrogen has been added as fertilizer to pure grass swards on china clay waste, nitrogen mineralization can be almost negligible (Bradshaw *et al.* 1975).

TABLE 4. Comparison of nitrogen content and compartmentation between major ecosystem pools in china clay wastes and semi-natural temperate ecosystems (Marrs *et al.* 1982)

Ecosystem	Total nitrogen (kg ha^{-1})	% Shoots	% Roots	% Litter	% Soil	Sampling depth (cm)	Reference
Semi-natural temperate ecosystems							
Heathland	5644	3 (shoots + roots)		1	96	20	Robertson & Davies (1965)
Grassland	10 460	0·7	0·5	0·3	98·5	30	Perkins (1978)
Oak woodland	7940	4·9 (shoots + roots)		0·9	94·2	30	Duvigneaud & Denaeyer de Smet (1970)
Naturally colonized china clay wastes							
Pioneer (*Lupinus arboreus*)	291	37	1	7	56	21	Marrs *et al.* (1981)
Pioneer (*Calluna vulgaris*/*Ulex europaeus*)	823	13	5	3	79	21	Marrs *et al.* (1981)
Intermediate (*Salix atrocinerea*)	981	8	18	6	68	21	Marrs *et al.* (1981)
Woodland (*Betula pendula*/*Rhododendron ponticum*/*Quercus robur*)	1770	30	3	0	67	21	Marrs *et al.* (1981)
Reclaimed china clay wastes							
Sand tips	211	11	59	nd	30	21	Roberts *et al.* (1980)
Mica dam walls	441	8	61	nd	31	21	Roberts *et al.* (1980)

nd, no data.

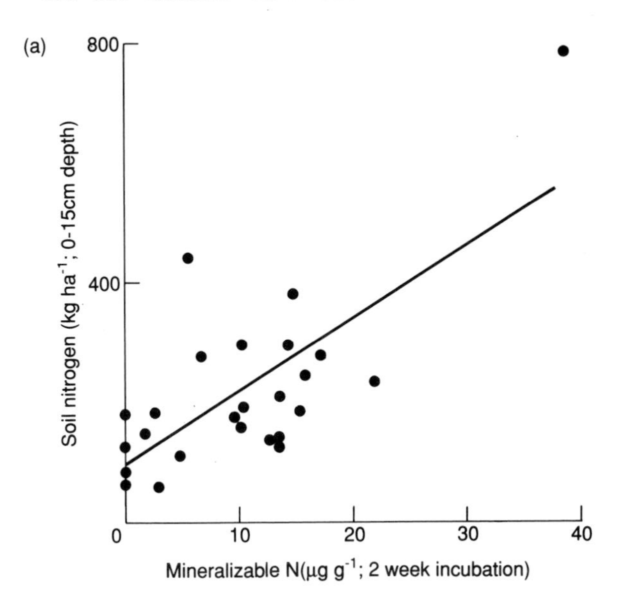

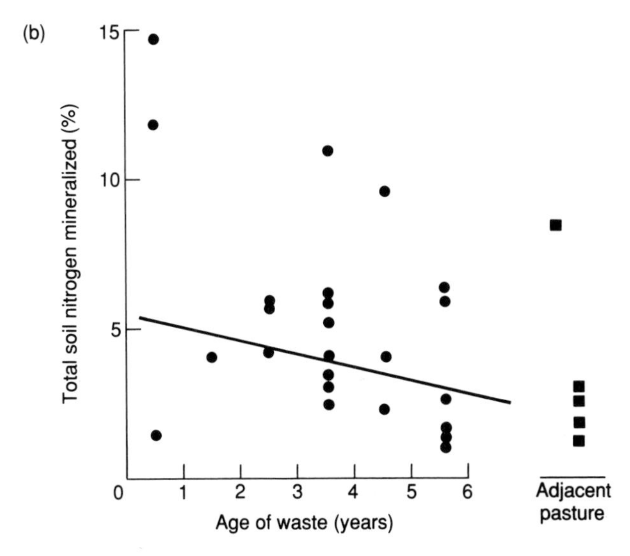

FIG. 5. Mineralization of soil nitrogen in reclaimed china clay sand wastes: (a) mineralization rate against total nitrogen; and (b) index of nitrogen turnover (mineralization expressed as a % of total nitrogen) against age since reclamation compared with adjacent pasture (Skeffington & Bradshaw 1981).

On the sites where the roots are the main nitrogen pool, decomposition of the roots is clearly a rate-limiting step. For china clay wastes we have very limited information on plant decomposition processes, but a first approximation suggested that 30% decomposed annually (Lanning & Williams 1979), with the main limiting factor being the adverse carbon : nitrogen ratio. The only ways of stimulating root decomposition is to redress the carbon : nitrogen ratio, either by adding fertilizer or by using plants with a low carbon : nitrogen ratio, usually nitrogen-fixing species.

A further site of nitrogen blockage in some grassland swards is in the above-ground biomass. An easy way to remove this blockage is to increase the turnover by grazing or mowing. Grazing, for example, tends to remove very little nitrogen from the system, only about 5% of grazed nitrogen (Newbould & Floate 1977; Perkins 1978; Bülow-Olsen 1980), but transfers nitrogen as faeces and urine to the soil pools. As a result, much more rapid cycling of the accumulated capital can be achieved (Floate 1981), and on reclaimed china clay wastes a minimum of 27% of the nitrogen taken up into the above-ground pool could be returned by a managed grazing regime (Marrs *et al.* 1980). Moreover, much of the above-ground nitrogen occurs in dead plant material, which can also be put back into effective circulation (Marrs *et al.* 1980). Presumably, the addition or encouragement of soil animals would also encourage this process, but little is known about their effects, although colonization of restored colliery wastes by some faunal groups can be very rapid (<2 years), with *Collembola* in particular having an important influence on decomposition (Hutson 1980a, b).

The importance of other nutrients

There is an immense literature showing the importance of various nutrient additions, particularly phosphorus, for the successful development of ecosystems on man-made wastes (for reviews see Hutnik & Davies 1973; Schaller & Sutton 1978; Bradshaw & Chadwick 1980).

It could be argued that these extra nutrients are required only for the establishment and maintenance of highly productive species, with a consequent high nutrient demand. This, however, begs the question as to where a primary succession terminates. Perhaps we should consider that termination occurs at or near a 'climax' community, when, under temperate conditions, some species at least are likely to be nutrient-demanding. Second, the development of species characteristic of the early stages of primary succession are sometimes limited by nutrients other than nitrogen. Marked responses have also been obtained in some restoration schemes where other nutrient elements are added.

If we consider the growth of nitrogen-fixing species, even those such as *Sarothamnus scoparius* and *Ulex europaeus* which are normally considered to be tolerant to low levels of phosphorus and calcium, then marked responses to these elements are found (Roberts *et al.* 1982) (Fig. 6a). Other species, for example *Trifolium repens*, show even more spectacular responses to calcium and phosphorus (Fig. 6b). As calcium is leached rapidly, and phosphorus is either leached or fixed, such species have a recurrent need for these elements.

In many developing ecosystems, calcium is likely to be lost through leaching, and it is possible that calcicoles will be favoured in the early stages, but disappear

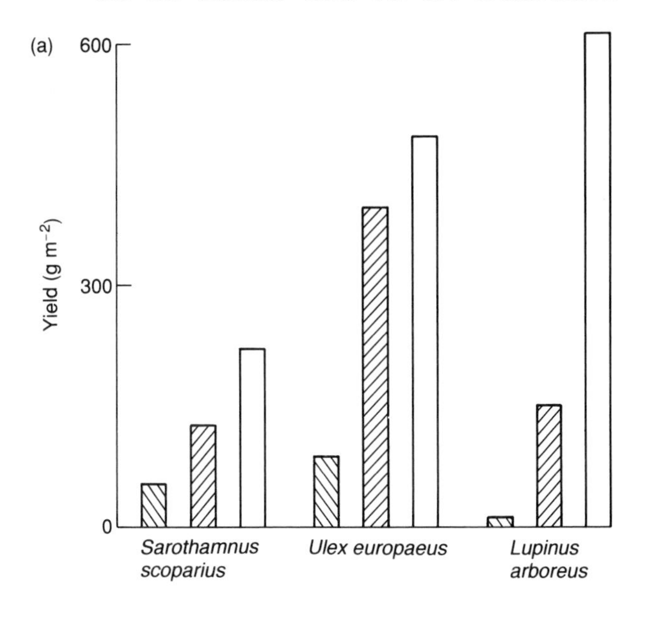

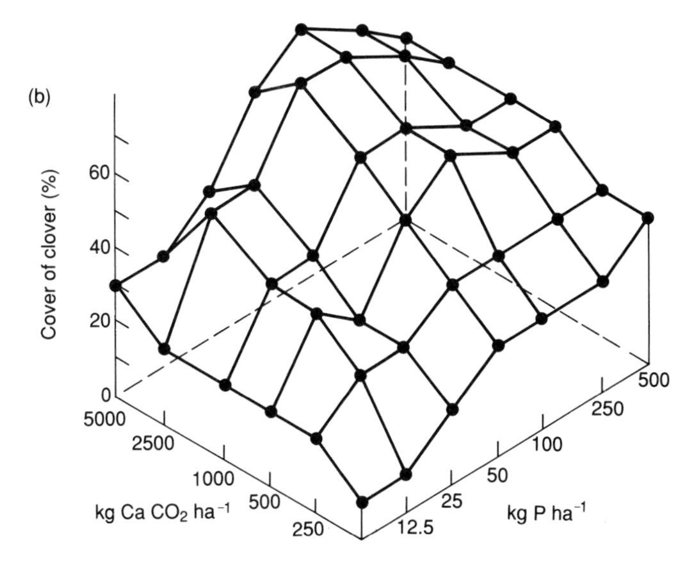

FIG. 6. Effects of lime and phosphorus on the yield of (a) three species of legume; ▧ with 2 t limestone ha^{-1} and 5 kg P ha^{-1}; ▨ with 50 kg P ha^{-1}; □ with 2 t limestone ha^{-1} and 50 kg P ha^{-1} (Roberts *et al.* 1982), and (b) a 2-year-old *Trifolium repens* sward (Bradshaw *et al.* 1978).

later. Phosphorus loss through leaching is smaller with most being fixed into sparingly soluble forms in the short term, although it is likely to increase in both total and available forms in the longer term (Knabe 1973). These two opposing processes, coupled with their potential direct interaction with nitrogen fixation, could have interesting ecological effects. Unfortunately, these effects have not been studied in detail.

THE SCIENTIFIC VALUE OF STUDYING PRIMARY SUCCESSIONS ON MAN-MADE WASTES

In the final part of this chapter we illustrate where we believe studies of man-made wastes can allow us to make significant progress in understanding the dynamics of soil formation and ecosystem development. Four examples are used: the first is a hypothetical one illustrating the enormous theoretical potential of man-made wastes, the second is a case-study where an attempt has been made to realize a subset of this potential, and the last two case-studies illustrate the value that lysimeter experiments may have as an investigative tool.

Testing Jenny's 'clorpt' equation for ecosystem genesis

In his classic equation developed to explain soil and ecosystem formation Jenny (1941, 1980) expressed ecosystem properties as a function of state variables (equation 1):

$$\underset{\text{(system properties)}}{l, v, a, s} = \underset{\text{(state factors)}}{f(cl, \phi, r, p, t, \ldots, m)} \qquad (1)$$

where the system properties:

l = total system; e.g. total carbon content and system respiration
v = vegetation; e.g. biomass, communities
a = animals; e.g. work performed, reproduction and health
s = soils; e.g. pH, texture, bacterial counts, humus content

are linked in multiple interactions with the state factors:

cl = climate
ϕ = pool of flora and fauna
r = topography (or relief), water-table etc.
p = parent material
t = age
$\ldots$ = any unknown factors
m = an additional factor added here to represent the influence of human beings.

Jenny (1980) points out that these state variables **can be** independent (his emphasis), uncorrelated and orthogonal, although in practice for most ecosystems this ideal proves to be impossible. He therefore developed a further series of equations with dominant and subordinate factors to separate major relationships

TABLE 5. Series of equations derived by Jenny (1980) to investigate the relationship between major functions (**bold**), subordinate functions and ecosystem development

l,v,a,s = f(**cl,ø,r,p,t,...**)	climofunction or climosequence
l,v,a,s = f(**ø,cl,r,p,t,...**)	biofunction or biosequence
l,v,a,s = f(**r,cl,ø,p,t,...**)	topofunction or toposequence
l,v,a,s = f(**p,cl,ø,r,t,...**)	lithofunction or lithosequence
l,v,a,s = f(**t,cl,ø,r,p,...**)	chronofunction or chronosequence
l,v,a,s = f(**...,cl,ø,r,p,t**)	dot-function or dot-sequence

between certain state factors and ecosystem development (Table 5). For most studies of plant succession the chronosequence model is used, where time is the dominant factor, and the other factors subordinate with their effects assumed to be zero. A chronosequence approach is therefore used where there may be a great deal of noise in the data, which may reflect unmeasured variability in the unknown factors. We argue that artificially accelerated successions on man-made wastes offer a real opportunity to investigate Jenny's equations using properly designed factorial experiments. Theoretically all state factors can be varied independently, for example:

1 Climate. Usually the climate for a given area is fixed for a given chronosequence; however, to misquote Francis Bacon, 'if the climate will not come to the hill, the hill will go to the climate'. As the wastes are the results of industrial processes the wastes can be dumped anywhere, and there is no reason why large volumes of waste cannot be moved to areas with a different climate. However, if the study has a low budget, lysimeter cores could be transplanted in different climatic regimes! Waste material and lysimeters have been moved to different areas for experimental convenience (Sheldon & Bradshaw 1977; Marrs & Bradshaw 1980), and there is no reason why the effects of climate *per se* could not be studied.

2 Parent material. The variability in physico-chemical properties of the parent material is a major problem in studying succession (Jenny 1980). However, many man-made wastes are produced by sieving, grinding and/or flotation/separation operations, which are designed to maintain uniform particle sizes. Thus, it is possible to start with a parent material of known and uniform composition. Studies of mineral weathering and its interaction with biotic processes can be made with more confidence than under natural systems. Moreover, it is possible to produce mixtures with wastes which have different properties (see case-study C). In many cases these waste mixtures may have a practical reclamation value.

3 Topography. Individual sites can, within reason, be tailored to test, for example, the effects of slope and compaction.

4 Biotic pool. This is the main factor to be modified during artificially accelerated successions, but studies could be extended to manipulating the effects of:

(a) herbivory, which has major impacts on both plant production and species composition during secondary succession; and

(b) the soil flora and fauna, which play a major part in nutrient cycling.

5 Humans. Planned management is an important feature of accelerating primary

successions during land restoration and the effects of fertilizer input, grazing, cutting, burning and crop offtake can be assessed and measured.

Clearly, there is much scope for designing experiments on man-made wastes to test the effects of all state variables alone and in combination. We believe such process-orientated studies would produce valuable information which would be of relevance to natural primary successions.

Case-study A: Testing the effects of varying biota, fertilizer inputs and grazing

Davis *et al.* (1985) report, a good example of an experimental study of the early phases of an artificially accelerated primary succession on a limestone quarry floor, where three factors limiting ecosystem development were alleviated. These factors were:

1 A low input or retention of propagules, alleviated by the addition of seeds of *Brachypodium sylvaticum* and *Lotus corniculatus*.
2 A low nutrient availability, alleviated by fertilizer addition.
3 Severe herbivore activity (mainly rabbits), alleviated by excluding the rabbits.

Effectively this study related vegetation development (**v**) to three state variables (equation (2), using Jenny's approach of bold type to identify main factors):

$$\mathbf{v} = f(\mathbf{t}, \boldsymbol{\o}, \mathbf{m}, p, r, cl, \ldots) \quad (2)$$

where the biotic pool ($\boldsymbol{\o}$) = f($\mathbf{v}_{seed}$, $\mathbf{a}_{grazing}$). In this study the subordinate factors remained unknown, but their effects on the interpretation of the data were minimized through the use of a randomized block ($n = 4$) experimental design, with factorial combinations of all treatments. Spatial variability in the subordinate factors across the site was to some extent constrained by block placement, and their effects within the blocks would be randomly associated with individual treatments.

The objective of this particular study was to develop methods for creating species-rich communities with a high vegetation cover. Poor colonization was a major limiting factor, because seeded plots had greater cover values than those that were unseeded, but the other two factors also limited vegetation development. A cover of 83% was found after 4 years where all three constraints were alleviated, at least double all other treatments (Fig. 7a). However, detailed assessments of flowering performance and rooted frequencies of individual species showed that the optimal treatments for providing both species diversity and cover was to combine seed addition with caging or fertilizer use, but not both (Fig. 7b).

Case study B: The use of lysimeters in studies of ecosystem development on man-made wastes

In studies of developing ecosystems on man-made wastes, field studies can often be augmented by meaningful experiments using monolith-type lysimeters.

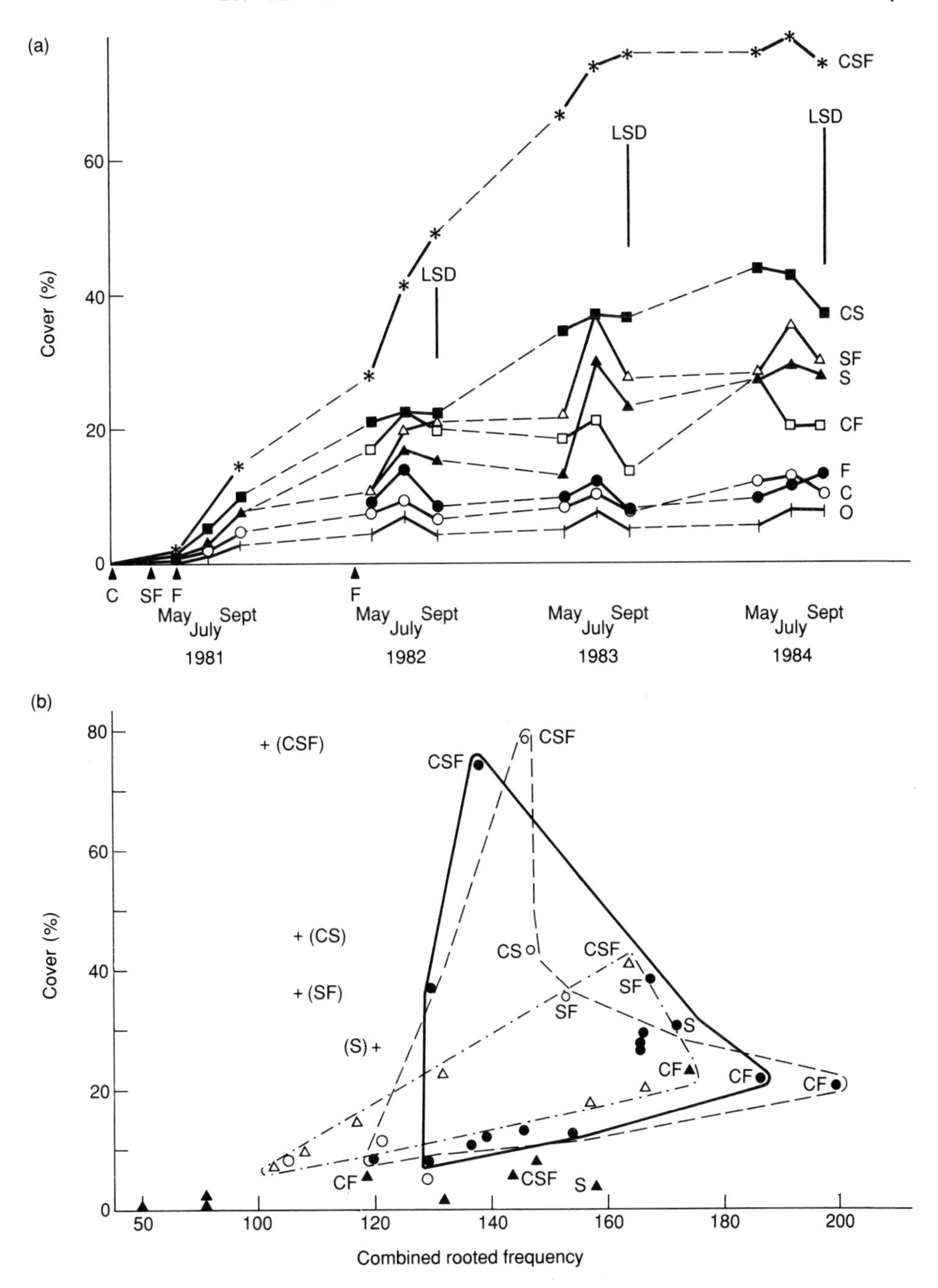

FIG. 7. Vegetation response to fertilizer (F), seed addition (S) and caging to exclude rabbits (C) in factorial combination on a limestone quarry floor (Davis *et al.* 1985): (a) the response of each treatment over a 4-year period (vertical bars indicate LSDs, $P<0{\cdot}05$); and (b) the relationship between ground cover and rooted frequency in July (▲) 1981, (Δ) 1982, (●) 1983, (○) 1984, (+) 1984 minus sown species.

Lysimeter studies on developed ecosystems often run into three main technical problems:

1 The difficulties in sampling intact soil cores.

2 The problems in maintaining the existing soil profiles, their horizons and biota.

3 The edge : area ratio of the lysimeter; if the lysimeter is too small, it is possible that edge flow may seriously affect the results.

As far as most man-made wastes are concerned, the first two problems are irrelevant: it is perfectly valid to fill a lysimeter tube from the top to simulate the normal dumping procedure, with horizons developing through time. Even where early successional vegetation is present, soil horizons often have not developed and it is still reasonably easy to take intact cores. The third problem is certainly true for small lysimeters, but even this is surmountable if large lysimeters could be filled by bulldozers.

Simple monolith lysimeters were used by Marrs & Bradshaw (1980) to simulate accelerated successions on china clay sand wastes. Two broadly defined scenarios were studied: first the vegetation establishment phase when seed and fertilizer were added to raw wastes, and second the effects of fertilizer addition to newly created grass swards. The leakiness of these two types of ecosystem differed: cations and phosphorus were lost rapidly from both ecosystems, but nitrogen losses were lower where vegetation was already present. For all elements the loss was greatest immediately after application. These results imply that as vegetation develops during primary successions nutrient cycling tends to become much tighter for nitrogen, but less so for other elements (Table 6).

TABLE 6. Comparison of nutrients leached from the lysimeters in relation to measured inputs ($kg\,ha^{-1}$) (Marrs & Bradshaw 1980)

	N	P	K	Ca	Mg
Developing vegetation on raw waste					
Fertilizer input	104	32	12	120	56
Rainfall input	9	2	7	8	6
Total input	113	34	19	128	62
Leaching losses	19	4	37	70	42
% leached	17	12	195	55	68
Vegetated wastes					
Fertilizer input	87	27	9	120	50
Rainfall input	9	2	7	8	6
Total input	96	29	15	128	62
Leaching losses	5	19	42	99	45
% leached	5	65	280	77	72

Case-study C: The use of lysimeters to determine the influence of soil parent materials in ecosystem development on china clay wastes

Using the same type of lysimeter, set up in the waste-heap environment, Smith (1985) investigated the influence of various amendments on ecosystem development on coarse sand wastes (silt and clay fractions <2%; Bradshaw *et al.* 1975). It is obvious from the previous discussion that these raw wastes not only contain very low levels of the major nutrients, but also that their retention of both water and nutrients is poor.

The range of amendments given and their effects are shown in Table 7. A basic application of fertilizer (50 kg N ha^{-1} and 40 kg P ha^{-1} plus appropriate additions of potassium, calcium and magnesium) was given at the start, and the experiment was run over the summer growing period (May–September). Where the coarse sand was ameliorated with finer materials there was a marked improvement in growth, and water and nitrogen retention. This experiment is an analogue of the effects of rock weathering in soils during primary succession and illustrates the importance of changes in the soil physico-chemical environment on ecosystem development. The exact mechanisms involved required further study. The improved

TABLE 7. The effects of various amendments to coarse china clay sand waste on vegetation growth, nutrient loss and retention in a lysimeter experiment (Smith 1985). Of particular interest is the mica amendment, which is also a man-made waste material, but with a very low and uniform particle size (>90% in the silt and clay fractions)

(a) *Plant response and leaching losses*

Treatment	Application (cm)	Leachate (mm)	Harvest ($g m^{-2}$)	Leaching losses (kg ha^{-1})		
				NO_3–N	NH_4–N	P
Nil	—	252_a	74_a	35_a	$2{\cdot}1_a$	$1{\cdot}9_a$
Mica	10	130_c	245_b	14_{bc}	$0{\cdot}09_b$	$1{\cdot}4_{ab}$
Mica	20	64_d	304_{bc}	12_c	$0{\cdot}07_b$	$1{\cdot}1_b$
Overburden	10	194_b	315_{bc}	14_{bc}	$0{\cdot}20_b$	$1{\cdot}4_{ab}$
Overburden	20	174_b	469_c	15_{bc}	$0{\cdot}15_b$	$1{\cdot}2_b$
Topsoil	5	191_b	253_b	18_b	$0{\cdot}17_b$	$1{\cdot}3_b$
Fertilizer input	—	—	—	25	25	40
Precipitation input	—	444	—	2·52	0·35	0·28

Treatment means not followed by the same letter are significantly different ($P < 0{\cdot}05$).

(b) *Retention recoveries (%)*

Treatment	Application	Precipitation	N	P
Nil	—	57	29·9	95·2
Mica	10	29	73·4	96·5
Mica	20	14	76·5	97·3
Overburden	10	44	72·5	96·4
Overburden	20	39	71·8	97·2
Topsoil	5	43	66·4	96·8

water retention, may in itself control the amount of nutrients leached, either directly because of reduced water throughput, or indirectly via improved plant growth and hence greater nutrient uptake (Smith 1985).

CONCLUSIONS

The singular fascination of primary succession is that it is the process by which mature ecosystems have developed from their rudimentary beginnings, covering a range of abiotic and biotic processes, which can have widely different origins and interactions. While it is not difficult to suggest the range of processes involved, it is much more difficult to give them priority and relevance. We argue that this can only come from a combination of: (a) logical theoretical analysis; and (b) critical experimentation. The first step of theoretical analysis can to some extent come from observation of chronosequences and clear thinking; proof, however, can only come from experimentation. As natural primary successions are restricted to a few rather specialized environments, man-made wastes offer an ideal analogue for testing our hypotheses in both the field and laboratory.

In our man-made analogues all the processes can be studied with relative ease. Such work has the added stimulus that demands for reclamation generate needs to isolate the important constraints. Thus, our concentration on the intermediate development phase is to some extent a response to reclamation demands that mature ecosystems be developed. Except where ecosystem development is curtailed by severe climatic extremes (e.g. Arctic/Alpine environments) this intermediate phase involves the accumulation of nutrient resources (particularly nitrogen), and the development of biological nutrient cycling. Accumulation and cycling of nitrogen are key factors in this phase, and these processes either take a very long time to develop when the only inputs come from the atmosphere, or they can be accelerated if nitrogen-fixing species colonize. Other resource nutrients such as phosphorus, are important and are accumulated in organic forms and made available at differing rates by biological cycling. Facilitation (Connel & Slatyer 1977) is therefore an important principle within the intermediate development phase, because if the resource supply has not developed then mid- and late-successional species will be unable to persist to maturity.

Compared with resource acquisition, we would argue that other factors are less important, although particular stress factors such as pH, drought or low temperature, may determine that only certain species will be able to establish. As long as species adapted to these factors have evolved, growth is then determined by resource levels. This conclusion is borne out by plant growth on metal spoils, where growth was related to phosphorus addition (nitrogen was not examined) and not metal level (Smith & Bradshaw 1979). Plants can overcome metal toxicity by evolution, but there appears to be a limit to the levels of resource deficiency that they can overcome by this means.

In all this work, distributional and descriptive data provide the initial evidence. But the critical evidence comes from field nutrient additions and lysimeter studies.

In nearly all cases, the experiments have been aimed at determining how best to produce a vegetation cover. The species involved are not necessarily those that would occur together in nature and unusual mixtures are possible. As far as we are aware no studies have been done to investigate the relationships between resource levels and species interaction/replacement within the intermediate development phase. There is an almost unlimited opportunity to design such experiments on man-made wastes.

We believe that the work we have described emphasizes one major point often overlooked, that the most critical part of primary succession is the intermediate development phase and the acquisition of nutrients.

ACKNOWLEDGMENTS

We thank NERC and English China Clays Ltd for financial support during the fieldwork phase of this project, our colleagues Drs R. D. Roberts and R. A. Skeffington, whose contribution to this project was considerable, and Dr P. J. Grubb for critical comment and advice.

REFERENCES

Allen, E.B. & Allen, M.F. (1980). Natural re-establishment of vesicular-arbuscular mycorrhizae following stripmine reclamation in Wyoming. *Journal of Applied Ecology*, **17**, 139–47.

Allen, S.E., Grimshaw, H.M., Parkinson, J.A. & Quarmby, C. (1974). *Chemical Analysis of Ecological Materials*. Blackwell Scientific Publications, Oxford.

Ashenden, T.W. & Bell, S.A. (1989). Rural concentrations of nitrogen dioxide pollution throughout Wales. *Environmental Pollution*, **58**, 179–94.

Bobbink, R. & Willems, J.H. (1987). Increasing dominance of *Deschampsia pinnatum* (L.) Beauv. in chalk grasslands: a threat to a species-rich ecosystem. *Biological Conservation*, **40**, 301–14.

Bloomfield, H.E., Handley, J.F. & Bradshaw, A.D. (1982). Nutrient deficiencies and the aftercare needs of reclaimed derelict land. *Journal of Applied Ecology*, **19**, 151–8.

Bradshaw, A.D. (1983). The reconstruction of ecosystems. *Journal of Applied Ecology*, **20**, 151–8.

Bradshaw, A.D. & Chadwick, M.J. (1980). *The Restoration of Land*. Blackwell Scientific Publications, Oxford.

Bradshaw, A.D., Dancer, W.S., Handley, J.S. & Sheldon, J.C. (1975). Biology of land revegetation and reclamation of china clay wastes. *The Ecology of Resource Degradation and Renewal* (Ed. by M.J. Chadwick & G.T. Goodman), pp. 363–84. Blackwell Scientific Publications, Oxford.

Bradshaw, A.D., Humphries, R.N., Johnson, M.S. & Roberts, R.D. (1978). The restoration of vegetation on derelict land produced by industrial activity. *The Breakdown and Restoration of Ecosystems* (Ed. by M.W. Holdgate & M.J. Woodman), pp. 249–74. Plenum, New York.

Brimblecombe, P. & Stedman, D.H. (1982). Historical evidence for a dramatic increase in the nitrate component of acid rain. *Nature*, **298**, 460–1.

Bülow-Olsen, A. (1980). Nutrient cycling in a grassland dominated by *Deschampsia flexuosa* (L.) Trin. and grazed by nursing cows. *Agro-Ecosystems*, **6**, 209–20.

Chapin III, F.S. (1980). The mineral nutrition of wild plants. *Annual Review of Ecology and Systematics*, **11**, 236–60.

Connel, J.H. & Slatyer, R.D. (1977). Mechanisms of succession in natural communities and their role in community stability and organization. *American Naturalist*, **111**, 1119–44.

Crocker, R.L. & Major, J. (1955). Soil development in relation to surface age at Glacier Bay, Alaska. *Journal of Ecology*, **43**, 427–48.

Daft, M.J. & Hacskaylo, E. (1976). Arbuscular mycorrhizas in the anthracite and bituminous coal wastes of Pennsylvania. *Journal of Applied Ecology*, **13**, 523–31.

Dancer, W.S. (1975). Leaching losses of ammonium and nitrate in the reclamation of sand spoils in Cornwall. *Journal of Environmental Quality*, **4**, 499–504.

Dancer, W.S., Handley, J.F. & Bradshaw, A.D. (1977). Nitrogen accumulation in kaolin mining wastes in Cornwall. II. Forage legumes. *Plant and Soil*, **48**, 303–14.

Davies, B.N.K., Lakhani, K.H., Brown, M.C. & Park, D.G. (1985). Early seral communities in a limestone quarry: an experimental study of treatment effects on cover and richness of vegetation. *Journal of Applied Ecology*, **22**, 473–90.

During, H.J. & Willems, J.H. (1986). The impoverishment of the bryophyte and lichen flora of the Dutch chalk heaths over the thirty years 1953–1983. *Biological Conservation*, **36**, 143–58.

Dutton, R.A. & Bradshaw, A.D. (1982). *Land Reclamation in Cities*. HMSO, London.

Duvigneaud, D. & Denaeyer de Smet, S. (1970). Biological cycling of minerals in temperate deciduous forests. *Temperate Forests* (Ed. by D.F. Reichle), pp. 199–225. Springer-Verlag, Berlin.

Elias, C.O., Morgan, A.L., Palmer, J.P. & Chadwick, M.J. (1982). *The Establishment, Maintenance and Management of Vegetation on Colliery Spoil Sites*. Derelict Land Research Unit, University of York.

Epstein, E. (1972). *Mineral Nutrition of Plants: Principles and Perspectives*. Wiley, New York.

Floate, M.J. (1981). Effects of grazing by large herbivores on nitrogen cycling in agricultural ecosystems. *Terrestrial Nitrogen Cycles* (Ed. by F.E. Clark & T. Rosswall), pp. 585–601. Swedish Natural Science Research Council, Stockholm.

Fowler, D. (1987). Rain cloud chemistry and acid deposition on mountains. *Annual Report of the Institute of Terrestrial Ecology, 1986–87*, pp. 61–3. NERC, Swindon.

Friedel, H. (1938a). Die Pflanzenbesiedlung im Vorfelde des Hintereisfereners. *Zeitschrift Gletscherkunde*, **26**, 215–39.

Friedel, H. (1938b). Bodem und Vegetations entwicklung im Vorfelde des Rhonegletschers. *Berichte des Geobotanischen Forsch. Institutes Rübel, Zurich 1937*, pp. 65–76.

Gadgil, R.L. (1971a). The nutritional role of *Lupinus arboreus* in coastal sand dune forestry. I. The potential influence of undamaged lupin plants and nitrogen uptake. *Plant and Soil*, **34**, 357–67.

Gadgil, R.L. (1971b). The nutritional role of *Lupinus arboreus* in coastal sand dune forestry. III. Nitrogen distribution in the ecosystem before tree planting. *Plant and Soil*, **35**, 113–26.

Grime, J.P. & Hunt, R. (1975). Relative growth rate: its range and adaptive significance in a local flora. *Journal of Ecology*, **63**, 393–422.

Grubb, P.J. (1986). The ecology of establishment. *Ecology and Design in Landscape* (Ed. by A.D. Bradshaw, D.A. Goode & E. Thorp), pp. 83–98. Blackwell Scientific Publications, Oxford.

Grubb, P.J. (1987). Some generalizing ideas about colonization and succession in green plants and fungi. *Colonization, Succession and Stability* (Ed. by A.J. Gray, M.J. Crawley & P.J. Edwards), pp. 80–102. Blackwell Scientific Publications, Oxford.

Haeberli, W. (Ed.) (1985). *Fluctuations of Glaciers (1975–80)*, vol. IV. IAHS(ICSI)-UNESCO, Paris.

Handley, J.F., Dancer, W.S., Sheldon, J.C. & Bradshaw, A.D. (1978). The nitrogen problems in derelict land reclamation with special reference to the British china clay industry. *Environmental Management of Mineral Wastes* (Ed. by G.T. Goodman & M.J. Chadwick), pp. 215–36. Sijthoff & Noordhoff, Alphen aan den Rijn.

Heil, G. & Diemont, W.H. (1983). Raised nutrient levels change heathland into grassland. *Vegetatio*, **53**, 113–20.

Hirose, T. (1986). Nitrogen uptake and plant growth. II. An empirical model of vegetation growth and partitioning. *Annals of Botany*, **58**, 487–96.

Hutnik, R.J. & Davies, G. (Eds) (1973). *Ecology and Reclamation of Devastated Land*, 2 vols. Gordon & Breach, New York.

Hutson, B.R. (1980a). The influence of soil development on the invertebrate fauna colonizing industrial reclamation sites. *Journal of Applied Ecology*, **17**, 277–86.

Hutson, B.R. (1980b). Colonization of industrial reclamation sites by Acari, Collembola and other invertebrates. *Journal of Applied Ecology*, **17**, 255–75.

Jefferies, R.A., Bradshaw, A.D. & Putwain, P.D. (1981a). Growth, nitrogen accumulation and nitrogen transfers by legume species established on mine spoils. *Journal of Applied Ecology*, **18**, 945–56.

Jefferies, R.A., Willson, K. & Bradshaw, A.D. (1981b). The potential of legumes as a nitrogen source for the reclamation of derelict land. *Plant and Soil*, **59**, 175–7.

Jenny, H. (1941). *Factors of Soil Formation*. McGraw-Hill, New York.

Jenny, H. (1980). *The Soil Resource*. Springer-Verlag, Berlin.

Jenny, H., Gessel, S.P. & Bingham, F.T. (1949). Comparative study of decomposition rates of organic matter in temperate and tropical ecosystems. *Soil Science*, **68**, 419–432.

Keeney, D.R. & Bremner, J.M. (1967). Determination and isotope-ratio analysis of different forms of nitrogen in soils, 6. Mineralizable nitrogen. *Proceedings of the Soil Science Society of America*, **31**, 34–9.

Knabe, W. (1973). Investigations of soils and tree growth in five deep-mine refuse piles in the hard coal region of the Ruhr. *Ecology and Reclamation of Devastated Land*, Vol. 1 (Ed. by R.J. Hutnik & G. Davis), pp. 307–24. Gordon & Breach, New York.

Lambert, D.H. & Cole, H. (1980). Effects of mycorrhiza on establishment and performance on forage species in mine spoil. *Agronomy Journal*, **72**, 257–60.

Lanning, S. & Williams, S.T. (1979). Nitrogen in revegetated china clay waste. 1. Decomposition of plant material. *Environmental Pollution*, **20**, 147–59.

Lawrence, D.B., Schoenike, R.E., Quispel, A. & Bond, G. (1967). The role of *Dryas drummondii* in vegetation development following ice recession at Glacier Bay, Alaska, with special reference to its nitrogen fixation by root nodules. *Journal of Ecology*, **55**, 793–813.

Leisman, G.A. (1957). A vegetation and soil chronosequence on the Mesabi iron range spoil banks, Minnesota. *Ecological Monographs*, **27**, 221–45.

Marrs, R.H. & Bradshaw, A.D. (1980). Ecosystem development on reclaimed china clay wastes. III. Leaching of nutrients. *Journal of Applied Ecology*, **17**, 727–36.

Marrs, R.H. & Bradshaw, A.D. (1982). Nitrogen accumulation, cycling and the reclamation of china clay wastes. *Journal of Environmental Management*, **15**, 139–57.

Marrs, R.H., Granlund, I.H. & Bradshaw, A.D. (1980). Ecosystem development on reclaimed china clay wastes. IV. Recycling of above-ground plant nutrients. *Journal of Applied Ecology*, **17**, 803–13.

Marrs, R.H., Roberts, R.D., Skeffington, R.A. & Bradshaw, A.D. (1981). Ecosystem development on naturally-colonized china clay wastes. II. Nutrient compartmentation. *Journal of Ecology*, **69**, 163–70.

Marrs, R.H., Roberts, R.D., Skeffington, R.A. & Bradshaw, A.D. (1983). Nitrogen and the development of ecosystems. *Nitrogen as an Ecological Factor* (Ed. by J.A. Lee, S. McNeill & I.H. Rorison), pp. 113–36. Blackwell Scientific Publications, Oxford.

Miles, J. (1979). *Vegetation Dynamics*. Chapman & Hall, London.

Miles, J. (1987). Succession: past and present perceptions. *Colonization, Succession and Stability* (Ed. by A.J. Gray, M.J. Crawley & P.J. Edwards), pp. 1–30. Blackwell Scientific Publications, Oxford.

Müller, F. (Ed) (1977). *Fluctuations of Glaciers (1970–1975)*, Vol. III. IAHS(ICSI)-UNESCO, Paris.

Newbould, P. & Floate, M.J. (1977). Agro-ecosystems in the United Kingdom. *Agro-Ecosystems*, **4**, 33–69.

Olsen, J.S. (1958). Rates of succession and soil changes on southern Lake Michigan sand dunes. *Botanical Gazette*, **199**, 125–70.

Olsen, J.S. (1963). Energy storage and the balance of producers and decomposers in ecological systems. *Ecology*, **44**, 322–31.

Palaniappan, V.M., Marrs, R.H. & Bradshaw, A.D. (1979). The effect of *Lupinus arboreus* on the nitrogen status of china clay wastes. *Journal of Applied Ecology*, **16**, 825–33.

Perkins, D.F. (1978). The distribution and transfer of nutrients in the *Agrostis–Festuca* grassland ecosystem. *Production Ecology of British Moors and Montane Grasslands* (Ed. by O.W. Heal & D.F. Perkins), pp. 375–96. Springer-Verlag, Berlin.

Reuss, J.O. & Innis, G.S. (1977). A grassland nitrogen flow simulation model. *Ecology*, **58**, 379–88.

Richard, J.L. (1968). Les groupment vegetaux de la reserve d'Aletsch. *Beitraege Geobotanischen Landesaufnahme der (Schweiz)*, **51**, 305.

Roberts, R.D., Marrs, R.H. & Bradshaw, A.D. (1980). Ecosystem development on reclaimed china clay wastes. II. Nutrient compartmentation and nitrogen mineralization. *Journal of Applied Ecology*, **17**, 153–61.

Roberts, R.D., Marrs, R.H., Skeffington, R.A. & Bradshaw, A.D. (1981). Ecosystem development on naturally-colonized china clay wastes. I. Vegetation changes and overall accumulation of organic matter and nutrients. *Journal of Ecology*, **69**, 153–62.

Roberts, R.D., Marrs, R.H., Skeffington, R.A., Bradshaw, A.D. & Owen, L.D.C. (1982). The importance of plant nutrients in the restoration of derelict land and mine wastes. *Transactions of the Institution of Mining and Metallurgy*, **91A**, 42–4.

Robertson, R.A. & Davies, G.E. (1965). Quantities of plant nutrients in heather ecosystems. *Journal of Applied Ecology*, **2**, 211–9.

Robertson, G.P. & Vitousek, P.M. (1981). Nitrification potentials in primary and secondary successions. *Ecology*, **62**, 376–86.

Robinson, D. (1986). Compensatory changes in the partitioning of dry matter in relation to nitrogen uptake and optimal variation in growth. *Annals of Botany*, **58**, 841–8.

Schaller, F.W. & Sutton, P. (Eds) (1978). *Reclamation of Drastically Disturbed Lands*. American Society of Agronomy, Madison, Wisconsin.

Sheldon, J.C. & Bradshaw, A.D. (1977). The development of a hydraulic seeding technique for unstable sand slopes. I. Effects of fertilizers, mulches and stabilizers. *Journal of Applied Ecology*, **14**, 905–18.

Skeffington, R.A. & Bradshaw, A.D. (1980). Nitrogen fixation by plants growing on reclaimed china clay wastes. *Journal of Applied Ecology*, **17**, 469–77.

Skeffington, R.A. & Bradshaw, A.D. (1981). Nitrogen accumulation in kaolin mining wastes of Cornwall. IV. Sward quality and the development of a nitrogen cycle. *Plant and Soil*, **62**, 438–51.

Smith, A.F. (1985). *Amerlioration of china clay mining wastes for vegetation establishment*. PhD Thesis, University of Liverpool.

Smith, R.A. & Bradshaw, A.D. (1979). The use of metal tolerant plant populations for the reclamation of metalliferous wastes. *Journal of Applied Ecology*, **16**, 595–612.

Sprent, J.I. (1987). *The Ecology of the Nitrogen Cycle*. Cambridge University Press, Cambridge.

Stahl, P.D., Williams, S.E. & Christensen, M. (1988). Efficacy of native vesicular-arbuscular mycorrhizal fungi after severe soil disturbance. *New Phytologist*, **110**, 347–54.

Tilman, D. (1985). The resource-ratio hypothesis of plant succession. *American Naturalist*, **125**, 827–52.

Tilman, D. (1986). Nitrogen-limited growth in plants from successional stages. *Ecology*, **67**, 555–63.

Viereck, L.A. (1966). Plant succession and soil development on gravel outwash of the Muldrow Glacier, Alaska. *Ecological Monographs*, **36**, 181–99.

Vitousek, P.M. & Walker, L.R. (1987). Colonization, succession and resource availability: ecosystem-level interactions. *Colonization, Succession and Stability* (Ed. by A.J. Gray, M.J. Crawley & P.J. Edwards), pp. 207–23. Blackwell Scientific Publications, Oxford.

Williams, J.D.H. & Walker, T.W. (1969). Fractionation of phosphate in a maturity sequence of New Zealand basaltic soil profiles. 2. *Soil Science*, **107**, 213–19.

Nitrogen fixers and species replacements in primary succession

L. R. WALKER
Department of Biological Sciences, University of Nevada Las Vegas, 4505 Maryland Parkway, Box 454004, Las Vegas, Nevada 89154–4004, USA

SUMMARY

1 The impact of vascular plants with symbiotic nitrogen fixers (termed 'nitrogen fixers') on species' replacements in primary succession was examined by surveying 150 studies from the ecological literature.
2 Nitrogen fixers were present in 77% of all seres but were a dominant part of the vegetation in only 28%. They were most abundant on glacial moraines and mudflows, of intermediate abundance on mine tailings, landslides, floodplains, and dunes, and least abundant on volcanoes and rock outcrops.
3 No correlation emerged in cross-site comparisons between the presence of dominant nitrogen fixers and total nitrogen accumulation in surface soils. Nitrogen fixers provide the principal inputs of nitrogen but little is known about the relative magnitude of inputs and outputs in primary succession.
4 Nitrogen fixers facilitate growth of associated species in some cases (e.g. mine tailings). However, generalizations about their facilitative role are misleading because nitrogen fixers often compete successfully for resources with other colonizers, thereby inhibiting growth of associated species and slowing species turnover.

INTRODUCTION

Primary succession can be defined as the replacement through time of one group of species by another on substrates with no prior soil development. Descriptions of this process are as old as ecology but a comprehensive explanation for it is still lacking. Early paradigms suggesting the process is similar to the development of an organism (Clements 1916) have been discarded in favour of a reductionist approach (McIntosh 1980) that examines mechanisms responsible for species change. Despite this emphasis, however, extrapolations beyond a particular study are still tentative, suggesting we are still far from a truly representative theory of succession.

The observation that nitrogen limits plant growth in the early stages of primary succession has focused considerable attention on the possible effect nitrogen sources may have on species composition. Precipitation may be the only source of nitrogen in recently disturbed habitats (Vitousek & Walker 1989), although other atmospheric inputs of nitrogen can include arthropods (Edwards *et al.* 1987), salt-spray (Morris *et al.* 1974), or perhaps volcanic gases (Griggs

1933; Eggler 1963). Bacteria and blue-green algae fix nitrogen and can be important free-living colonizers of new substrates or have symbiotic associations with fungi or higher plants. However, most of the nitrogen added to primary seres can be attributed to legumes and non-leguminous plants which have bacterial symbionts that fix nitrogen in root nodules (Marrs *et al.* 1983).

What is the impact that vascular plants with symbiotic nitrogen fixers have on species replacements in the low nitrogen environment of primary succession? Do nitrogen fixers facilitate species change in primary succession, as is widely assumed? This chapter reviews the evidence for facilitation by vascular plants with symbiotic nitrogen fixers (henceforward referred to as 'nitrogen fixers') and examines the possibility that nitrogen fixers also have an inhibitory role in primary succession. The abundance, life-form and impacts of nitrogen fixers on soil nitrogen, growth of associated species and species replacements in succession will be examined in seven naturally occurring seres (glacial moraines, mudflows, landslides, floodplains, dunes, volcanoes and rock outcrops) and naturally occurring succession on one type of human disturbance (mine tailings). These seres include classic primary succession on substrates where initial nitrogen levels are quite low (dunes, volcanoes, rock outcrops), as well as seres where nitrogen content is higher (floodplains, landslides, mudflows). These latter seres can be considered the middle of a continuum from low nutrient primary seres to secondary seres (Vitousek & Walker 1987). The chapter considers successional changes in the early stages of regrowth following severe disturbances where nitrogen is considered most limiting, but does not address the dynamics of colonization (see Grubb 1986, 1987) or the special conditions of very long-term seres where phosphorus may become more limiting than nitrogen (Walker & Syers 1976).

SOURCES

Abundance and life-forms of nitrogen fixers were determined from a survey of 150 studies describing 141 examples of primary succession (Table 1) that included species lists or some discussion of species composition. The majority of the studies offered no experimental evidence of mechanisms affecting species replacements, although patterns of abundance, changes in life-forms and species replacements can offer clues to underlying mechanisms. In the section on effects of nitrogen fixers on successional change, I reference those studies that were most helpful in determining the role of nitrogen fixers. A subset of 25 studies or closely related ones (found in parentheses in Table 1) also presented data on nitrogen accumulation. These data are discussed in the section on effects of nitrogen fixers on soil nitrogen. The seres are arranged throughout the tables and text in descending order by the abundance of nitrogen fixers (see Table 2).

ABUNDANCE

Nitrogen fixers were present in 77% of all primary seres (Table 2) that were investigated but were dominant in only 36% of the seres where they were present, or 28% of all seres. Nitrogen fixers are most likely to dominate glacial moraines, mudflows and mine tailings, and are least likely to dominate rock outcrops and volcanoes. On landslides, floodplains and dunes, nitrogen fixers are equally likely to be scarce or dominant. Plants with the ability to fix nitrogen may have a competitive advantage when soil nitrogen levels are low (Vitousek & Walker 1987). However, there does not appear to be either a strong negative or positive correlation between initial total nitrogen pools (see below) and nitrogen fixer

Table 1. Sources for the survey of abundance and life-form of nitrogen fixers in primary succession

Sere type and location	Description*	References
Glacial moraines		
Klutlan, British Columbia	AD2	Birks 1980 (Jacobsen & Birks 1980)†
Yoho, British Columbia	AD2	Bray & Struik 1963
Glacier Bay, Alaska	CD2; CD2	Cooper 1923; Crocker & Major 1955†
Mount Robson, British Columbia	AD2; AD2	Cooper 1916; Tisdale *et al.* 1966
Herbert, British Columbia	AD2	Crocker & Dickson 1957†
Mendenhall, British Columbia	AD2	Crocker & Dickson 1957†
Athabaska, British Columbia	AD2	Heusser 1956
Columbia, British Columbia	AD2	Heusser 1956
Freshfield, British Columbia	AD2	Heusser 1956
Saskatchewan, British Columbia	AD2	Heusser 1956
Lyell, British Columbia	AD2	Heusser 1956
Storbreen, Norway	CD2	Matthews 1979
Muldrow, Alaska	AD2	Viereck 1966†
Mudflows		
Mount St Helens, Washington	CE, BE, AD2	Dale 1986; del Moral & Clampitt 1985; Halpern & Harmon 1983 (seed traps, bioassays)
Shasta, California	AD2	Dickson & Crocker 1953†
Mount Rainier, Washington	BD2, CD2	Frehner 1957; Frenzen *et al.* 1988†
N Trondelag, Norway	AD2	Gimingham & Boggie 1957 (clay)
N Trondelag, Norway	AD2	Gimingham & Boggie 1957 (sand)
Mount Lassen, California	AD2	Heath 1967
Mines		
Several (20), Pennsylvania	AD2 (coal)	Bramble & Ashley 1955
Cornwall, UK	AD2 (clay)	Dancer *et al.* 1977†
Clipsham Quarry, UK	BE (limestone)	Davis *et al.* 1985 (factorial expt.)
Several (8), Maryland, West Virginia	AD2 (coal)	Hardt & Forman 1989
North Bohemia, Czechoslovakia	BD2 (coal)	Hejkal 1985
Sussex, UK	CD2 (chalk)	Hope-Simpson 1940
Mesabi Range, Minnesota	AE (iron)	Leisman 1957† (fertilization)
Middlesex, UK	CD1 (chalk)	Locket 1945
Several (8), Nauru Island	AD2 (phosphate)	Manner *et al.* 1984
Several (7), Florida	AD2 (phosphate)	McClanahan 1986
Kuala Lumpur, Malaysia	AD2 (tin)	Palaniappan 1974
Maggie Pie, UK	AD2 (clay)	Palaniappan *et al.* 1979†

TABLE 1. *Contd.*

Sere type and location	Description*	References
Several (2), Pennsylvania	BE (coal)	Schramm 1966 (seeding)
Several (10), Pennsylvania	AD2 (coal)	Schuster & Hutnik 1987
Landslides		
Mount St Helens, Washington	CE	Dale 1986 (seed traps)
Several (29), New Hampshire	AD2	Flaccus 1959
Several, (4), Virginia	AD2	Hull & Scott 1982
Massanutten Mount, Virginia	BD2	Hupp 1983
Elk Mountains, Colorado	AD2	Langenheim 1956
Fiordland, New Zealand	AD2	Mark *et al.* 1964
Cascade Mountains, Oregon	AD2	Miles & Swanson 1986
Queen Charlotte Island, British Columbia	AD2	Smith *et al.* 1986
Luquillo, Puerto Rico	BD2	Varela *et al.* 1983
Valdivian Andes, Chile	AD2	Veblen & Ashton 1978
Luquillo, Puerto Rico	BE	L. Walker, unpublished (factorial expt.)
Floodplains		
Colville River, Alaska	AD2	Bliss & Cantlon 1957
San Antonio River, Texas	AD2	Bush & Van Auken 1984
Vancouver Island, British Columbia	AD2	Clement 1985
Indiana River, Yukon	AD1	Crampton 1987
Kvikkjokk River, Sweden	CD2	Dahlskog 1982
Kluane River, Yukon	AD2	Douglas 1987
L. Missouri River, North Dakota	AD2	Everitt 1968
Hoh River, Washington	AD2, AD2	Fonda 1974; Luken & Fonda 1983[†]
Raritan River, New Jersey	CD2	Frye & Quinn 1979
Blaeberry River, British Columbia	AD2	Fyles & Bell 1986
Mackenzie River, Northwest Territories	AD2	Gill 1973
Nile River, Sudan	AD2	Halwagy 1963
Passage Creek, Virginia	AD2	Hupp & Osterkamp 1985
Missouri River, North Dakota	AD2	Johnson *et al.* 1976
Atchafalaya, Louisiana	BD2	Johnson *et al.* 1985
Mount St Helens, Washington	BD2	Kiilsgaard *et al.* 1986
Tanana River, Alaska	BD2, AD2, AD2	Krasny *et al.* 1984;[†] Van Cleve *et al.* 1971;[†] Walker 1989
Wabash & Tippecanoe, Indiana	AD2	Lindsey *et al.* 1961
Dry Creek, Louisiana	AD2	McBride & Strahan 1984
Susitna River, Alaska	AD2	McKendrick *et al.* 1982
Hudson River, New York	AD1	McVaugh 1947
Several (5), Wisconsin	AD2	Menges 1986
Savannah River, South Carolina	AD2	Muzika *et al.* 1987
Beatton River, British Columbia	AD2	Nanson & Beach 1977
Trout Creek, Arizona	AD2	Reichenbacher 1984
Potomac River, Washington DC	AD1	Sigafoos 1964
Amazon River, Peru	CD2	Terborgh 1985
Chena River, Alaska	AD2	Viereck 1970[†]
Missouri River, South Dakota	AD2	Wilson 1970[†]

TABLE 1. *Contd.*

Sere type and location	Description*	References
Dunes		
Mediterranean Sea, Egypt	AD2 (coastal)	Ayyad 1973[†]
Olduvai Gorge, Tanzania	AD2 (inland)	Belsky & Amundson 1986
Fremont County, Idaho	AD2 (inland)	Chadwick & Dalke 1965
Atlantic Ocean, New Jersey	AE (coastal)	Charette & Shisler 1985 (sand fences, fertilization and plantings)
Point Reyes, California	AD2 (coastal)	Holton & Johnson 1979
Pacific Ocean, Oregon	AD2 (coastal)	Kumler 1962
Voorne, The Netherlands	AD2 (coastal)	van der Laan 1979
Lake Vogelmeer, The Netherlands	AD2 (inland)	Londo 1971
Voorne, The Netherlands	CD2 (coastal)	van der Maarel *et al.* 1985
Algoa Bay, South Africa	AD2 (coastal)	McLachlan *et al.* 1987
El Morro de la Mancha, Mexico	AD2 (coastal)	Moreno-Casasola 1986
Atlantic Ocean, New Jersey	AD1 (coastal)	Morris *et al.* 1974
Lake Michigan, Indiana	AD2 (inland)	Olson 1958[†]
Baltic Sea, Poland	AD1 (coastal)	Piotrowska 1988
Pacific Ocean, California	AD1 (coastal)	Purer 1936
Newborough, Wales	AD2 (coastal)	Ranwell 1960
Pacific Ocean, South Carolina	AD2 (coastal)	Rayner & Batson 1976
Lake Michigan, Indiana	AE, AE (inland)	Robertson 1982; (Robertson & Vitousek 1981[†]) (soil incubations)
Gibraltar Pt, UK	AD1 (coastal)	Stewart & Pearson 1967
Krakatoa, Indonesia	AD2 (coastal)	Tagawa *et al.* 1985
Atlantic Ocean, Iceland	AD2 (coastal)	Tuxen 1970
Several (2), Virginia	AD2 (coastal)	Tyndall & Levy 1976
Cooloola, Australia	AD2 (coastal)	Walker *et al.* 1981
Grand Bend, Ontario	AD2 (inland)	Yarranton & Morrison 1974
Namib Desert, Namibia	AD2 (inland)	Yeaton 1988
Volcanoes		
Mount St Helens, Washington	BD2, CD2	Anderson & MacMahon 1985; del Moral & Wood 1988 (ash)
Tarawera, Philippines	AD1	Aston 1916 (ash)
Mauna Loa and Kilauea, Hawaii	AD2	Atkinson 1970 (ash)
Mauna Loa and Kilauea, Hawaii	AD2	Atkinson 1970 (pahoehoe)
Soufrière, St Vincent	AD1	Beard 1945 (ash)
Surtsey, Iceland	AD1, BD2, BD2	Brock 1973; Fridriksson 1970; (Henriksson & Rodgers 1978[†]) (lava, ash)
Taal, Philippines	AD2, AD1	Brown *et al.* 1917;[†] Gates 1914 (ash)
Krakatoa, Indonesia	AD1, AD2, CD2	Campbell 1909; Tagawa *et al.* 1985; Whittaker *et al.* 1989 (lava, ash)
Snake River Plains, Idaho	AD2	Eggler 1941 (lava)
Parícutin, Mexico	CD1, CD2	Eggler 1959, 1963 (playas)
Parícutin, Mexico	CD1, CD2	Eggler 1959, 1963 (a'a)
Parícutin, Mexico	CE, CD2	Eggler 1959, 1963 (ash; sown seeds)
Jorullo, Mexico	CD1, CD2	Eggler 1959, 1963 (playas)

TABLE 1. *Contd.*

Sere type and location	Description*	References
Jorullo, Mexico	CD1, CD2	Eggler 1959, 1963 (a'a)
Jorullo, Mexico	CE, CD2	Eggler 1959, 1963 (ash; sown seeds)
Several (16), Hawaii	AD1	Eggler 1971 (a'a and pahoehoe)
Soufrière, Guadeloupe	AE, AD1	Fritz-Sheridan & Portecop 1987; Howard *et al.* 1980 (ash; soil incubations)
Katmai, Alaska	AD1	Griggs 1933 (ash)
Isla Fernadina, Galápagos	CD2	Hendrix 1981 (ash)
Mount Fuji, Japan	AD2, AD2	Nakamura 1985; Ohsawa 1984 (lava)
Several (2), Greece	AD1	Raus 1988 (lava and ash)
Gunong Guntur, Java	AD1	Schimper 1903 (from Brown *et al.* 1917)
Carrizozo, New Mexico	AD1	Shields & Crispin 1956 (lava)
Papua, New Guinea	AD1	Taylor 1957 (ash)
Oshima, Japan	AD2	Tezuka 1961† (lava and ash)
Usu, Japan	BD2	Tsuyuzaki 1989 (ash)
Several (20), Hawaii	AD2	Uhe 1988a (a'a and pahoehoe)
Yasour, New Hebrides	AD2	Uhe 1988b (ash)
Mauna Loa and Kilauea, Hawaii	AE	Vitousek *et al.* 1983† (ash; soil incubation)
Mauna Loa and Kilauea, Hawaii	AE	Vitousek *et al.* 1983† (pahoehoe; incubation)
Rock outcrops		
Piedmont Plateau, Georgia	BD2, AE	Burbanck & Platt 1964; Houle & Phillips 1988, 1989 (buried seed bank, species additions/removals)
James Peak, Colorado	AD1	Cox 1933
Rocky Mountains National Park, Colorado	AD1	Griggs 1956
Randolph County, Alabama	AD1	Harper 1939
Alexander County, North Carolina	AD1	Keever *et al.* 1951
Several (20), South-east USA	AD1	McVaugh 1943
Several, North Carolina	AD1	Oosting & Anderson 1939
Lake Hjalmaren, Norway	CD2	Rydin & Borgegard 1988
Panola Mountains, Georgia	BD2	Shure & Ragsdale 1977
Several (90), Oklahoma	AE	Uno & Collins 1987 (buried seed bank)
Llano County, Texas	AD1	Whitehouse 1933
Shawnee Hills, Illinois	AD2	Winterringer & Vestal 1956
Lake Opinicon, Ontario	AD2	Woolhouse *et al.* 1985

* A = one-time comparison of different successional communities; B = study that followed succession seasonally or annually for <5 years; C = study that followed succession for >5 years; D1 = qualitative descriptive study; D2 = quantitative descriptive study (e.g. including density, cover, etc.); E = experimental study (e.g. manipulations, such as fertilization). Actual experiments follow the references with this descriptive code.
† These studies were also used for discussing soil nitrogen accumulation. Parentheses indicate they were not used for abundance or life-form summaries.

TABLE 2. Distribution and abundance of nitrogen fixers in primary succession

Abundance category	Type of succession*								
	Mo	Mu	Mi	La	Fl	Du	Vo	Ro	Total
Absent†	0‡	0	2	1	6	5	14	5	33
Present	1	1	3	3	9	9	7	6	39
Abundant	5	2	3	6	6	2	5	1	30
Dominant	7	3	6	1	8	9	4	1	39
Total	13	6	14	11	29	25	30	13	141
Per cent abundant and dominant	92	83	64	63	48	44	30	15	49

* Mo = glacial moraine, Mu = mudflow, Mi = mine tailing, La = landslide, Fl = floodplain, Du = dune, Vo = volcano, Ro = rock outcrop.
† Absent (not mentioned in species list); present (mentioned in species list, <5% cover; abundant (in several stages and >5% cover); dominant (one of three species with the highest cover or importance value). When several nitrogen fixers were present, only the highest abundance category was used.
‡ Number of seres.

abundance. Other factors that might determine the abundance of nitrogen fixers in primary succession include substrate stabilization (Moreno-Casasola 1986), tolerance of flooding (Gill 1973) or salt-spray (Morris *et al.* 1974), ability to survive grazing pressures (Dahlskog 1966) or droughts and dispersal abilities (Walker *et al.* 1986).

Increases in temperature with decreasing latitude can increase nitrogen availability and may affect nitrogen fixer abundance. Table 3 indicates that nitrogen fixers are more likely to be abundant or dominant in boreal than tropical seres. However, this apparent pattern does not hold within sere types and probably reflects the dominance of nitrogen fixers on boreal moraines and the relative absence of nitrogen fixers on tropical volcanoes.

This survey suggests that symbiotic nitrogen fixers are not always present in primary succession and are not always dominant when they are present. Their presence is not correlated with soil nitrogen status or latitude. However, they are

TABLE 3. Distribution of nitrogen fixers in primary succession by latitude

Latitude	Type of succession								
	Mo	Mu	Mi	La	Fl	Du	Vo	Ro	Total
Boreal	1*/12†	0/2	–/–	1/1	5/6	1/1	2/–	–/1	10/23
Temperate	–/–	1/3	3/9	3/6	9/7	10/9	7/6	11/1	44/41
Tropical	–/–	–/–	2/–	–/–	1/1	3/1	12/3	–/–	18/5

* Number of seres where nitrogen fixers were 'absent' or 'present'.
† Number of seres where nitrogen fixers were 'abundant' or 'dominant'.
See Table 2 for definitions of abundance categories.

more likely to occur in certain habitats (e.g. glacial moraines, mines and mudflows) than in others (e.g. dunes and volcanoes).

LIFE-FORMS

Life-form and life-history of individual species can be important in determining colonizing ability (Grubb 1987) and competitive outcomes (Tilman 1988) among plants. Legumes, the most common symbiotic nitrogen fixers present in primary succession, comprise 59% of the 185 recorded occurrences of nitrogen fixers in primary succession (Table 4). Both woody and herbaceous legumes are least likely to occur on moraines, mudflows or floodplains where other nitrogen fixers dominate. Woody nitrogen fixers (leguminous or not) predominate on glacial moraines, mudflows and volcanoes; herbaceous (leguminous) fixers predominate only on dunes. Both woody and herbaceous legumes are less likely to be dominant, however, than non-leguminous fixers such as *Myrica*, *Alnus*, and *Dryas* (Fig. 1). This may in part be due to the ability of these three genera to form dense thickets that exclude other species (Heusser 1956; Luken & Fonda 1983; Walker & Chapin 1986; Walker & Vitousek 1991). Short-lived nitrogen fixers that do not form dense canopies (e.g. herbaceous legumes) may have a less inhibitory effect on associated, non-fixing plants than relatively long-lived canopy dominants such as *Alnus*.

EFFECTS ON SOIL NITROGEN

Nitrogen fixers are the principal contributors to nitrogen accumulation in soils of primary succession on floodplains (Van Cleve, *et al.* 1971; Luken & Fonda 1983), mine tailings (Bradshaw 1983), volcanoes (Vitousek & Walker 1989) and glacial moraines (Crocker & Major 1955). Characteristic nitrogen accumulation rates can vary from 27 to 163 kg N ha^{-1} $year^{-1}$ (Luken & Fonda 1983; Marrs *et al.* 1983). I compared 25 primary seres (see Table 1) to determine if nitrogen fixers significantly altered the rate or magnitude of nitrogen accumulation (kg N ha^{-1} $year^{-1}$; determined from bulk density × soil depth × p.p.m. N/10) in surface mineral soils

TABLE 4. Number and distribution of nitrogen-fixing in species primary succession by life-form

Life-form	Mo	Mu	Mi	La	Fl	Du	Vo	Ro	Total
Herbaceous legumes	8	5	9	8	15	18	9	4	76
Woody legumes	0	2	10	1	6	3	8	3	34
Other woody species	19	7	1	6	21	7	8	2	75

All abundance categories (see Table 2) were combined. When several nitrogen fixers were present in one sere, each was included.

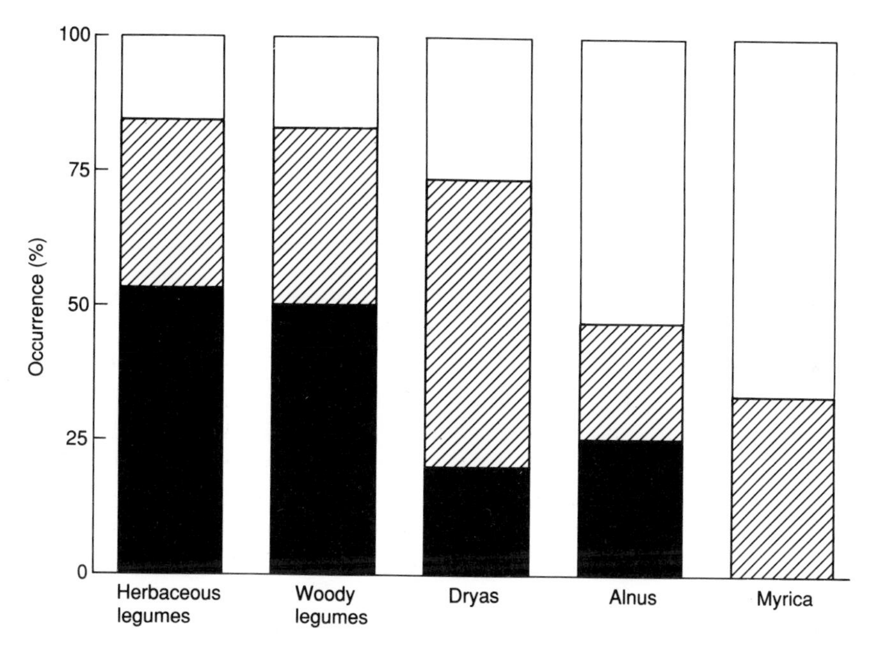

FIG. 1. Abundance of herbaceous and woody legumes and three other woody nitrogen fixers (percentage of times found in each of three abundance categories): (shaded) present; (hatched) abundant; (white) dominant. Abundance categories are defined in Table 2.

to depths of 10–25 cm. Surface mineral soils can contain 70–84% of the nitrogen in young developing ecosystems (Bradshaw 1983).

The general pattern of nitrogen accumulation is one of rapid increases for the first 50–200 years and then an asymptote at 2000–5000 kg N ha^{-1} (Fig. 2). Many seres (e.g. dunes; Fig. 2d) appear to reach this asymptote only after 1000 years (Olson 1958) while others (e.g. mine tailings and floodplains; Fig. 2b, c) reach it much sooner. Large variations in accumulation rates occur within types of seres whether nitrogen fixers are dominant (mine tailings; Fig. 2b) or absent (dunes; Fig. 2d). In some cases, the presence of a dominant stand of nitrogen fixers did not alter either the basic pattern or rate of nitrogen accumulation. For example, two floodplains (Fig. 2c) without dominant nitrogen fixers (Viereck 1970; Wilson 1970) accumulated more nitrogen at a faster rate than two *Alnus*-dominated floodplains (Luken & Fonda 1983; Walker 1989). Similarly, although all glacial moraines were dominated by nitrogen fixers, the five examples in Fig. 2a had widely divergent amounts and rates of nitrogen accumulation.

No clear correlation emerges between the presence of dominant stands of nitrogen fixers and the accumulation of nitrogen in mineral soils during primary succession. Inputs, predominately from symbiotic nitrogen fixers, are known for a few systems, but losses of nitrogen are less well-understood. Variations in the relative magnitude of inputs and losses of soil nitrogen make across-site comparisons difficult. Nitrogen availability may be more important to successional development

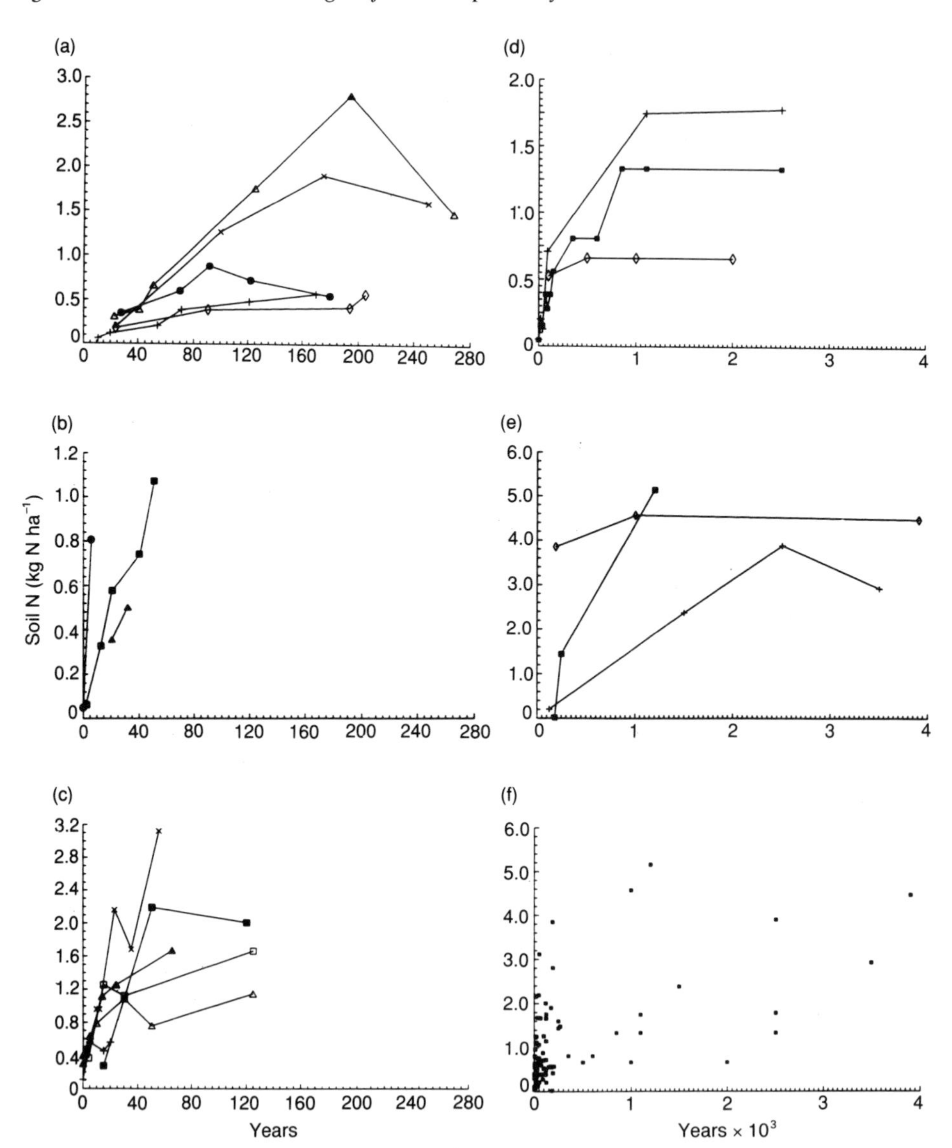

FIG. 2. Total soil nitrogen in surface mineral soil to depths of 10–25 cm from six types of primary succession (kg N ha^{-1}). Sources followed by an asterisk had dominant stands of nitrogen fixers. (a) Glacial moraines: (●) Crocker & Major 1955,* (△) Jacobsen & Birks 1980,* (x) Viereck 1966,* (+) Crocker & Dickson 1957 – Mendenhall Glacier,* (◇) Crocker & Dickson 1957 – Herbert Glacier.* (b) Mine tailings: (■) Leisman 1957,* (●) Palaniappan *et al.* 1979,* (▲) Dancer *et al.* 1977.* (c) Floodplains: (x) Wilson 1970, (■) Viereck 1970, (▲) Luken & Fonda 1983,* (□) Krasny *et al.* 1984,* (△) Walker 1989,* (+) Van Cleve *et al.* 1971.* (d) Dunes: (+) Robertson & Vitousek 1981, (■) Olson 1958, (◇) Ayyad 1973. (e) Volcanoes: (■) Tezuka 1961,* (◇) Vitousek *et al.* 1983 – ash, (+) Vitousek *et al.* 1983 – pahoehoe. (f) All seres: a combination of all the previous graphs.

than total pools of nitrogen (Vitousek & Walker 1987) but it has rarely been measured in primary succession (cf. Robertson & Vitousek 1981; Vitousek *et al.* 1983; Walker 1989).

EFFECTS ON ASSOCIATED SPECIES AND SUCCESSIONAL CHANGE

The impact of nitrogen fixers is considered beneficial in terms of long-term soil and vegetation development (Marrs *et al.* 1983). This has led to the assumption that nitrogen fixers also have a beneficial effect on the growth of associated species. However, examination of both direct and indirect interactions between nitrogen fixers and associated species reveals competitive as well as facilitative interactions (Moore 1982; Walker & Chapin 1987; Morris & Wood 1989; del Moral 1993; Walker & Vitousek 1991). The life-history and life-form of the nitrogen fixer, as well as when it occurs during succession are also clearly important in determining the nature of interactions with other species. These interactions are discussed below, beginning with primary seres where nitrogen fixers are most often dominant (e.g. glacial moraines and mine tailings) and presumably important to successional change. The discussion ends with primary seres where symbiotic nitrogen fixers are generally absent (e.g. volcanoes and rock outcrops). Most of the literature is descriptive and field experiments are rare. Therefore, conclusions regarding the effect of nitrogen fixers on successional change must be tentative.

Glacial moraines

Symbiotic nitrogen fixers were abundant or dominant on all but one of the glacial moraines surveyed (see Table 2). *Dryas* and *Alnus* were the most common nitrogen fixers. *Dryas* forms dense mats on many moraines which may inhibit colonization by other species (Heusser 1956). However, the *Dryas* mat develops outward from a centre which dies after about 20 years. Establishment of other species such as *Alnus* (Cooper 1923), *Salix* and *Populus* (Birks 1980), or conifers (Heusser 1956) in the central portion of the mat may eventually cause the death of the remaining mat by shading (Birks 1980). *Sheperdia*, another nitrogen fixer that is abundant on glacial moraines (Tisdale *et al.* 1966; Viereck 1966) may also inhibit other species by shading.

The most cited example of facilitative effects of nitrogen fixers in primary succession is from Glacier Bay, Alaska. Crocker & Major (1955) established a correlation between nutrient accumulation from symbiotic nitrogen fixation by alder and vegetation development at Glacier Bay. Lawrence and colleagues (Schoenike 1958; Lawrence *et al.* 1967) experimentally demonstrated that *Dryas* plants and *Alnus* litter increased growth of associated *Populus* in early succession at Glacier Bay. There is, however, no evidence to support the general assumption that *Alnus* also facilitates growth of *Picea*, the species that dominates later successional stages (Chapin & Walker 1990). In fact, Cooper (1923) reported that *Picea* grew as well in *Salix* thickets as in thickets of *Alnus*. Data from another boreal site (Walker & Chapin 1986) demonstrated that *Alnus* inhibited *Picea* germination and seedling growth. Until more experimental evidence is presented on the actual interactions of nitrogen fixers with other species at Glacier Bay, it seems premature to state that *Alnus* facilitates growth of later successional species. Generalizing from one boreal example to other primary seres may also be unrealistic.

Despite the general assumption that glacial moraines are classic examples of how nitrogen fixers facilitate vegetation development in primary succession, the available evidence suggests that the direct interactions of nitrogen fixers with other species are more competitive than facilitative. More experimental data, particularly from glacial moraines outside North America, are needed to test this observation.

Mudflows

Nitrogen fixers are also abundant on mudflows (see Table 2), despite much higher nitrogen levels than glacial moraines in the initial substrates. Succession on mudflows is strongly influenced by organic debris, and live and standing dead plant material that may be embedded in the substrate and can provide sites for seed entrapment and seedling establishment (Frenzen *et al.* 1988). For example, root sprouts from surviving *Lupinus* on Mount St Helens trapped windblown seeds of other species, and may facilitate germination and establishment by providing shade and increased soil moisture (Dale 1986). This kind of facilitation by a nitrogen fixer may be independent of its nitrogen-fixing abilities.

Alnus dominates many mudflows (Frehner 1957; Halpern & Harmon 1983) and can form dense thickets that are replaced by later successional species like *Picea* (Gimingham & Boggie 1957) only after the alder dies. *Purshia* and *Ceanothus*, other non-leguminous fixers, were considered moderately important in soil development on mudflows in California (Dickson & Crocker 1953) but neither their actual impact on soil development nor their effects on other species were described. Despite relatively high nutrient levels, nutrients may still limit establishment on mudflows (del Moral & Clampitt 1985), although microenvironments and substrate appear to be more important limiting factors than nutrients.

Mine tailings

Studies of primary succession on mine tailings provide some of the most detailed experimental evidence for species interactions in primary succession. Mine tailings can present a wide variety of substrates for plant colonization (Bradshaw 1983) but are usually of known age and origin. Nitrogen fixers are in most cases abundant or dominant, but in other cases can be unimportant components of the vegetation (Manner *et al.* 1984; Hejkal 1985). Unlike moraines and mudflows, herbaceous and woody legumes are the most common type of nitrogen fixer (see Table 4). Evidence of facilitation by nitrogen fixers comes from the observations by Palaniappan *et al.* (1979) that non-leguminous plants colonized beneath *Lupinus* and apparently benefited from it. Leisman (1957) compared various substrates and suggested trees in lean ore mines with few nitrogen fixers did not grow as quickly as trees on stripping banks with more abundant nitrogen fixers. However, both sites were nitrogen-limited so favourable microsites and soil texture were also considered important for tree seed germination and growth.

Legumes can be the only colonizers of mine tailings (Schramm 1966) and can dominate early succession (Hope-Simpson 1940; Bramble & Ashley 1955). However, legumes are usually short-lived and are generally replaced by non-nitrogen fixers such as grasses (Locket 1945), *Rhododendron* (Dancer *et al.* 1977), or later successional tree species. Occasionally legumes persist; *Trifolium* dominated mine tailings between 13 and 51 years in Minnesota (Leisman 1957). Replacement may also be due to competitive interactions, drought or frost intolerance (Locket 1945). Experimental additions of nitrogen fixers to mine tailings in England (Davis *et al.* 1985) did not increase cover of other species unless rabbit grazing was excluded, underscoring the multiple causes for plant establishment and growth. Alternatively, *Prunus* trees increased in importance and cover when nitrogen-fixing *Robinia* trees were experimentally added to mine tailings in Ohio (Larson 1984) and Pennsylvania (Schuster & Hutnick 1987).

Landslides

On landslides, nitrogen fixers can be abundant (Table 2) on less severe habitats such as basal debris fans (Smith *et al.* 1986), but are generally absent on slip faces (Mark *et al.* 1964). The vegetation on landslides may closely resemble the vegetation of the surrounding pre-disturbance vegetation (Veblen & Ashton 1978), although distinctive early successional communities are more common (Flaccus 1959; Varela *et al.* 1983). Both legumes (Langenheim 1956; Hull & Scott 1982) and other nitrogen fixers like *Alnus*, *Ceanothus* and *Coriaria* (Veblen & Ashton 1978; Dale 1986; Miles & Swanson 1986) occur (Table 4) but their effects on other species are poorly documented and are probably less important than organic debris, substrate stability, availability of water and protection from exposure (Flaccus 1959).

Floodplains

Non-leguminous plants such as *Alnus* or *Myrica* typically dominate mid-successional stages on floodplains (e.g. Dahlskog 1982; McKendrick *et al.* 1982; Clement 1985; Muzika *et al.* 1987). However, nitrogen fixers can also be the first colonizers on floodplains (Bliss & Cantlon 1957; Kiilsgaard *et al.* 1986) or only arrive very late in succession (Terborgh 1985; Fyles & Bell 1986). Legumes are less abundant (Wilson 1970; Frye & Quinn 1979; Bush & Van Auken 1984), unimportant or even absent (Lindsey *et al.* 1961; Sigafoos 1964; Johnson *et al.* 1976; Menges 1986;

Crampton 1987). Allogenic factors such as sedimentation, flooding, soil texture and drainage are generally considered more important in determining succession than nitrogen added by nitrogen fixation, although experimental tests of these assumptions are uncommon (cf. Walker & Chapin 1986).

Non-leguminous fixers such as *Alnus* may form thickets that inhibit germination and seedling establishment of other species (Luken & Fonda 1983; Walker & Chapin 1986) until their death releases understorey tree seedlings. In some cases, shade may facilitate the establishment of herbs that are only able to colonize after the thicket has developed (Clement 1985). Alternatively, species may colonize at the same time as the *Alnus* thicket but grow more slowly (Van Auken & Bush 1985). Nitrogen additions by *Alnus* facilitated growth of seedlings of the late successional *Picea* in greenhouse experiments in Alaska (Walker & Chapin 1986). However, the *Alnus* canopy inhibited growth of *Picea* seedlings by shade and root competition, providing a net negative effect. Nitrogen fixers can therefore facilitate or inhibit establishment and growth of other species, although facilitation by nitrogen additions on floodplains has not been documented under field conditions. Inhibition by the formation of thickets is not limited to nitrogen fixers but can be a characteristic of other species on floodplains including *Populus* (Nanson & Beach 1977) or *Tamarix* (Halwagy 1963). Sometimes thickets of *Alnus* may be prolonged by the higher palatibility of earlier or later species to herbivores (Dahlskog & Cramer 1985; Bryant & Chapin 1986).

Dunes

Nitrogen fixers may be present throughout dune succession (Tuxen 1970; Ayyad 1973; Rayner & Batson 1976), may be early colonizers (Moreno-Casasola 1986), or may be present only in intermediate (Morris *et al.* 1974; Tagawa *et al.* 1985) or final (Chadwick & Dalke 1965; Londo 1971) stages of succession. Yet the importance of nitrogen fixers to the nutrition of other dune plants has not been determined (Barbour *et al.* 1985). Succession on dunes may be primarily driven by the accumulation of organic matter and allogenic factors such as shifting sand or the availability of water.

Some early-colonizing nitrogen fixers such as *Chamaecrista* and *Hippophae* may do best in shifting sand(Moreno-Casasola 1986), and successful early colonizers can often establish monospecific stands, regardless of their nitrogen-fixing abilities. *Ammophila* stands may resist invasion by the nitrogen-fixing *Myrica* for as much as 150 years (Morris *et al.* 1974) and dense patches of the nitrogen-fixing *Hippophae* resist invasion by other species thereby decreasing species diversity (Moore 1982). On coastal dunes in Virginia, Tyndall & Levy (1976) demonstrated that nitrogen fixers may undergo a form of cyclic succession with other species. On the Virginia dunes, neither *Myrica* nor *Pinus* was dependent on the other for establishment but the taller *Pinus* succeeded the shade-intolerant *Myrica* (*Myrica* vigour declined under *Pinus*). However, *Pinus* was less salt tolerant and died when too exposed, leaving *Myrica* dominant again, perhaps to be replaced eventually by *Quercus* (Tyndall & Levy 1976). Long-term studies of dunes show site deterioration and retrogressive succession (Olson 1958; Walker *et al.* 1981) where phosphorus and other nutrients become more limiting than nitrogen.

Volcanoes

Symbiotic nitrogen fixers are more often absent than abundant or dominant on volcanoes (Table 2). However, some of the volcanic seres have non-vascular nitrogen fixers that may be important sources of nitrogen. Blue-green algae were common colonizers of ash on Krakatoa and may have facilitated fern growth (Tagawa *et al.* 1985). They also grew on Parícutin in Mexico (Eggler 1963) and as epiphytes on surviving trees on La Soufrière in Guadeloupe (Fritz-Sheridan & Portecop 1987). Lichens are common on ash on La Soufrière (Fritz-Sheridan & Portecop 1987), on lava but not ash on Surtsey (Brock 1973), on a'a lava of Parícutin (Eggler 1963), and on ash and pahoehoe lava in Hawaii (Atkinson 1970; Smathers & Mueller-Dombois 1974). The contribution of these non-vascular fixers to total nitrogen accumulation is generally small compared with other sources such as precipitation (Vitousek & Walker 1989) that may support growth of early colonizers in the absence of nitrogen fixation (Eggler 1959, 1963; Tezuka 1961).

Herbaceous legumes are not generally important on volcanic substrates although *Desmodium* occurred on lava flows in Papua, New Guinea (Taylor 1957) and Taal, Philippines (Brown *et al.* 1917) and *Lupinus* was a pioneer on a variety of volcanic substrates on Mount St Helens (del Moral & Wood 1988). Woody legumes and other non-leguminous species can form an important mid-successional stage that often invades an earlier grass-dominated stage (Aston 1916; Brown *et al.* 1917; Tagawa *et al.* 1985; del Moral & Wood 1988; Uhe 1988b). This invasion may be due to the ability of the invader to resprout after fire (*Acacia* on Taal; Brown *et al.* 1917), or from vegetative reproduction via stem suckers (*Alnus* on Fiji; Tezuka 1961). These stands of nitrogen fixers may in turn resist invasion of later successional trees by forming dense thickets (*Coriaria* on Taal; Aston 1916) or may represent the final stage of succession (*Alnus* on Katmai, Alaska; Griggs 1936). Nitrogen fixers may facilitate establishment of other species (del Moral & Wood 1988), especially if they form an open canopy (Tezuka 1961). However, even in the low nitrogen soils of volcanic origin, nitrogen fixers have not been shown to facilitate growth of other species via nitrogen additions. Instead, the nitrogen fixers often form successful stands, capture the resources and resist invasion by other species until they are unable to regenerate (e.g. *Casuarina* on Krakatoa; Tagawa *et al.* 1985; Whittaker *et al.* 1989). *Myrica*, when introduced to volcanic soils in Hawaii that previously lacked vascular nitrogen fixers, also formed dense stands that excluded other species (Mueller-Dombois & Whiteaker 1990; Walker & Vitousek 1991).

Rock outcrops

Vascular nitrogen fixers are rarely abundant on rock outcrops and when present, are not an important part of succession. Lichens are present on most rock seres and may be important to soil development (Shure & Ragsdale 1977), although mosses are perhaps more important than lichens (Oosting & Anderson 1939; McVaugh 1943). Herbaceous legumes (Whitehouse 1933; Harper 1939; McVaugh 1943; Griggs 1956) are most common on rock outcrops, but woody legumes (*Robinia*; Keever *et al.* 1951) and other nitrogen fixers (Cox 1933) are also found. One long-term study on rocky islands in Norway (Rydin & Borgegard 1988) suggests *Alnus* may affect other species by shading, but found species interactions were generally weak. Soil development and water retention, perhaps interacting with competition (Sharitz & McCormick 1972; Houle & Phillips 1989) are probably more important to successional development on rock outcrops than nitrogen additions by nitrogen fixers.

CONCLUSIONS

A majority (77%) of the primary seres reviewed had vascular plants associated with symbiotic nitrogen-fixing bacteria. Of those that did not, 54% (12% of total) reported lichens or blue-green algae present in the early stages of succession. In the remaining studies (11%), non-vascular nitrogen fixers may have been present but not reported. The results thereby suggest that nitrogen fixers are almost always associated with primary succession. What is less clear and harder to determine is the effect these plants have on other species and the ecosystem.

Many studies have demonstrated the large additions that nitrogen fixers make to the total ecosystem store of nitrogen. Actually determining the impact such nitrogen additions have on succession requires detailed nitrogen budgets (Marrs *et al.* 1983) over long periods of time. Comparisons across study sites, however, indicate no correlation between the presence of dominant stands of nitrogen fixers and the rate or pattern of ecosystem nitrogen accumulation over successional time. The process may be too complex to isolate the impact of a single factor such as nitrogen fixation alone.

Vascular species with nitrogen-fixing symbionts are often not the first colonizers in primary succession (Sprent 1993), but tend to invade after stands of early ruderals. This is more likely due to poor dispersal ability of heavy seeds (e.g. legumes) or insufficient nitrogen to allow development of nodules (Grubb 1986) than an inability to tolerate harsh conditions (Wood & del Moral 1987). Nitrogen fixers generally do not invade later stages of succession and many have characteristics of early successional plants, including low shade tolerance, rapid growth and short life spans (Chapin 1993).

Nitrogen fixers can facilitate the arrival and establishment of later colonizers by providing perches for birds, traps for wind-blown seeds or a favourably shady, moist microenvironment. Nitrogen fixation may not be involved in this form of facilitation. In some cases (e.g. mine tailings), nitrogen additions by nitrogen fixers have been shown to facilitate growth of associated species in primary succession. Also, the nitrogen additions in early succession are likely to provide long-term benefits to later successional species by improving soil development. Facilitation may alter the rate more than the pattern of succession (Dean & Hurd 1980; Walker & Chapin 1987). However, facilitation is difficult to demonstrate, perhaps because many nitrogen fixers can be nitrogen-limited themselves or may not add nitrogen to their environment until they decay (Sprent 1987; Sprent & Sprent 1989).

Nitrogen fixers can be very successful in capturing resources, and often form dense thickets that inhibit colonization by other species. Not all nitrogen fixers form thickets, however, and herbaceous legumes, the most common form of nitrogen fixer, may not inhibit successional change at all. When thickets do form they are sometimes invaded by longer-lived or taller-growing plants that may then shade out the nitrogen fixer. However, the nitrogen fixer often disappears only after it reaches its full life-span and fails to regenerate under its own shade. The capability of sprouting vegetatively may prolong the thicket stage. The basic pattern of many seres can thus be explained by successive replacement of short-lived by longer-lived species. Competition, like facilitation, may alter the rate more than the pattern of succession. The outcome of competition may result in species displacements that increase the rate of successional change or may result in resistance to species changes that slows succession.

A realistic appraisal of the role of nitrogen fixers in primary succession includes all aspects of their ecology. Their particular ability to fix atmospheric nitrogen, a nutrient generally limiting plant growth, has led to unrealistic generalizations about their facilitative role in primary succession. By providing the limiting nutrient in the early stages of primary succession, nitrogen fixers certainly contribute to the process of ecosystem development. Yet the relationships between nitrogen fixers and other species are also characterized by a full range of competitive interactions.

ACKNOWLEDGMENTS

Comments by G. Aplet, J. Matthews, R. del Moral, E. Powell, R. Sanford, J. Sprent and P. Vitousek improved the manuscript. Financial assistance for writing was provided by NSF grant BSR 87–18004 and by the Summer Initiatives Program of the University of Puerto Rico. Development of the ideas presented here occurred while I was supported by NSF grants BSR 84–15821, BSR 84–05269, and BSR 88–1789.

REFERENCES

Anderson, D.C. & MacMahon, J.A. (1985). Plant succession following the Mount St Helens volcanic eruption — Facilitation by a burrowing rodent, *Thomomys talpoides*. *American Midland Naturalist*,

114, 62–9.

Aston, B.C. (1916). The vegetation of the Tarawera Mountain, New Zealand, *Journal of Ecology*, **4**, 18–26.

Atkinson, I.A.E. (1970). Successional trends in the coastal and lowland forest of Mauna Loa and Kilauea Volcanoes, Hawaii. *Pacific Science*, **24**, 387–400.

Ayyad, M.A. (1973). Vegetation and environment of the western Mediterranean coastal land of Egypt. *Journal of Ecology*, **61**, 509–23.

Barbour, M.G., de Jong, T.M. & Pavlik, B.M. (1985). Marine beach and dune plant communities. *Physiological Ecology of North American Plant Communities* (Ed. by B.F. Chabot & H.A. Mooney), pp. 296–322. Chapman & Hall, New York.

Beard, J.S. (1945). The progress of plant succession on the Soufrière of St Vincent. *Journal of Ecology*, **33**, 1–9.

Belsky, A.J. & Amundson, R.G. (1986). Sixty years of successional history behind a moving sand dune near Olduvai Gorge, Tanzania. *Biotropica*, **18**, 231–5.

Birks, H.J.B. (1980). The present flora and vegetation of the moraines of the Klutlan Glacier, Yukon Territory, Canada: A study in plant succession. *Quaternary Research*, **14**, 60–86.

Bliss, L.C. & Cantlon, J.E. (1957). Succession on river alluvium in northern Alaska. *American Midland Naturalist*, **58**, 452–69.

Bradshaw, A.D. (1983). The reconstruction of ecosystems. *Journal of Applied Ecology*, **20**, 1–17.

Bramble, W.C. & Ashley, R.H. (1955). Natural revegetation of spoil banks in central Pennsylvania. *Ecology*, **36**, 417–23.

Bray, J.R. & Struik, G.J. (1963). Forest growth and glacial chronology in eastern British Columbia, and their relation to recent climatic trends. *Canadian Journal of Botany*, **41**, 1245–71.

Brock, T.D. (1973). Primary colonization of Surtsey, with special reference to the blue-green algae. *Oikos*, **24**, 239–43.

Brown, W.H., Merrill, E.D. & Yates, H.S. (1917). The revegetation of Volcano Island, Luzon, Philippine Islands, since the eruption of Taal Volcano in 1911. *Philippine Journal of Science*, **12**, 177–243.

Bryant, J.P. & Chapin III, F.S. (1986). Browsing–woody plant interactions during a boreal forest plant succession. *Forest Ecosystems in the Alaskan Taiga; A Synthesis of Structure and Function* (Ed. by K. Van Cleve, F.S. Chapin III, P.W. Flanagan, L.A. Viereck & C.T. Dyrness), pp. 213–25. Springer-Verlag, New York.

Burbanck, M.P. & Platt, R.B. (1964). Granite outcrop communities of the Piedmont plateau in Georgia. *Ecology*, **45**, 292–306.

Bush, J.K. & Van Auken, O.W. (1984). Woody-species composition of the Upper San Antonio River gallery forest. *Texas Journal of Science*, **36**, 139–48.

Campbell, D.H. (1909). The new flora of Krakatoa. *American Naturalist*, **43**, 449–60.

Chadwick, H.W. & Dalke, P.D. (1965). Plant succession on dune sands in Fremont County, Idaho. *Ecology*, **46**, 765–80.

Chapin III, F.S. (1993). Physiological controls over plant establishment in primary succession. *Primary Succession on Land* (Ed. by J. Miles & D.W.H. Walton), pp. 161–78. Blackwell Scientific Publications, Oxford.

Chapin III, F.S. & Walker, L.R. (1990). The importance of Glacier Bay to tests of current theories of plant succession. *Proceedings of the Second Glacier Bay Science Symposium, Glacier Bay National Park, Gustavus, Alaska*, pp. 136–9.

Charette, D.J. & Shisler, J.K. (1985). Comparison of the plant communities in natural and managed foredunes along the Atlantic coast of New Jersey. *Gambling with the Shore*. Proceedings of the Ninth Annual Conference of the Coastal Society. 1984. Atlantic City, New Jersey.

Clements, F.E. (1916). *Plant Succession: An Analysis of the Development of Vegetation*. Carnegie Institution of Washington Publication **242**.

Clement, C.J.E. (1985). Floodplain succession on the west coast of Vancouver Island. *Canadian Field Naturalist*, **99**, 34–9.

Cooper, W.S. (1916). Plant succession in the Mount Robson region, British Columbia. *Plant World*, **19**, 11–238.

Cooper, W.S. (1923). The recent ecological history of Glacier Bay, Alaska. II. The present vegetation cycle. *Ecology*, **4**, 223–46.

Cox, C.F. (1933). Alpine plant succession on James Peak, Colorado. *Ecological Monographs*, **3**, 299–372.

Crampton, C.B. (1987). Soils, vegetation and permafrost across an active meander of Indian River, Central Yukon, Canada. *Catena*, **14**, 157–63.

Crocker, R.L. & Dickson, B.A. (1957). Soil development on the recessional moraines of the Herbert and Mendenhall Glaciers, south-eastern Alaska. *Journal of Ecology*, **45**, 169–85.

Crocker, R.L. & Major, J. (1955). Soil development in relation to vegetation and surface age at Glacier Bay, Alaska. *Journal of Ecology*, **43**, 427–48.

Dahlskog, S. (1966). Sedimentation and vegetation in a Lapland mountain delta. *Geografiska Annaler*, **48A**, 86–101.

Dahlskog, S. (1982). Shrub canopy establishment in primary successions, *Meddelanden från Väextekologiska Institutionen Uppsala*, **3**, 63–8.

Dahlskog, S. & Cramer, W. (1985). Development of *Alnus crispa* thickets in the Kvikkjokk Delta, Swedish Lapland. *Meddelanden från Väextekologiska Institutionen Uppsala*, **2**, 21.

Dale, V.H. (1986). Plant recovery on the debris avalanche at Mount St Helens. *Mount St Helens: Five Years Later* (Ed. by S.A.C. Keller), pp. 208–14. Eastern Washington Press, Cheney, Washington.

Dancer, W.S., Handley, J.F. & Bradshaw, A.D. (1977). Nitrogen accumulation in kaolin mining wastes in Cornwall. I. Natural communities. *Plant and Soil*, **48**, 153–67.

Davis, B.N.K., Lakhani, K.H., Brown, M.C. & Park, D.G. (1985). Early seral communities in a limestone quarry — an experimental study of treatment effects on cover and richness of vegetation. *Journal of Applied Ecology*, **22**, 473–90.

Dean, T.A. & Hurd, L.E. (1980). Development in an estuarine fouling community: The influence of early colonists on later arrivals. *Oecologia*, **46**, 295–301.

del Moral, R. & Clampitt, C.A. (1985). Growth of native plant species on recent volcanic substrates from Mount St Helens. *American Midland Naturalist*, **114**, 374–83.

del Moral, R. & Wood, D.M. (1988). Dynamics of herbaceous vegetation recovery on Mount St Helens, Washington, USA, after a volcanic eruption. *Vegetatio*, **74**, 11–27.

del Moral, R. (1993). Mechanisms of primary succession on volcanoes: A view from Mount St Helens. *Primary Succession on Land* (Ed. by J. Miles & D.W.H. Walton), pp. 79–100. Blackwell Scientific Publications, Oxford.

Dickson, B.A. & Crocker, R.L. (1953). A chronosequence of soils and vegetation near Mt Shasta, California. II. The development of the forest floors and the carbon and nitrogen profiles of the soils. *Journal of Soil Science*, **4**, 142–55.

Douglas, D.A. (1987). Growth of *Salix setchelliana* on a Kluane River point bar, Yukon Territory, Canada. *Arctic and Alpine Research*, **19**, 35–44.

Edwards, J.E., Crawford, R.L., Sugg, P.M. & Peterson, M. (1987). Arthropod colonization in the blast zone of Mount St Helens. *Mount St Helens: Five Years Later* (Ed. by S.A.C. Keller), pp. 329–34. Eastern Washington Press, Cheney, Washington.

Eggler, W.A. (1941). Primary succession on volcanic deposits in southern Idaho. *Ecological Monographs*, **11**, 277–98.

Eggler, W.A. (1959). Manner of invasion of volcanic deposits by plants, with further evidence from Parícutin and Jorullo. *Ecological Monographs*, **29**, 267–84.

Eggler, W.A. (1963). Plant life of Parícutin Volcano, Mexico. Eight years after activity ceased. *American Midland Naturalist*, **69**, 38–68.

Eggler, W.A. (1971). Quantitative studies of vegetation on 16 young lava flows on the Island of Hawaii. *Tropical Ecology*, **12**, 66–100.

Everitt, B.L. (1968). Use of the cottonwood in an investigation of the recent history of a flood plain. *American Journal of Science*, **266**, 417–39.

Flaccus, E. (1959). Revegetation of landslides in the White Mountains of New Hampshire. *Ecology*, **40**, 692–703.

Fonda, R.W. (1974). Forest succession in relation to river terrace development in Olympic National Park, Washington, *Ecology*, **55**, 927–42.

Frehner, H.K. (1957). *Development of soil and vegetation on Kautz Creek flood deposit in Mount Rainier National Park*. MS thesis, University of Washington, Seattle.

Frenzen, P.M., Krasny, M.E. & Rigney, L.P. (1988). Thirty-three years of plant succession on the Kautz Creek mudflow, Mount Rainier National Park, Washington. *Canadian Journal of Botany*, **66**, 130–7.

Fridrikksson, S. (1970). *The colonization of vascular plants on Surtsey in 1968*. Surtsey Research Progress Report V. The Surtsey Research Society, Reykjavik.

Fritz-Sheridan, R.P. & Portecop, J. (1987). Nitrogen fixation on the tropical volcano, La Soufrière (Guadaloupe): I. A survey of nitrogen fixation by blue-green algal microepiphytes and lichen endophytes. *Biotropica*, **19**, 194–9.

Frye, R.J. & Quinn, J.A. (1979). Forest development in relation to topography and soils on a floodplain of the Raritan River, New Jersey. *Bulletin of the Torrey Botanical Club*, **106**, 334–45.

Fyles, J.W. & Bell, M.A.M. (1986). Vegetation colonizing river gravel bars in the Rocky Mountains of southeastern British Columbia. *Northwest Science*, **60**, 8–14.

Gates, F.C. (1914). The pioneer vegetation of Taal Volcano. *Philippine Journal of Science*, **9**, 391–434.

Gill, D. (1973). Floristics of a plant succession sequence in the Mackenzie Delta, Northwest Territories. *Polarforschung*, **43**, 55–65.

Gimingham, C.H. & Boggie, R. (1957). Stages in the recolonization of a Norwegian clay-slide. *Oikos*, **8**, 38–64.

Griggs, R.F. (1933). The colonization of the Katmai ash, a new and inorganic 'soil'. *American Journal of Botany*, **20**, 92–113.

Griggs, R.F. (1936). The vegetation of the Katmai district. *Ecology*, **17**, 380–417.

Griggs, R.F. (1956). Competition and succession on a rocky mountain fellfield. *Ecology*, **37**, 8–20.

Grubb, P.J. (1986). The ecology of establishment. *Ecology and Landscape Design* (Ed. by A.D. Bradshaw, D.A. Goode & E. Thorp), pp. 83–98. Symposium of the British Ecological Society, 24. Blackwell Scientific Publications, Oxford.

Grubb, P.J. (1987). Some generalizing ideas about colonization and succession in green plants and fungi. *Colonization, Succession and Stability* (Ed. by A.J. Gray, M.J. Crawley & P.J. Edwards), pp. 81–102. Symposium of the British Ecological Society, 26. Blackwell Scientific Publications, Oxford.

Halpern, C.B. & Harmon, M.E. (1983). Early plant succession on the Muddy River mudflow, Mount St Helens, Washington. *American Midland Naturalist*, **110**, 97–106.

Halwagy, R. (1963). Studies on the succession of vegetation on some islands and sand banks in the Nile near Khartoum, Sudan. *Vegetatio*, **11**, 217–34.

Hardt, R.A. & Forman, R.T.T. (1989). Boundary form effects on woody colonization of reclaimed surface mines. *Ecology*, **70**, 1252–60.

Harper, R.M. (1939). Granite outcrop vegetation in Alabama. *Torreya*, **39**, 153–9.

Heath, J.P. (1967). Primary conifer succession, Lassen Volcanic National Park. *Ecology*, **48**, 270–5.

Hejkal, J. (1985). The development of a carabid fauna (Coleoptera, Carabidae) on spoil banks under conditions of primary succession. *Acta Entomologica Bohemoslovaca*, **82**, 321–46.

Hendrix, L.B. (1981). Post-eruption succession on Isla Fernandina, Galapagos. *Madroño*, **28**, 242–54.

Henriksson, L.E. & Rodgers, G.A. (1978). *Further studies in the nitrogen cycle of Surtsey, 1974–1976*. Surtsey Research Progress Report VIII. The Surtsey Research Society, Reykjavik.

Heusser, C.J. (1956). Post-glacial environments in the Canadian Rocky Mountains. *Ecological Monographs*, **26**, 263–302.

Holton, B. Jr & Johnson, A.F. (1979). Dune scrub communities and their correlation with environmental factors at Point Reyes National Seashore, California. *Journal of Biogeography*, **6**, 317–28.

Hope-Simpson, J.F. (1940). Studies of the vegetation of the English chalk. VI. Late stages in succession leading to chalk grassland. *Journal of Ecology*, **28**, 386–402.

Houle, G. & Phillips, D.L. (1988). The soil seed bank of granite outcrop plant communities. *Oikos*, **52**, 87–93.

Houle, G. & Phillips, D.L. (1989). Seed availability and biotic interactions in granite outcrop plant communities. *Ecology*, **70**, 1307–16.

Howard, R.A., Portecop, J. & de Montaignac, P. (1980). The post-eruptive vegetation of La Soufrière, Guadaloupe, 1977–1979. *Journal of the Arnold Arboretum*, **6**, 749–64.

Hull, J.C. & Scott, R.C. (1982). Plant succession on debris avalanches of Nelson County, Virginia.

Castanea, **47**, 158–76.

Hupp, C.R. (1983). Seedling establishment on a landslide site. *Castanea*, **48**, 89–98.

Hupp, C.R. & Osterkamp, W.R. (1985). Bottomland vegetation distribution along Passage Creek, Virginia, in relation to fluvial landforms. *Ecology*, **66**, 670–81.

Jacobsen, G.L. Jr & Birks, H.J.B. (1980). Soil development on recent end moraines of the Klutlan Glacier, Yukon Territory, Canada. *Quaternary Research*, **14**, 87–100.

Johnson, W.B., Sasser, C.E. & Gosselink, J.G. (1985). Succession of vegetation in an evolving river delta, Atchafalaya Bay, Louisiana. *Journal of Ecology*, **73**, 973–86.

Johnson, W.C., Burgess, R.L. & Keammerer, W.R. (1976). Forest overstory vegetation and environment on the Missouri River floodplain in North Dakota. *Ecological Monographs*, **46**, 59–84.

Keever, C., Oosting, H.J. & Anderson, L.E. (1951). Plant succession on exposed granite of Rocky Face Mountain, Alexander County, North Carolina. *Bulletin of the Torrey Botanical Club*, **78**, 401–21.

Kiilsgaard, C.W., Green, S.E., Stafford, S.G. & McKee, W.A. (1986). Recovery of riparian vegetation in the northeastern region of Mount St Helens. *Mount St Helens: Five Years Later* (Ed. by S.A.C. Keller), pp. 222–30. Eastern Washington Press, Cheney, Washington.

Krasny, M.E., Vogt, K.A. & Zasada, J.C. (1984). Root and shoot biomass and mycorrhizal development of white spruce seedlings naturally regenerating in interior Alaskan floodplain communities. *Canadian Journal of Forest Research*, **14**, 554–8.

Kumler, M.L. (1962). Plant succession on the sand dunes of the Oregon coast. *Ecology*, **50**, 695–704.

Langenheim, J.H. (1956). Plant succession on a subalpine earthflow in Colorado. *Ecology*, **37**, 301–17.

Larson, M.M. (1984). Invasion of volunteer tree species on strip mine plantations in east-central Ohio. *Ohio State University. Agricultural Research and Development Center Research Bulletin* No. 1158.

Lawrence, D.B., Schoenike, R.E., Quispel, A. & Bond, G. (1967). The role of *Dryas drummondii* in vegetation development following ice recession at Glacier Bay, Alaska, with special reference to its nitrogen fixation by root nodules. *Journal of Ecology*, **55**, 793–813.

Leisman, G. (1957). A vegetation and soil chronosequence on the Mesabi Iron Range spoil banks, Minnesota. *Ecological Monographs*, **27**, 221–44.

Lindsey, A.A., Pretty, R., Sterling, D. & Van Asdall, W. (1961). Vegetation and environment along the Wabash and Tippecanoe Rivers. *Ecological Monographs*, **31**, 105–56.

Locket, G.H. (1945). Observations on the colonization of bare chalk. *Journal of Ecology*, **33**, 205–9.

Londo, G. (1971). Successive mapping of dune slack vegetation. *Vegetatio*, **29**, 51–64.

Luken, J.O. & Fonda, R.W. (1983). Nitrogen accumulation in a chronosequence of red alder communities along the Hoh River, Olympic National Park, Washington. *Canadian Journal of Forest Research*, **13**, 1228–37.

Manner, H.I., Thaman, R.R. & Hassall, D.C. (1984). Phosphate mining induced vegetation changes on Nauru Island. *Ecology*, **65**, 1454–65.

Mark, A.F., Scott, G.A.M., Sanderson, F.R. & James, P.W. (1964). Forest succession on landslides above Lake Thomson, Fiordland. *New Zealand Journal of Botany*, **2**, 60–89.

Marrs, R.H., Roberts, R.D., Skeffington, R.A. & Bradshaw, A.D. (1983). Nitrogen and the development of ecosystems. *Nitrogen as an Ecological Factor* (Ed. J.A. Lee, S. McNeill & I.H. Rorison), pp. 113–36. Symposium of the British Ecological Society, 22. Blackwell Scientific Publications, Oxford.

Matthews, J.A. (1979). The vegetation of the Storbreen gletschervorfeld, Jotunheimen, Norway. I. Introduction and approaches involving classification. *Journal of Biogeography*, **6**, 17–47.

McBride, J.R. & Strahan, J. (1984). Establishment and survival of woody riparian species on gravel bars of an intermittent stream. *American Midland Naturalist*, **112**, 235–45.

McClanahan, T.R. (1986). The effect of a seed source on primary succession in a forest ecosystem. *Vegetatio*, **65**, 175–8.

McIntosh, R.P. (1980). The relationship between succession and the recovery process in ecosystems. *The Recovery Process in Damaged Ecosystems* (Ed. by J. Cairns), pp. 11–62. Ann Arbor Science, Ann Arbor, MI.

McKendrick, J., Collins, W., Helm, D., McMullen, J. & Koranda, J. (1982). *Alaska Power Authority, Susitna Hydroelectric Project, Environmental Studies. Subtask 7·12, Plant Ecology Studies, Phase I*

of Final Report. University of Alaska Agricultural Experiment Station, Palmer, Alaska.
McLachlan, A., Ascaray, C. & du Toit, P. (1987). Sand movement, vegetation succession and biomass spectrum in a coastal dune slack in Algoa Bay, South Africa. *Journal of Arid Environments*, **12**, 9–25.
McVaugh, R. (1943). The vegetation of the granitic flat-rocks of the southeastern United States. *Ecological Monographs*, **13**, 121–66.
McVaugh, R. (1947). Establishment of vegetation on sand-flats along the Hudson River, New York. *Ecology*, **28**, 189–93.
Menges, E.S. (1986). Environmental correlates of herb species composition in five southern Wisconsin floodplain forests. *American Midland Naturalist*, **115**, 106–17.
Miles, D.W.R. & Swanson, F.J. (1986). Vegetation composition on recent landslides in the Cascade Mountains of western Oregon. *Canadian Journal of Forest Research*, **16**, 739–44.
Moore, P. (1982). Struggles among the dunes. *Nature*, **296**, 805–6.
Moreno-Casasola, P. (1986). Sand movement as a factor in the distribution of plant communities in a coastal dune system. *Vegetatio*, **65**, 67–76.
Morris, M., Eveleigh, D.E., Riggs, S.C. & Tiffney, W.N. Jr (1974). Nitrogen fixation in the bayberry (*Myrica pensylvanica*) and its role in coastal succession. *American Journal of Botany*, **61**, 867–70.
Morris, W.F. & Wood, D.M. (1989). The role of lupine in succession on Mount St Helens: Facilitation or inhibition? *Ecology*, **70**, 697–703.
Mueller-Dombois, D. & Whiteaker, L.D. (1990). Plants associated with *Myrica faya* and two other pioneer plants on a recent volcanic surface in Hawaii Volcanoes National Park. *Phytocoenologia*, **19**, 29–41.
Muzika, R.M., Gladden, J.B. & Haddock, J.D. (1987). Structural and functional aspects of succession in southeastern floodplain forests following a major disturbance. *American Midland Naturalist*, **117**, 1–9.
Nakamura, T. (1985). Forest succession in the subalpine region of Mt Fuji, Japan. *Vegetatio*, **64**, 15–27.
Nanson, G.C. & Beach, H.F. (1977). Forest succession and sedimentation on a meandering river floodplain, northeast British Columbia, Canada. *Journal of Biogeography*, **4**, 229–51.
Ohsawa, M. (1984). Differentiation of vegetation zones and species strategies in the subalpine region of Mt Fuji. *Vegetatio*, **57**, 15–52.
Olson, J.S. (1958). Rates of succession and soil changes on southern Lake Michigan sand dunes. *Botanical Gazette*, **119**, 125–70.
Oosting, H.J. & Anderson, L.E. (1939). Plant succession on granite rock in eastern North Carolina. *Botanical Gazette*, **100**, 750–68.
Palaniappan, V.M. (1974). Ecology of tin tailing areas: plant communities and their succession. *Journal of Applied Ecology*, **11**, 133–50.
Palaniappan, V.M., Marrs, R.H. & Bradshaw, A.D. (1979). The effect of *Lupinus arboreus* on the nitrogen status of china clay wastes. *Journal of Applied Ecology*, **16**, 825–31.
Piotrowska, H. (1988). The dynamics of the dune vegetation on the Polish Baltic coast. *Vegetatio*, **77**, 169–175.
Purer, E.A. (1936). Studies of certain coastal sand dune plants of southern California. *Ecological Monographs*, **6**, 1–64.
Ranwell, D. (1960). Newborough Warren, Anglesey. II. Plant associes and succession cycles of the sand dune and dune slack vegetation. *Journal of Ecology*, **48**, 117–41.
Raus, Th. (1988). Vascular plant colonization and vegetation development on sea-born volcanic islands in the Aegean (Greece). *Vegetatio*, **77**, 139–47.
Rayner, D.A. & Batson, W.T. (1976). Maritime closed dunes vegetation in South Carolina. *Castanea*, **41**, 58–70.
Reichenbacher, F.W. (1984). Ecology and evolution of southwestern riparian plant communities. *Desert Plants*, **6**, 15–22.
Robertson, G.P. (1982). Factors regulating nitrification in primary and secondary succession. *Ecology*, **63**, 1561–73.
Robertson, G.P. & Vitousek, P.M. (1981). Nitrification potentials in primary and secondary succession. *Ecology*, **62**, 376–86.

Rydin, H. & Borgegard, S. (1988). Primary succession over sixty years on hundred-year old islets in Lake Hjalmaren, Sweden. *Vegetatio*, **77**, 159–68.

Schimper, A.F.W. (1903). *Plant Geography Upon a Physiological Basis*. English translation by W.R. Fisher. Clarendon Press, Oxford.

Schramm, J.R. (1966). Plant colonization studies on black wastes from anthracite mining in Pennsylvania. *Transactions of the American Philosophical Society*, **56**, 1–194.

Schuster, W.S. & Hutnik, R.J. (1987). Community development on 35-year-old planted minespoil banks in Pennsylvania. *Reclamation and Revegetation Research*, **6**, 109–20.

Schoenike, R.E. (1958). Influence of mountain avens (*Dryas drummondii*) on growth of young cottonwoods (*Populus trichocarpa*) at Glacier Bay, Alaska. *Proceedings of the Minnesota Academy of Science*, **25–26**, 55–8.

Sharitz, R.R. & McCormick, J.F. (1972). Population dynamics of two competing annual plant species. *Ecology*, **54**, 723–40.

Shields, L.M. & Crispin, J. (1956). Vascular vegetation of a recent volcanic area in New Mexico. *Ecology*, **37**, 341–51.

Shure, D.J. & Ragsdale, H.L. (1977). Patterns of primary succession on granite outcrop surfaces. *Ecology*, **58**, 993–1006.

Sigafoos, R. (1964). Botanical evidence of floods and flood-plain deposition. *Geological Survey Professional Paper* **485-A.**

Smathers, G.A. & Mueller-Dombois, D. (1974). Invasion and recovery of vegetation after a volcanic eruption in Hawaii. *Hawaii National Park Service Monograph*, No. 5.

Smith, R.B., Commandeur, P.R. & Ryan, M.W. (1986). Soils, vegetation, and forest growth on landslides and surrounding logged and old-growth areas on the Queen Charlotte Islands. *British Columbia Ministry of Forests Land Management Report* No. 41.

Sprent, J.I. (1987). *The Ecology of the Nitrogen Cycle*. Cambridge University Press, Cambridge.

Sprent, J.I. (1993). The role of nitrogen fixation in primary succession on land. *Primary Succession on Land* (Ed. by J. Miles & D.W.H. Walton), pp. 209–19. Blackwell Scientific Publications, Oxford.

Sprent, J.I. & Sprent, P. (1989). *Nitrogen Fixing Organisms – Pure and Applied Aspects*. Chapman & Hall, London.

Stewart, W.D.P. & Pearson, M.C. (1967). Nodulation and nitrogen-fixation by *Hippophae rhamnoides* in the field. *Plant and Soil*, **26**, 348–59.

Tagawa, H., Suzuki, E., Partoihardjo, T. & Suriadarma, A. (1985). Vegetation and succession on the Krakatoa Islands, Indonesia. *Vegetatio*, **60**, 131–46.

Taylor, B.W. (1957). Plant succession on recent volcanoes in Papua. *Journal of Ecology*, **45**, 233–43.

Terborgh, J. (1985). Habitat selection in Amazonian birds. *Habitat Selection in Birds* (Ed. by M.L. Cody), pp. 311–38. Academic Press, New York.

Tezuka, Y. (1961). Development of vegetation in relation to soil formation in the volcanic island of Oshima, Izu, Japan. *Japan Journal of Botany*, **17**, 371–402.

Tilman, D. (1988). *Plant Strategies and the Dynamics and Structure of Plant Communities*. Princeton University Press, Princeton, New Jersey.

Tisdale, E.W., Fosberg, M.A. & Poulton, C.E. (1966). Vegetation and soil development on a recently glaciated area near Mount Robson, British Columbia. *Ecology*, **47**, 517–23.

Tsuyuzaki, S. (1989). Analysis of revegetation dynamics on the Volcano Usu, Northern Japan, deforested by 1977–1978 eruptions. *American Journal of Botany*, **76**, 1468–77.

Tuxen, R. (1970). Pflanzensoziologische beobachtungen an Islandischen dunengesellschaften. *Vegetatio*, **20**, 251–78.

Tyndall, R.W. & Levy, G.F. (1976). Vegetational zonation and succession within interdunal depressions on the barrier dune system of southeastern Virginia. *Virginia Journal of Science*, **27**, 60.

Uhe, G. (1988a). The composition of the plant communities inhabiting the recent volcanic deposits of Maui and Hawaii, Hawaiian Islands. *Tropical Ecology*, **29**, 26–47.

Uhe, G. (1988b). The composition of the plant communities inhabiting the recent volcanic ejecta of Yasour Tanna, New Hebrides. *Tropical Ecology*, **29**, 48–54.

Uno, G.E. & Collins, S.L. (1987). Primary succession on granite outcrops in southwestern Oklahoma. *Bulletin of the Torrey Botanical Club*, **114**, 387–92.

Van Auken, O.W. & Bush, J.K. (1985). Secondary succession on terraces of the San Antonio River. *Bulletin of the Torrey Botanical Club*, **112**, 158–66.

Van Cleve, K., Viereck, L.A. & Schlentner, R.L. (1971). Accumulation of nitrogen in alder (*Alnus*) ecosystems near Fairbanks, Alaska. *Arctic and Alpine Research*, **3**, 101–14.

van der Laan, D. (1979). Spatial and temporal variation in the vegetation of dune slacks in relation to the ground water regime. *Vegetatio*, **39**, 43–51.

van der Maarel, E., de Cock, N. & de Wildt, E. (1985). Population dynamics of some major woody species in relation to long-term succession on the dunes of Voorne. *Vegetatio*, **61**, 209–19.

Varela, C., Ortíz, D., Berrios, A. & Alvarez, H.J. (1983). Estudio de la colonización en areas de derrumbres en el bosque experimental de Luquillo. *Los Bosques de Puerto Rico* (Ed. by A.E. Lugo), pp. 110–22. US Fish and Wildlife Service and the Dept of Natural Resources.

Veblen, T.T. & Ashton, D.H. (1978). Catastrophic influences on the vegetation of the Valdivian Andes, Chile. *Vegetatio*, **36**, 149–67.

Viereck, L.A. (1966). Plant succession and soil development on gravel outwash of the Muldrow Glacier, Alaska. *Ecological Monographs*, **36**, 181–91.

Viereck, L.A. (1970). Forest succession and soil development adjacent to the Chena River in interior Alaska. *Arctic and Alpine Research*, **2**, 1–26.

Vitousek, P.M. & Walker, L.R. (1987). Colonization, succession and resource availability: ecosystem-level interactions. *Colonization, Succession and Stability* (Ed. by A.J. Gray, M.J. Crawley & P.J. Edwards), pp. 207–24. Symposium of the British Ecological Society, 26. Blackwell Scientific Publications, Oxford.

Vitousek, P.M. & Walker, L.R. (1989). Biological invasion by *Myrica faya* in Hawaii: Plant demography, nitrogen fixation, and ecosystem effects. *Ecological Monographs*, **59**, 247–65.

Vitousek, P.M., Van Cleve, K., Balakrishnan, N. & Mueller-Dombois, D. (1983). Soil development and nitrogen turnover in montane rainforest soils on Hawaii. *Biotropica*, **15**, 268–74.

Walker, J., Thompson, C.H., Fergus, I.F. & Tunstall, B.R. (1981). Plant succession and soil development in coastal sand dunes of subtropical eastern Australia. *Forest Succession* (Ed. by D.C. West, H.H. Shugart & D.B. Botkin), pp. 107–31. Springer-Verlag, New York.

Walker, L.R. (1989). Soil nitrogen changes during primary succession on a floodplain in Alaska, USA. *Arctic and Alpine Research*, **21**, 341–9.

Walker, L.R. & Chapin III, F.S. (1986). Physiological controls over seedling growth in primary succession on an Alaskan floodplain. *Ecology*, **67**, 1508–23.

Walker, L.R. & Chapin III, F.S. (1987). Interactions among processes controlling successional change. *Oikos*, **50**, 131–5.

Walker, L.R. & Vitousek, P.M. (1991). An invader alters germination and growth of a native dominant tree in Hawaii. *Ecology*, **72**, 1449–55.

Walker, L.R., Zasada, J.C. & Chapin III, F.S. (1986). The role of life history processes in primary succession on an Alaskan floodplain. *Ecology*, **67**, 1243–53.

Walker, T.W. & Syers, J.K. (1976). The fate of phosphorus during pedogenesis. *Geoderma*, **15**, 1–19.

Whitehouse, E. (1933). Plant succession on central Texan granite. *Ecology*, **14**, 391–405.

Whittaker, R.J., Bush, M.B. & Richards, K. (1989). Plant recolonization and vegetation succession on the Krakatau Islands, Indonesia. *Ecological Monographs*, **59**, 59–123.

Wilson, R.E. (1970). Succession in stands of *Populus deltoides* along the Missouri River in southeast South Dakota. *American Midland Naturalist*, **83**, 330–42.

Winterringer, G.S. & Vestal, A.G. (1956). Rock-ledge vegetation in southern Illinois. *Ecological Monographs*, **26**, 105–30.

Wood, D.M. & del Moral, R. (1987). Mechanisms of early primary succession in subalpine habitats on Mount St Helens. *Ecology*, **68**, 780–90.

Woolhouse, M.E.J., Harmsen, R. & Fahrig, L. (1985). On succession in a saxicolous lichen community. *Lichenologist*, **17**, 167–72.

Yarranton, G.A. & Morrison, R.G. (1974). Spatial dynamics of a primary succession: Nucleation. *Journal of Ecology*, **62**, 417–28.

Yeaton, R.I. (1988). Structure and function of the Namib dune grasslands: characteristics of the environmental gradients and species distributions. *Journal of Ecology*, **76**, 744–58.

Soil organisms in coastal foredunes involved in degeneration of *Ammophila arenaria*

W. H. VAN DER PUTTEN
Netherlands Institute of Ecology, PO Box 40, 6666 ZG, Heteren, The Netherlands

SUMMARY

1 Biomass production of *Ammophila arenaria* is reduced when grown in sand originating from its own rooting zone. This is due to harmful soil organisms, probably both nematodes and fungi.
2 Vigour of *A. arenaria* depends on its constantly colonizing windblown sand. As this sand does not contain harmful organisms, such colonization probably enables *A. arenaria* to escape from harmful organisms.
3 If *A. arenaria* produces new roots in windblown sand, the soil pathogens colonize the new root zone within one growing season.
4 In stable dunes, degeneration of *A. arenaria* seems to be caused by a combination of harmful soil organisms that attack the root system and abiotic stress, such as drought and high soil temperatures.
5 Harmful soil organisms may be involved in primary succession in coastal foredunes. The relationship between soil pathogens and vegetation succession, however, needs to be established.

INTRODUCTION

Primary succession of plant species in coastal foredunes is related to the formation and development of sand-dunes, which results in changes in abiotic soil factors. On the beach, salt-tolerant plant species (e.g. *Elymus farctus, Atriplex hastata, Salsola kali* and *Cakile maritima*) check the wind so that drifting sand grains accumulate. Subsequently, less salt-tolerant species (e.g. *Honkenya peploides* and *Ammophila arenaria*) colonize such raised sites on the beach (Salisbury 1952). *Ammophila arenaria* (marram grass) is a perennial grass species that dominates the coastal foredunes of north-western Europe and the Mediterranean (Huiskes 1979). It is introduced in Australia and North America. Special properties enable it to tolerate burial by windblown sand up to 1 m year^{-1} (Ranwell 1958; Huiskes 1979). The North American *Ammophila breviligulata* (American beachgrass) (Maun & Lapierre 1984) also has this property. Sand drift is a key factor in primary succession on foredunes. Both *Ammophila* spp. are vigorous if buried regularly by windblown sand from the beach, whereas they degenerate after sand deposition ceases (Marshall 1965; Hope-Simpson & Jefferies 1966; Huiskes 1979; Eldred & Maun 1982). Possible explanations for the relationship between sand deposition and vigour of *Ammophila* spp. are:

1 Windblown sand acts as fertilizer. Windblown sand supplies *Ammophila* spp. with nutrients, e.g. P, K and $CaCO_3$, whereas decaying debris is a source of N (van Dieren 1934; Lux 1964).
2 Windblown sand eliminates competing species. Specialization of *Ammophila* spp. to survive sand burial presumably resulted in loss of competitive ability (Huiskes 1979). Hence in stabilized dunes, invading species expel *Ammophila* spp. from the vegetation by interspecific competition (Halwagy 1953, cited by Marshall 1965; Huiskes 1979).
3 Windblown sand prevents plants from physiological ageing. Roots of *Ammophila* spp. have a limited life-span and are formed on nodes which can, however, produce only a limited number of roots. The formation of new nodes depends on the deposition of fresh, windblown sand (Gemmell *et al.* 1953; Marshall 1965).
4 Colonizing windblown sand enables *Ammophila* to escape from soil pathogens. Recently it has been shown that coastal foredunes contain soil organisms which are harmful to *A. arenaria* and which could be involved in the degeneration of natural stands of this species (van der Putten *et al.* 1988). This chapter reviews that investigation and subsequent research on the possible involvement of harmful soil organisms in degeneration of *A. arenaria*. The effect of the organisms on growth of *Calammophila baltica* is also discussed. The latter is known as Baltic marram grass, a sterile hybrid of *A. arenaria* and *Calamagrostis epigejos*, and it occurs in coastal foredunes of north-western Europe (Westergaard 1943).

HARMFUL SOIL ORGANISMS IN THE ROOT ZONE OF *A. ARENARIA*

In The Netherlands, coastal foredunes have to conform to minimum dimensions because of flood danger. The foredune ridge at Voorne Island was much too weak. From 1985 to 1988 the foredune was raised and strengthened with $6 \cdot 10^6\,m^3$ of sand. There was some controversy as to whether dredged sea sand could be used instead of sand from the local dune area. The problem was that after raising the foredune, *A. arenaria* had to be re-established. This species is supposed to benefit from soil micro-organisms occurring in dune sand, e.g. vesicular-arbuscular mycorrhiza, which would not be present in virgin sea sand. Therefore, seedlings of *A. arenaria* were grown in pots with sand from both sea-floor and foredune origin. Sea sand was also mixed with sand from the foredune. All pots were supplied weekly with Hoagland nutrient solution without N, P and K. Three groups of pots received different amounts of NPK, i.e. (a) 0, (b) 1/8, and (c) 1/2 strength of Hoagland-NPK (van der Putten *et al.* 1988).

After 14 weeks growth in a greenhouse, the biomass production in sea sand was significantly higher ($P < 0{\cdot}05$) than in rhizosphere sand if plants were supplied with NPK (Fig. 1). The addition of rhizosphere sand to the sea sand had a negative effect on biomass production. Hence, foredune rhizosphere sand contained a factor that reduced the growth of *A. arenaria*.

An experiment was done to examine the cause of growth reduction. Sand was

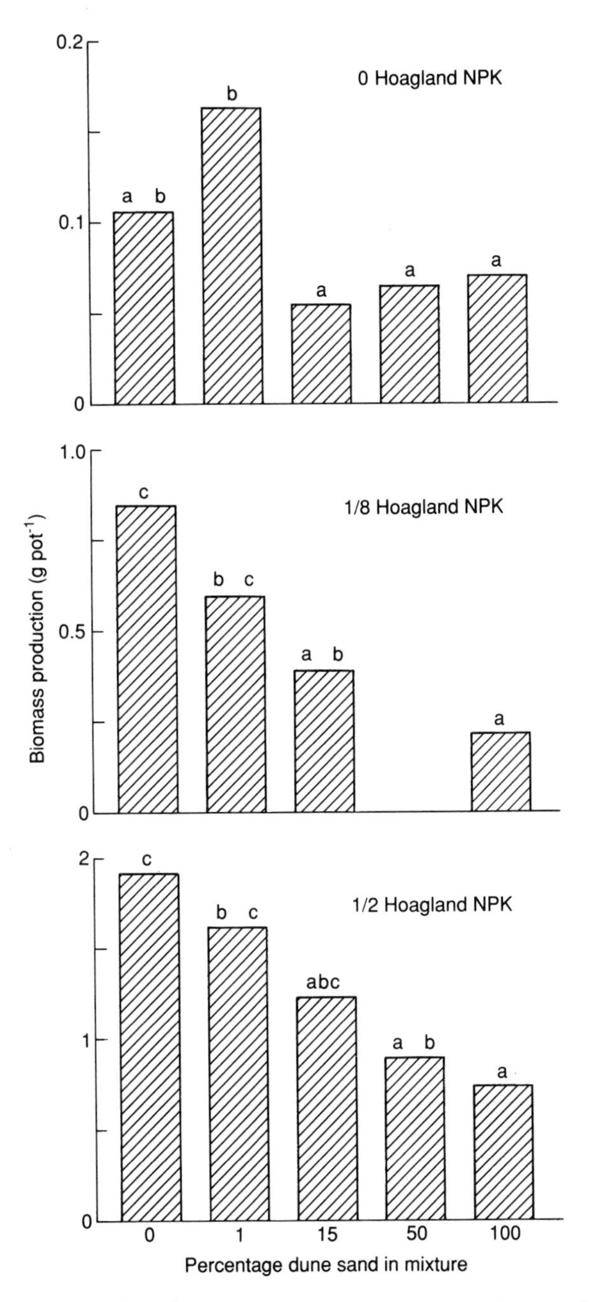

FIG. 1. Biomass production (g pot^{-1}) of *A. arenaria* seedlings in five mixtures of sea and dune sand grown at three nutrient levels: 0 Hoagland NPK, 1/8 Hoagland NPK, and 1/2 Hoagland NPK (all 100 ml pot^{-1} $week^{-1}$). In each plot significant differences ($P < 0{\cdot}05$) are indicated by different letters (data from van der Putten *et al.* 1988).

collected from three extreme stages of foredune succession: the beach (*A. arenaria* absent), the mobile dune (with vigorous *A. arenaria*) and the stable dune (with degenerated *A. arenaria*). Half of each sand type was sterilized by means of gamma irradiation (2·5 Mrad; Oremus & Otten 1981), put in pots and planted with seedlings of *A. arenaria*. The plants were grown in a greenhouse. Artefacts, e.g. due to nutrient flushes caused by soil sterilization, were avoided by supplying ample Hoagland nutrient solution (van der Putten & Troelstra 1990).

After 44 days, biomass production was significantly lower in unsterilized sand from the mobile and stable dunes than in unsterilized sand from the beach (Fig. 2). Soil sterilization did not affect biomass production in beach sand; however, in sand from the mobile and stable dunes soil sterilization caused a significant increase of biomass production. Thus the root zone of the mobile as well as of the stable dunes contained harmful soil organisms, whereas these organisms did not occur on the beach.

CHARACTERIZATION OF HARMFUL SOIL ORGANISMS IN COASTAL FOREDUNES

The harmful soil organisms occurring in sand from the root zone of *A. arenaria* were characterized by applying biocides to eliminate or inactivate groups of soil organisms as selectively as possible. In a greenhouse, seedlings of *A. arenaria* were grown in sand that was mixed with bactericides, fungicides or a nematicide and, as a control, in untreated and in sterilized sand. Results of two experiments

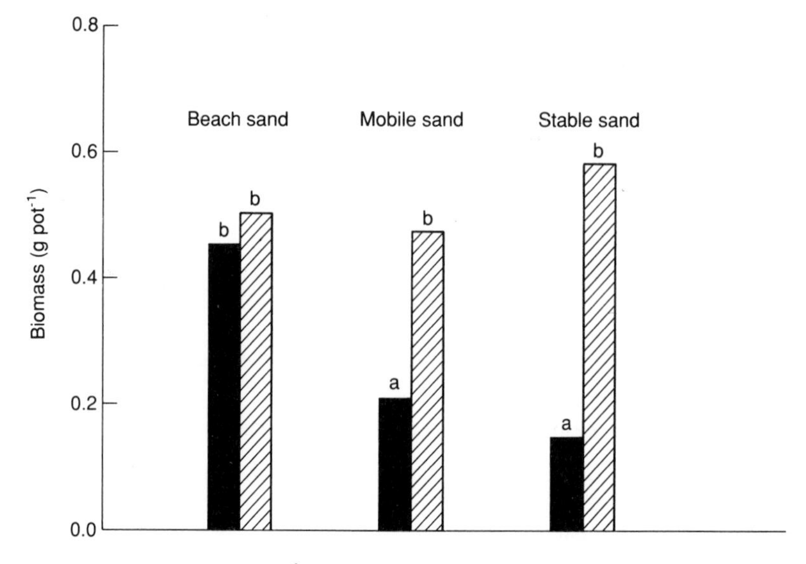

FIG. 2. Biomass production (g pot^{-1}) of *A. arenaria* seedlings in sterilized (black) and unsterilized (hatched) sand samples originating from the beach, the mobile and the stable dunes. Significant differences ($P<0{\cdot}05$) are marked by different letters (data from van der Putten & Troelstra 1990).

TABLE 1. Biomass production of *A. arenaria* seedlings as an indicator of the growth reducing effect of harmful organisms in the soil. Soil organisms were eliminated by: (a) gamma-sterilization; (b) bactericides; (c) a fungicide; and (d) a nematicide. − indicates poor growth; +, significant growth increase; ++, very significant growth increase. Sand was examined from (a) a mobile dune; and (b) a stable dune (with vigorous and degenerated *A. arenaria*, respectively)

	Sand origin	
Soil treatment	Mobile dune	Stable dune
Untreated	−	−
Gamma-sterilization	++	++
Bactericide (streptomycin + penicillin)	−	−
Fungicide (benomyl)	+	+
Nematicide (oxamyl)	+	+

(van der Putten *et al.* 1990) are summarized qualitatively. Bactericides did not stimulate growth (Table 1). On the other hand, in sand containing the nematicide oxamyl, as well as in that with the fungicide benomyl, biomass production was only slightly lower than in sterilized soil. The applied nematicide was supposed to be very specific and prevented roots from being colonized by endoparasitic nematodes (*Pratylenchus* sp., *Heterodera avenae* group and *Meloidogyne maritima*) (van der Putten *et al.* 1990). However, the fungicide benomyl was not specific, as it also eliminated nematodes. Therefore, with these experiments it could not be shown that only nematodes were involved in growth reduction of *A. arenaria* or that soil fungi were also responsible for this reduction.

Other experiments showed that even if no plant parasitic nematodes occurred in sand from the root zone, growth of *A. arenaria* could be improved by soil sterilization (van der Putten *et al.* 1989). From inoculation experiments it appeared that soil fungi collected from the root zone of a foredune could reduce growth of *A. arenaria* and that this reduction was higher if fungi were inoculated in combination with nematodes (van der Putten 1989). Possibly, soil fungi are also involved in growth reduction; however, the identification of the harmful soil organisms requires more investigation.

IMPACT OF HARMFUL SOIL ORGANISMS UNDER SEMI-NATURAL CONDITIONS

The effect of harmful soil organisms on plant growth under semi-natural conditions was examined by means of outdoor experiments. Sand was collected from the same dune sere as was examined before (see experiment of Fig. 2). Only sand from the stable dune was sterilized and containers of 50 litres were filled with sand from: (a) beach; (b) mobile dune; (c) stable dune; and (d) sterilized sand from the stable dune. Every container was planted with five cuttings of either *A. arenaria* or *Calammophila baltica* (May) and supplied with Osmocote (24–6–6 NPK, 3·25 g pot^{-1}, 9–12 months active at 21°C) slow-release fertilizer (van der Putten & Troelstra 1990).

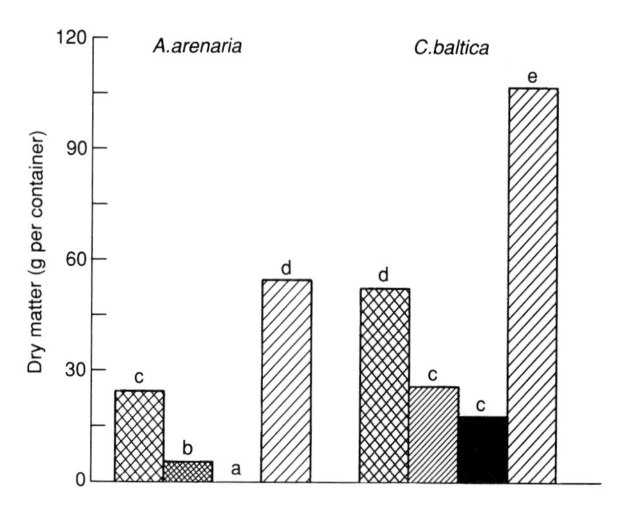

FIG. 3. Biomass production (g container^{-1}) of *A. arenaria* and *Calammophila baltica* cuttings in sand from the beach (cross hatched), the mobile dune (finely cross hatched), the stable dune (black) and in sterilized sand from the stable dune (single hatching). Significant differences ($P < 0{\cdot}05$) are marked by different letters (data from van der Putten & Troelstra 1990).

After 22 weeks, biomass production of *Calammophila baltica* was significantly higher than of *A. arenaria* in every sand type (Fig. 3). Both species produced the highest amount of biomass in the sterilized sand and then (in descending order) in sand from the beach, the mobile and the stable dunes. The difference between beach sand and sterilized sand could have been partly due to a nutrient flush occurring after soil sterilization (van der Putten & Troelstra 1990). It is obvious, however, that both species produced significantly less biomass in dune sand than in beach sand. In dune sand, as opposed to beach sand, biomass production of *A. arenaria* was more severely reduced when compared with *C. baltica*. Moreover, in unsterilized sand from the stable dune 44% of the *A. arenaria* plants died, whereas no mortality occurred in the containers with *Calammophila baltica*.

EFFECT OF HARMFUL SOIL ORGANISMS ON ROOT MORPHOLOGY OF *A. ARENARIA*

To quantify the effect of harmful soil organisms on root morphology, both roots and sand were collected from the soil profile of: (a) a mobile foredune (sand accretion of about 20 cm year^{-1}); and (b) a stable foredune (no sand accretion) at depths of 0–10, 10–20, 20–30, 30–40, 40–50, 50–60 and 60–100 cm below the surface. In a greenhouse, seedlings of *A. arenaria* were grown in both sterilized and unsterilized sand samples of each layer (van der Putten *et al.* 1989). After 8 weeks, plants were harvested and roots were examined. The length of roots covered with hairs was estimated as a percentage of the total root length. In the same way, disintegration of the cortex was quantified.

TABLE 2. Influence of various growth conditions (in the greenhouse and in the field) on the occurrence of root hairs and cortex disintegration (expressed as a percentage of total root length) on roots of *A. arenaria* (data from van der Putten *et al.* 1989)

		Mobile site			Stable site		
			Pot experiment			Pot experiment	
Layer	Depth (cm)	Field material	Unsterilized sand	Sterilized sand	Field material	Unsterilized sand	Sterilized sand
Root hairs (%)							
1	0–10	82·5	37·5	99·5	17·5	92·5	97·5
2	10–20	82·5	2·5	99·5	2·5	92·5	97·5
3	20–30	72·5	22·5	99·5	2·5	57·5	87·5
4	30–40	17·5	2·5	97·5	0	67·5	97·5
5	40–50	2·5	17·5	99·5	0	52·5	87·5
6	50–60	2·5	42·5	97·5	0	67·5	92·5
7	60–100	2·5	12·5	97·5	0	32·5	47·5
Cortex disintegration (%)							
1	0–10	7·5	7·5	0	22·5	2·5	0
2	10–20	17·5	12·5	0	72·5	2·5	0
3	20–30	12·5	2·5	0	82·5	2·5	7·5
4	30–40	72·5	2·5	0	100	2·5	0
5	40–50	92·5	2·5	0	100	12·5	2·5
6	50–60	87·5	2·5	0·5	100	7·5	2·5
7	60–100	87·5	2·5	0·5	100	12·5	2·5

Hairs were abundant on the roots collected from the upper part of the mobile dune (Table 2). In the stable dune only roots in the upper 10 cm had some hairs, whereas in all lower layers, the roots were virtually hairless. In the greenhouse, root hairs were abundant in sterilized sand, as well as in unsterilized sand from the stable dune. However, in unsterilized sand from the mobile dune the cover with hairs was usually low.

Cortex disintegration in the field increased with depth (Table 2). At the stable dune severe disintegration occurred below 10 cm, whereas in the mobile dune the same percentage was found at a depth of 40 cm.

DISCUSSION AND CONCLUSIONS

Sand from the root zone of *A. arenaria* contains harmful soil organisms. This is shown by comparing growth of *A. arenaria* seedlings in sterilized and unsterilized sand. Enhanced growth in sterilized sand can be due to a flush of nutrients, being released from the dead micro-organisms (De Nooij *et al.* 1986). In order to compensate for such flushes all plants in the greenhouse experiments were supplied with Hoagland nutrient solution with a strength that was ample for maximal plant growth (van der Putten *et al.* 1988). Under these experimental circumstances the comparison of plant growth in sterilized and unsterilized sand could be used for examination of the presence of harmful soil organisms in sand from different foredune locations.

The harmful organisms, probably a combination of nematodes and soil fungi, were detected in stable dunes with degenerated *A. arenaria*, as well as in mobile dunes with vigorous plants. Burial by windblown sand, usually originating from the beach, maintains vigour of *A. arenaria* in mobile dunes (Marshall 1965; Huiskes 1979). Since beach sand does not contain harmful soil organisms, upward colonization of windblown sand by *A. arenaria* enables the plants to form roots in sand without these organisms. Therefore, it may be concluded that sand deposition enables *A. arenaria* to escape from harmful organisms. These organisms, however, colonize newly formed roots within a year (van der Putten *et al.* 1989), so that the plants in order to maintain vigour constantly require windblown sand.

In stable dunes, roots can only be formed in sand that already contains the harmful organisms. As a result these new roots are hairless, branched shortly (van der Putten *et al.* 1989) and have a low capacity to take up water and nutrients (van der Putten *et al.* 1988). In lower layers in the foredune the roots did not have a cortex, in opposition to the upper layers. This indicates that new roots are mainly formed in the upper soil layers (Marshall 1965; van der Putten *et al.* 1989). Here they are subject to stress by, e.g. drought and extremely high soil temperatures (Baldwin & Maun 1983). The susceptibility of the roots to stress will be increased if they are deformed by the harmful soil organisms. Therefore degeneration of *A. arenaria* in stable dunes may be due to a combination of biotic and abiotic stress.

Calammophila baltica is less susceptible to the harmful soil organisms than *A. arenaria*; however, its growth is also reduced by these organisms. Field observations might establish whether *Calammophila baltica* is relatively more abundant in stable dunes than *A. arenaria*; however, this has yet not been studied. Degeneration of another dune species, *Hippophaë rhamnoides* (sea buckthorn), has been related to the occurrence of nematodes in the soil (Oremus & Otten 1981). It seems, therefore, that harmful soil organisms are strongly involved in primary succession in coastal sand-dunes; however, this needs to be examined.

REFERENCES

Baldwin, K.A. & Maun, M.A. (1983). Microenvironment of Lake Huron sand dunes. *Canadian Journal of Botany*, **61**, 241–55.

Eldred, R.A. & Maun, M.A. (1982). A multivariate approach to the problem of decline in vigour of *Ammophila. Canadian Journal of Botany*, **60**, 1371–80.

De Nooij, M.P., Troelstra, S.R. & Wagenaar, R. (1986). Growth reduction in *Plantago lanceolata* in relation to biotic factors in the soil environment. *Oecologia*, **70**, 266–72.

Gemmell, A.R., Greig-Smith, P. & Gimingham, C.H. (1953). A note on the behaviour of *Ammophila arenaria* (L.) Link in relation to sand dune formation. *Transactions of the Botanical Society of Edinburgh*, **36**, 132–6.

Hope-Simpson, J.F. & Jefferies, R.L. (1966). Observations relating to vigour and debility in marram grass (*Ammophila arenaria* (L.) Link). *Journal of Ecology*, **54**, 271–5.

Huiskes, A.H.L. (1979). Biological flora of the British isles: *Ammophila arenaria* (L.) Link (*Psamma* *arenaria* (L.) Roem. et Schult.: *Calamagrostis arenaria* (L.) Roth). *Journal of Ecology*, **67**, 363–82.

Lux, H. (1964). Die biologischen Grundlagen der Strandhaferpflanzung und Silbergrasansaat im Dünenbau: eine Untersuchung uber die Möglichkeiten biologischer Dünenbau- und Dünenbefestigungsmassnahmen in den Sylter Dünengebieten bei vorhandener und fehlender Sandablagerung. *Angewandte Pflanzensoziologie*, **20**, 5–53.

Marshall, J.K. (1965). *Corynephorus canescens* (L.) P. Beauv. as a model for the *Ammophila* problem. *Journal of Ecology*, **53**, 447–65.

Maun, M.A. & Lapierre, J. (1984). The effects of burial by sand on *Ammophila breviligulata. Journal of Ecology*, **72**, 827–39.

Oremus, P.A.I. & Otten, H. (1981). Factors affecting growth and nodulation of *Hippophaë rhamnoides* L. ssp. *rhamnoides* in soils from two successional stages of dune formation. *Plant and Soil*, **63**, 317–31.

Ranwell, D.S. (1958). Movement of vegetated sand dunes at Newborough Warren, Anglesey. *Journal of Ecology*, **46**, 83–100.

Salisbury, E. (1952). *Downs and Dunes: Their Plant Life and its Environment*. Bell, London.

van der Putten, W.H. (1989). *Establishment, growth and degeneration of* Ammophila arenaria *in coastal sand dunes*. PhD thesis, Agricultural University, Wageningen.

van der Putten, W.H. & Troelstra, S.R. (1990). Harmful soil organisms in coastal foredunes involved in the degeneration of *Ammophila arenaria* and *Calammophila baltica. Canadian Journal of Botany*, **68**, 1560–8.

van der Putten, W.H., van Dijk, C. & Troelstra, S.R. (1988). Biotic soil factors affecting the growth and development of *Ammophila arenaria. Oecologia*, **76**, 313–20.

van der Putten, W.H., van der Werf-Klein Breteler, J.T. & van Dijk, C. (1989). Colonization of the root zone of *Ammophila arenaria* by harmful soil organisms. *Plant and Soil*, **120**, 213–23.

van der Putten, W.H., Maas, P.W.Th., van Gulik, W.J.M. & Brinkman, H. (1990). Characterization of soil organisms involved in the degeneration of *Ammophila arenaria. Soil Biology and Biochemistry*, **22**, 845–52.

van Dieren, J.W. (1934). *Organogene Dünenbildung: eine geomorphologische Analyse der Dünenlandschaft der West-Friesischen Insel Terschelling mit pflanzensoziologischen Methoden*. Martinus Nijhoff, The Hague.

Westergaard, M. (1943). Cyto-taxonomical studies on *Calamagrostis epigejos* (L.) Roth, *Ammophila arenaria* (L.) Link, and their hybrids (*Ammophila baltica* (Fluegge) Link). *Biologiske Skrifter*, **II**, 1–68.

Primary succession on land: Community development and wildlife conservation

M. B. USHER*
Department of Biology, University of York, York YO1 5DD, UK

INTRODUCTION

There is no absolute distinction between the processes of the colonization of a new environment, the establishment of the first species of plants and animals, the parallel developments of populations and of the abiotic character of the site and the subsequent development of a community. All are aspects, often defined in a fuzzy manner, of the same continuum, divided into categories to aid the reductionist approach to the study of a 'large' process. Primary succession on land can be considered as 'large' because, first, the process occurs more or less wherever new land is exposed on Earth (the possible exception being some very hot deserts, but see Vestal (1993)), and, second, the process occupies a very lengthy time-scale from the initial colonization events to a community in which there is no further directional change (although there may be cyclical change).

If succession is such a 'large' process, why have conservationists been so preoccupied with the communities of plants and animals that are characteristic of the end of the successional sequence? True conservation is the integration of measures to protect: (a) individual species; (b) the communities that they form; and (c) the ecological processes that relate these species and communities in both time and space. All of these facets of conservation are necessary for preserving biodiversity.

This chapter concentrates on a subject that is often not associated with research into successional communities: conservation biology. Why have conservation biologists focused their attention on climax, or apparently unchanging, communities? As ecosystems are damaged by human activity, it becomes more and more important that a whole suite of successional species are available, either for introduction or natural colonization, so that the damage can be repaired as speedily as possible. It is, therefore, likely that the early- and mid-successional species are among those in greatest need of conservation. The aim of this chapter is therefore to explore some of the concepts required for the conservation of successional species and communities, viewing conservation as an important applied aspect of the study of ecological succession. The examples quoted in this chapter are drawn from the work of the former Nature Conservancy Council (NCC), the government conservation body which covered the whole of Great Britain.

* Present address: Scottish Natural Heritage, 2 Anderson Place, Edinburgh EH6 5NP, UK.

SUCCESSION, COMMUNITIES AND CONSERVATION

Very few of the key sites listed by Ratcliffe (1977) are in areas undergoing primary succession; the few that qualify are coastal, where sand-dunes or shingle beaches are naturally accreting. In these situations the series of different successional stages, each with its own suite of species or community, is valued. An excellent example is Ratcliffe's description of Tentsmuir Point in Fife, in which he says 'the actively aggrading lime-poor dune system has full seral successions from embryo slacks and dunes to alder, birch and willow scrub'. Genuine primary successions are probably rare where naturalness is accepted as a necessary prerequisite for giving a large value to the conservation importance of a site, and may indeed be confined largely to coastal systems and Alpine and polar systems where glaciers are retreating.

However, Bradshaw (1977) estimated that 90% of Ratcliffe's (1977) key sites were, to a greater or lesser extent, influenced by human activity, especially by either forestry or agriculture. On many of these sites the maintenance of the conservation value is dependent upon resisting the changes that would be part of a successional process. The chalk grasslands are maintained as species-rich communities by controlled grazing, the hay meadows by appropriate manuring and hay cutting, and both upland and lowland heathlands by periodic fire. Without such management intervention, all such habitats would become invaded by scrub and, presumably, the sites would slowly develop towards some kind of woodland (see Marrs *et al.* (1986) for a discussion of successional loss of heathlands). The management of succession as a facet of conservation management is therefore important in a country such as the United Kingdom where very few sites are truly natural (Jefferson & Usher 1986).

In what instances are primary successions a feature of conservation biology in Britain? There are probably five main categories:

1 Sand-dunes and other coastal zones of sediment accretion (i.e. natural primary successions).
2 Landslips in areas of soft geological strata, especially cliff faces on boulder clay or sandy materials (i.e. also natural primary successions).
3 Quarries and gravel pits (though here the purpose of their creation was originally an extractive industry).
4 Spoil and waste heaps of various kinds (though the purpose of their creation was the disposal of industrial and mining wastes).
5 Local primary successions planned for particular conservation purposes.

Category 1 is well-reviewed in the literature (e.g. Ranwell 1972; van der Maarel in press). Category 2 is less well-understood; Usher & Jefferson (1991) have listed some of the sites that could be important for early successional insects, such as the Hymenoptera that nest in small, sandy cliff faces and *Melitaea cinxia* (Glanville fritillary butterfly) of the undercliff on the south coast of the Isle of Wight. Sites abandoned by the extractive industries, category 3, can gain wildlife interest

(Jefferson 1984), especially when the substrate being extracted is base-rich and moderately soft (Usher 1979). Greenwood & Gemmel (1978) and Gemmell (1982) have demonstrated that some sites with industrial waste, category 4, can develop considerable wildlife interest. Once again there is a strong relationship between this interest and the pH of a site, the alkaline wastes often developing interesting, species-rich floras, with associated rich invertebrate faunas. Categories 1, 3 and 4 are, therefore, reasonably well-documented for the conservation interest in their successional species and communities.

In relation to the development of communities of conservation interest during primary succession on land, it is category 5 that is of particular concern. A question such as 'Are primary successions ever started, as a management aim, on National Nature Reserves?' would probably generate the answer 'No' from most people. Conservation management generally cannot begin coastal accretion, generally does not encourage mineral extraction and generally prohibits dumping of spoil or industrial waste. However, in a very small way, primary successions are started on some National Nature Reserves (NNRs) for specific management purposes. The following section deals with the creation of habitats on NNRs.

HABITAT CREATION FOR ARTHROPODS ON NNRs

Usher & Jefferson (1991) reviewed the creation and re-creation of habitats for invertebrates on NNRs in Great Britain. All 15 regions of the NCC were contacted, and nearly 100 examples of habitat creation or re-creation were found. By far the commonest management activity was the re-creation of habitats that had been lost due to succession because of lack of active management. After re-creation of habitats, reserve management is now aiming to 'hold' communities at some predetermined point in a successional sequence; in general the earlier that point is in the successional sequence, the greater the management input required. Only 20 of the examples found in the NNR survey related to habitat creation; these are listed in Table 1.

The particular feature of Table 1 is that 16 of the 20 management activities listed relate to freshwater habitats. The other four activities relate to the creation of very open habitats, and hence probably to the conservation of insect species that are 'early successional' in their ecological requirements. Habitat creation, starting a primary succession, seems largely confined to small freshwater bodies, at least on British nature reserves; how speedily do these freshwater species and communities become established and therefore become important from a conservation point of view? Usher & Jefferson (1991) quoted examples of the dragonflies and damselflies of Loch Lomond and Rum NNRs. At the former, a single pond of *c*. 260 m^2, created in September 1986, had five breeding species of Odonata in 1987, and seven breeding species in 1988 and 1989. At Rum NNR, a sheltered pond dug in 1983 had seven breeding species by 1989, whereas a less-sheltered pond dug at the same time had only four or five breeding species by 1989.

TABLE 1. A survey of habitat creation for invertebrates on NNRs in Great Britain (taken from a survey of habitat creation and re-creation by Usher & Jefferson (1991)

NNR	Management activity
Ainsdale sand-dunes, England	Creation of permanent and seasonal pools in the dune system for freshwater invertebrates generally
Ariundle, Scotland	Creation of ponds for Odonata
Aston Rowant, England	Creation of open, stony ground for Hemiptera (especially *Hallodopus montadoni* and *Taphropeltus hamulatus*)
Braunton Burrows, England	Excavation of small ponds for Odonata
Chartley Moss, England	Creation of open water for freshwater invertebrates generally and *Neuronia clathrata* in particular
Cors-y-Llyn, Wales	Creation of pond for Odonata
Dorset Heaths (five NNRs), England	Creation of ponds for Odonata
Dyfi, Wales	Dam building to create wet raised mire and pools for Odonata
Glen Strathfarrar, Scotland	Creation of ponds for Odonata
Gordano Valley, England	Excavation of ditch system for Odonata
Loch Leven, Scotland	Creation of lagoons for reeds, freshwater invertebrates and wildfowl feeding
Loch Lomond, Scotland	Creation of pool for Odonata
Lower Derwent Valley, England	Creation of south facing sandy cliff face for Hymenoptera: Aculeata
Newham Fen, England	Creation of ponds for Odonata
Rum, Scotland	Creation of two ponds for Odonata and as possible introduction sites for common frog and/or toad
Shapwick Heath, England	Excavation of ditch system and small ponds for Odonata
Taynish, Scotland	Dam outflow of mires to create areas of open water for Odonata
The Lizard, England	Excavation of ponds and scrapes for Odonata
Thursley Common, England	Creation and maintenance of sandy patches within *Calluna* heath, and creation of 1.5 m high, south-facing sandy cliffs, for sand wasps (Hymenoptera)
Walberswick, England	Re-excavation of sand pits to create sandy cliff faces for Hymenoptera: Aculeata

Unfortunately, monitoring of the ponds on these NNRs has concentrated only on the target groups of organisms (Odonata) and has not recorded the development of the remainder of the invertebrate community.

Monitoring of a farm pond in Yorkshire created in February 1981 (Usher 1987) showed the development of a multi-species community. The pond had only six species of aquatic insect, or a total of 11 species of aquatic invertebrate, by Spring 1983. These figures had increased to 20 insect species, and nine other invertebrate species, by Spring 1986. Five years had seen the development of a reasonably species-rich community in which amphibians were also present. However, as fish had not yet colonized the pond, further changes in both species complement and species abundance could be expected.

PRIMARY SUCCESSIONS ON NNRs IN ONE REGION

The region selected for this example is the South-east England Region, where primary successions have been identified on nine NNRs (see Table 2). Naturally-occurring primary successions in this part of England are associated with marine deposits either of very fine particles (leading to salt-marshes rather than dunes) or of shells. No natural successions of the landslip, cliff edge formation, type have been identified on these NNRs. There are only two examples of industrial dereliction — one a quarry and one colliery spoil — both of which pre-date the designation of the site as an NNR. Again, management activities are largely related to the creation of ponds, though the example of stored chestnut coppice, arguably a primary succession, could prove important for saproxylic insects if they are able to colonize this area. Because of the vulnerability of this particular group of insects (Speight 1989), it would be important to monitor the colonization of the windblown chestnut as well as to consider the introduction of saproxylic species to aid the development of this community.

In compiling data on the South-east England Region, the management activities associated with primary successions were also investigated (see Table 3). In general the management costs are low; the cost of managing nine of the 12 successions was indicated as less than £500, although much work done by NCC wardens may not be included in these costs (as for example the zero cost indicated for Ham Street Woods NNR). Relatively little information can be gained from the data on time-scales other than (a) the majority of habitat creation or re-creation is recent, mostly dating from the 1980s; and (b) wardening staff generally felt that a climax community had developed extremely rapidly, for example one pond was assumed to have developed a climax freshwater community within 21 months, and scrub/thicket was referred to as climax!

More importantly, benefits and disadvantages were investigated. The list in Table 3 generally focuses on increased diversity of a site; this is achieved by managing the site so that species of all successional stages can coexist. It is interesting that, within this one region, there is no focus on any one group of species, since the replies from individual reserve wardens have mentioned plant

TABLE 2. Primary successions on the NNRs in the South-east England Region of the NCC (data provided in Autumn 1989 by the Regional Officer and wardens of the Region)

	NNR	Succession category*	Area involved†	Habitat and succession
1	Blean Woods	5	2	60-year-old stored chestnut (*Castanea sativa*) coppice, partially windblown in 1987, with policy of non-intervention
2	Ham Street Woods	5	1	Digging and removal of glade vegetation; monitoring of succession
3	Kingley Vale	5	2	Two new ponds created
4	Lullington Heath	5	1	Dry, derelict dewpond with scrub excavated and relined; constant macrophyte control needed
5a	Medway Marshes	1	4	Salt-marsh accretion resulting in formation of a salt-marsh succession
5b	Medway Marshes	1	2	Formation of small shell beaches on the edge of the salt-marsh
6a	Stodmarsh	5	1	Pond dug in reed swamp and clay-lined from local materials
6b	Stodmarsh	4	3	Approximately 60 years accumulation of colliery waste (prior to NNR declaration) dumped on an original habitat of wet grazing meadow; colonized by scrub
7a	The Swale	1	4	Salt-marsh accretion resulting in formation of a salt-marsh succession
7b	The Swale	1	3	Formation and accretion of shell spit
8	Swanscombe Skull Site	3	3	Gravel quarry that ceased working in 1947; gravel heaps and sand faces naturally developing grassland and mixed scrub woodland; central grassland area managed to remove any developing scrub species
9	Thursley	5	4	New ponds and peat diggings to restart successions of bog and wet heath communities

* The categories here relate to the text, where coastal accretion is 1, extractive industry is 3, industrial waste disposal is 4, and planned conservation management is 5; in this region there are no examples of category 2 (natural landslips, etc.).

† Area categories are 1 (<0·1 ha), 2 (0·2–1 ha), 3 (1·1–10 ha) and 4 (>10 ha).

Table 3. Management data for the primary successions on NNRs in the South-east England Region of the NCC. The numbering of reserves is the same as in Table 2

	Reserve benefits arising	Cost code*	Undertaker†	Time-scale	Comments
1	Comparison of different invertebrate and flora communities	1	1, 2	—	—
2	Comparison of successional and non-disturbed communities	0	2	Started 1989	—
3	Increased diversity of Mollusca, Odonata and Diptera	3	2	1976, 1989	Included in overall management of secondary successions by cutting, grazing and mowing
4	Breeding site for Odonata (eight spp.) and two newt spp. Drinking water for birds (e.g. finches, swallows) and mammals (e.g. fox, badger)	4	2, 3, 4, 5	1978–79; 5 years to develop	Considerable public use
5a	Increased diversity of salt-marsh flora; wildfowl and wader feeding areas	0	1	—	No management intervention required
5b	Early successional flora; tern breeding areas	0	1	—	No management intervention required
6a	Increasing diversity: separate pond and connecting ditch system communities	2	4	Started 1986	Management to prevent vegetation encroaching
6b	High, well-drained area, colonized by breeding nightjar (*Caprimulgus europeaus*) and butterflies (*Callophrys rubi* and *Pararge aegeria*)	1	2	Spoil disposal ceased 1968; now patchy scrub and bare rock	Educational value (demonstration of succession); habitat type not otherwise found locally
7a	Increased diversity and area for wildfowl feeding	0	1	—	No management intervention required
7b	Sequence from pioneer flora to stabilized shell materials; little tern (*Sterna albifrons*) breeding colony	0	1	—	No management intervention required
8	Diverse grassland flora; insect and bird communities of the scrub woodland	2	2	Natural since 1947	Annual management input to control scrub

TABLE 3. *Contd.*

	Reserve benefits arising	Cost code*	Undertaker†	Time-scale	Comments
9	Increase overall site diversity; retention of ephemeral species	4	2	Initial work 1963; larger scale from 1984; 15 years from dry heath to scrub if unmanaged	Part of more extensive work to control secondary succession since dry heath is readily invaded by *Betula* spp. and *Pinus sylvestris*. Bare, sandy areas are important for some insects

* Cost codes at 1989 values: 0, essentially no cost; 1, <£100; 2, £101–500; 3, £501–1000; 4, £1000.
† Undertakers: 1, natural processes; 2, NCC staff; 3, staff of other governmental agencies; 4, voluntary groups; 5, contractors.

diversity, birds, mammals, amphibians and a variety of invertebrate groups. Work to increase the natural diversity of these NNRs pre-dates the publication of E. O. Wilson's (1988) book, and the subsequent public and political interest in the subject of biodiversity.

DISCUSSION

These examples of habitat creation and re-creation demonstrate that conservation biologists in Britain are unlikely ever to be dealing with spatially large-scale primary successions, except in coastal habitats. However, the creation of new habitats is one of many methods of management that may be essential for the conservation of some of the species and communities associated with the earlier stages of a successional process. The analysis of rarity by Hodgson (1986) showed that the species in most urgent need of conservation are those of the most distributed habitats, frequently the early successional species.

Studies in primary succession have relevance for conservation management and the maintenance of biodiversity. It is interesting that many studies have been concerned with the development of communities in potentially 'extreme' habitats. Why do ecologists have such a fascination for these extreme environments? Is it that the primary successions have been described for many other geographical areas and that these are the only environments remaining to be studied? Or is the apparent simplicity of some of these successions (i.e. with rather fewer species) a genuine scientific advantage in understanding how a community develops and is organized? As a preliminary, it is perhaps more important to answer the question 'does it matter what sort of community we study in order to explore the ecological principles of succession?'.

Van der Putten (1993) investigated what, from a botanical point of view, is also a simple community with few species. He showed that the development of this community is dependent upon the interplay between plant root growth, nematode activity and fungal infection. This relationship between plants and soil animals is reminiscent of the effect of phytophagous insects on the early stages of a secondary succession (Brown 1982), or the complex interactions between higher plant growth, mycorrhizal infection of the plant's roots and the soil arthropod community demonstrated by Warnock *et al.* (1982).

An important conclusion from studies of natural primary successions is that succession should not be studied by a botanist or a zoologist in isolation. The ecological principles only become apparent when the plants, animals and microbes, both above and below ground, are all investigated simultaneously, demonstrating that each group is interacting with the others and modifying the course of the succession.

Genuine ecological principles tend to be elusive and, if found, tend to have relatively little ability to be predictive. Conversely, if a management aim can be defined, ameliorative treatments that allow that aim to be achieved can generally be applied with a reasonable probability of success. The literature on reclamation of industrial spoil tends to give the impression that the restoration of biological productivity, physical stability, lack of pollution potential or scenic amenity have a much higher priority than the replacement of the original species in a successional sequence. How does the conservation biologist react to such a view? Should species which are not good colonists, or which have become rare in the geographical area, be introduced to reclamation sites? In other words, should the creation of biodiversity be built in as a management aim? The study of the butterflies of a landfill site in Essex by Davis (1989) indicated that many species will colonize naturally, though a few will need to be introduced if the aim is to include them in the butterfly fauna. Can and should communities be multipurpose, tailor-made for their productivity, stability, aesthetic amenity and conservation of wildlife? Conversely, will conservation biologists remain purists, valuing 'naturalness' above all else, and insisting that all sites must progress through the full successional sequence? The complement of successional species will be restricted since not all the species will be able to colonize these new sites due to their rarity, the fragmentation of their populations or other reasons.

This discussion, with the examples from the NNRs, has developed two particularly important themes about the relationship between primary succession and wildlife conservation — the preservation of biodiversity in nature reserves and the inclusion of nature conservation objectives in managing the wider countryside. The planned creation or re-creation of small areas where a successional sequence can commence or recommence raises several issues. The most important is where the colonists, or species associated with later successional stages, will come from. Some of these plant species may exist on site in dormant seed banks; Jefferson & Usher (1987) demonstrated that there were a few species in soil seed banks that could not be found in the vegetation in the quarry sites they studied. In genuine primary successions, where there is no seed bank, the arrival of species is a stochastic process, the probability of arrival and establishment declining as progagules need to travel longer distances. The retention of the complete suite of species on site therefore becomes more important. True conservation will ensure the long-term survival of these successional species, many of which would, in any case, be required to remedy the occasional damage (windthrow, fire, flooding) that can naturally occur on sites.

In the wider countryside, conservation biologists should be more enthusiastic about creating complete communities. It is often accepted that planting trees is satisfactory, but why should the herbaceous and lower plants, invertebrates and soil biota all be expected to colonize the newly planted areas naturally? It would certainly be more beneficial to the nature resource of the country if as complete a community as possible could be established without complete reliance on the vagaries of the stochastic arrival and subsequent colonization of species.

Nature reserves are not divorced from the wider countryside. They may be used to provide species that can be used as colonists for other areas. However, if the national aim is to preserve biodiversity, conservation management must (a) retain all of the species that are associated with successions and (b) ensure an equitable spatial distribution of these species within their biogeographical range. This will ensure more rapid natural colonization of newly created sites and a smaller probability of chance extinction if one or two populations are destroyed by disease, development, or other agencies.

REFERENCES

Bradshaw, A.D. (1977). Conservation problems in the future. *Proceedings of the Royal Society of London, Series B*, **197**, 77–96.

Brown, V.K. (1982). The phytophagous insect community and its impact on early successional habitats. *Proceedings of the 5th International Symposium on Insect–Plant Relationships*, pp. 205–13. Pudoc, Wageningen.

Davis, B.N.K. (1989). Habitat creation for butterflies on a landfill site. *The Entomologist*, **108**, 109–22.

Gemmell, R.P. (1982). The origin and botanical importance of industrial habitats. *Urban Ecology* (Ed. by R. Bornkamm, J.A. Lee & M.R.D. Seawood), pp. 33–9. Blackwell Scientific Publications, Oxford.

Greenwood, E.F. & Gemmell, R.P. (1978). Derelict industrial land as a habitat for rare plants in South Lancs (V.C. 59) and West Lancs (V.C. 60). *Watsonia*, **12**, 33–40.

Hodgson, J.G. (1986). Commonness and rarity in plants with special reference to the Sheffield flora, part II: the relative importance of climate, soils and land use. *Biological Conservation*, **36**, 253–74.

Jefferson, R.G. (1984). Quarries and wildlife conservation in the Yorkshire Wolds, England. *Biological Conservation*, **29**, 363–80.

Jefferson, R.G. & Usher, M.B. (1986). Ecological succession and the evaluation of non-climax communities. *Wildlife Conservation Evaluation* (Ed. by M.B. Usher), pp. 69–91. Chapman & Hall, London.

Jefferson, R.G. & Usher, M.B. (1987). The seed bank in soils of disused chalk quarries in the Yorkshire Wolds, England: implications for conservation management. *Biological Conservation*, **42**, 287–302.

Marrs, R.H., Hicks, M.J. & Fuller, R.M. (1986). Losses of lowland heath through succession at four sites in Breckland, East Anglia, England. *Biological Conservation*, **36**, 19–38.

Ranwell, D.S. (1972). *Ecology of Salt Marshes and Sand Dunes*. Chapman & Hall, London.

Ratcliffe, D.A. (ed.) (1977). *A Nature Conservation Review*, vols 1 & 2. Cambridge University Press, Cambridge.

Speight, M.C.D. (1989). Saproxylic invertebrates and their conservation. *Council of Europe, Nature and Environment Series*, **42**, 1–82.

Usher, M.B. (1979). Natural communities of plants and animals in disused quarries. *Journal of Environmental Management*, **8**, 223–36.

Usher, M.B. (1987). Creation of a new pond, Hopewell House, North Yorkshire. *Conservation Monitoring and Management* (Ed. by R. Matthews), pp. 12–18. Countryside Commission, Cheltenham.

Usher, M.B. & Jefferson, R.G. (1991). Creating new and successional habitats for arthropods. *The Conservation of Insects and their Habitats* (Ed. by N.M. Collins & J.A. Thomas), pp. 263–91. Academic Press, London (in press).

van der Maarel, E. (in press). *Ecosystems of the World*, 2. *Dry Coastal Ecosystems*. Elsevier, Amsterdam (in press).

van der Putten, W.H. (1993). Soil organisms in coastal foredunes involved in degeneration of *Ammophila arenaria*. *Primary Succession on Land* (Ed. by J. Miles & D.W.H. Walton), pp. 273–81. Blackwell Scientific Publications, Oxford.

Vestal, J.R. (1993). Cryptoendolithic communities from hot and cold deserts: Speculation on microbial colonization and succession. *Primary Succession on Land* (Ed. by J. Miles & D.W.H. Walton), pp. 5–16. Blackwell Scientific Publications, Oxford.

Warnock, A.J., Fitter, A.H. & Usher, M.B. (1982). The influence of a springtail *Folsomia candida* (Insecta, Collembola) on the mycorrhizal association of leek *Allium porrum* and the vesicular-arbuscular mycorrhizal endophyte *Glomus fasciculatum*. *New Phytologist*, **90**, 285–92.

Wilson, E.O. (ed.) (1988). *Biodiversity*. National Academy Press, Washington, DC.

Primary succession revisited

J. MILES* AND D. W. H. WALTON[†]

* *Institute of Terrestrial Ecology, Hill of Brathens, Banchory, Kincardineshire AB31 4BY, and* [†] *British Antarctic Survey, Natural Environment Research Council, High Cross, Madingley Road, Cambridge CB3 0ET, UK*

INTRODUCTION

The aim of the symposium was to focus attention on the early stages of primary succession. It was hoped that, in reviewing the field of knowledge and concentrating on the major processes involved in primary succession, the symposium would, as well as identifying gaps in knowledge, identify more clearly than hitherto those features which distinguish primary from secondary succession.

In practice, gaps in knowledge appeared well before the symposium took place, as we tried to find speakers to cover the entire field of interest. Thus it proved quite impossible to find an aerobiologist to discuss long-distance dispersal of spores and other propagules in an ecological context. There also appeared to be an apparent lack of interest in the processes leading to the establishment of animal communities; Majer's (1989) recent volume *Animals in Primary Succession* throws little light on this. Dispersal has long been recognized by biologists as a key process, but one about which little is known for most species. Indeed, until recently, dispersal seemed commonly to be regarded by the major agencies which fund research as smacking of Victorian-era natural history. Interestingly, one consequence of current concerns about global climate change, and in the European Community about possible land-use changes (including the abandonment of agricultural land, which might result from reform of the Common Agricultural Policy), has been to persuade funding agencies of the need to support further work on processes of animal and plant dispersal and migration!

In what follows, we consider the symposium proceedings in the light of Bradshaw's analytical introduction, looking at succession: first, from the point of view of the plants and animals; second, from the point of view of their habitats; and third, considering species interactions and the development of succession. Interestingly, the steps which Bradshaw highlighted and discussed were all clearly identified early this century by Clements (1904, 1916).

ORGANISMS

Dispersal

Dispersal — the getting there — is clearly crucial to questions of colonization and

* Present address: The Scottish Office Central Research Unit, New St Andrew's House, Edinburgh EH1 3TG, UK.

succession. Smith's paper leads to the inference that there must be a large and widespread 'rain' of viable propagules of bryophytes in Antarctica. His studies have shown that ice and snowfields accumulate spores, and that one consequence of snowmelt is to ensure that recently exposed soils, even those remote from any established vegetation, contained often substantial pools of viable bryophyte propagules reflecting a diverse species assemblage. Similar rains of bryophyte propagules probably occur in other continents. Certainly, exposed soil in pots in glasshouses in temperate regions quickly becomes colonized by bryophytes, though these are usually ubiquitous pioneer species. However, in two experiments in upland Scotland to study soil change in which patches of mature birchwood were felled, the field layers killed by chemical herbicides, and the patches then planted with pre-established plants of *Calluna vulgaris* (heather), all the dozen or so mosses commonly associated with *Calluna* in nearby moorland were present after only 2–3 years (J. Miles, unpubl. data).

Little is known about the ability of most invertebrates to disperse over long distances, though the evidence correlating the spread of various European earthworm species across North America with the activities of freshwater anglers suggests that the dispersal powers of these are limited. Also in contrast to bryophytes, the dispersal powers of most flowering plants seem very limited. This poses few problems when sources of seeds or fruits are available in the vicinity. Thus, J. Miles (unpubl. data) examined the primary succession occurring on a 2-ha patch of calcareous mica-schist rock spoil which was deposited at the bottom of a hillside in Perthshire, Scotland, in the late 1950s during the construction of a long tunnel. The spoil was dumped at the conjunction of an expanse of *Calluna*-dominated moorland, a birchwood and a livestock farm. In 1974, the flora of the spoil heap contained most of the commonly occurring flowering plants of the adjacent moorland, woodland and reclaimed grasslands, and also a surprisingly large number of less common plants. The combination of the lime-rich and often finely crushed rock and an annual precipitation of about 1250 mm clearly provided a receptive substrate for what must have been a heavy local seed rain. However, relative isolation is probably more common. H. Ash (unpubl. data) has shown through experimental introductions that the characteristically poor floras of islands of industrial waste and spoil can be accounted for by their isolation from similar habitats where species grow which are adapted to the particular soil conditions. In this symposium, Whittaker & Bush's analysis of the re-colonization of the Krakatoa Islands after the 1883 eruption showed that even a century later the islands have very few tree species characteristic of inland areas, even though species characteristic of later stages of forest succession have continued to appear during the last 50 years. Most of the woodland cover is still provided by a few tree species characteristic of early successional stages. The effect of isolation in limiting succession on Krakatoa is also shown by the fact that most species colonizing are those commonly dispersed endozoically.

The studies of del Moral and his colleagues in the aftermath of the first Mount St Helens volcanic eruption in 1980 demonstrated yet again the crucial role of species accessibility in colonization and succession. Regrowth of vegetation at

Mount St Helens was quick only at sites with a species legacy. The conclusion is that the nature and direction of succession at Mount St Helens reflects the order of species arrival by dispersal, and nothing else. It would not be surprising if these stochastic effects persisted for decades. Indeed, studies such as those at Mount St Helens bear out the essential correctness of the views about the 'individualistic' nature of vegetation developed earlier this century independently by H. A. Gleason, F. Lenoble and L. G. Ramensky.

Establishment

Only Chapin's paper deals explicitly with establishment, and even he was forced in the main to make inferences from descriptive rather than experimental studies. Drawing largely on studies from north temperate and sub-Arctic regions, Chapin concluded that colonizers in primary succession commonly have traits which maximize dispersal but permit growth in dry, infertile sites. He concluded, like Whittaker & Bush, that woody species are important colonizers during primary succession, but also concluded that most have light, wind-dispersed seeds, which contrasts with the mainly endozoically dispersed tree flora that Whittaker & Bush found in Krakatoa. Species identified as common colonizers of primary substrates are characterized by low absolute growth rates, which contrasts with the generally high relative growth rates characteristic of pioneers in secondary successions. This highlights a crucial difference, indeed perhaps the most crucial difference, between primary and secondary successions, namely that the soils have a low content of organic matter and hence a low nitrogen content. The importance of acquiring a pool of nitrogen in soils on primary substrates to allow succession is stressed by Marrs & Bradshaw, and is also highlighted by Davy & Figueroa as a major determinant of establishment and success of strandline colonists.

Growth

Plants will only grow if they can acquire nutrients and water. Excluding the rather special cases of succession on beaches and sand dunes where the substrates may have been weathered and leached for millennia, primary successions commonly occur on relatively unweathered substrates. In these all the elements needed for plant nutrition may be present in abundance except for nitrogen and phosphorus, but especially the former. Most phosphorus comes directly from weathering of soil minerals, such as apatite. Its availability in primary successions is directly proportional to its abundance in the soil minerals. What is the origin of the nitrogen which accumulates during primary successions? Ammonia and nitric acid are now common atmospheric pollutants, though rates of deposition of these decline with increasing distance from source. Further, nearly all such inputs will be lost in run-off water unless an appreciable cover of plants is available to capture them. Wind-blown material should not be overlooked as a source of nitrogen; del Moral noted that organic matter fallout at Mount St Helens averaged $4\,mg\,m^{-2}\,day^{-1}$ during the growing season, and exceeded $10\,mg\,m^{-2}$

day^{-1} in places. Sprent noted that cyanobacteria are common on most sites, but that little is known about how much nitrogen they fix and how much of this is made available to subsequent stages in succession. However, fixation of nitrogen by cyanobacteria at maximum recorded levels of 15 kg N ha^{-1} year^{-1} could, together with other inputs, support some growth by flowering plants.

Both Gray and Chapin discuss characteristics of pioneer plants of primary succession. Such species do not show the sets of life-history traits usually thought to characterize pioneer plants of secondary succession. In contrast to secondary successions, the pioneer plants of primary successions are generally either long-lived iteroparous perennials or shrubs and trees. These are generally slow-growing, and commonly persist into the later stages of succession. Grubb (1986) commented on the remarkable variety of pioneer flowering-plant species found in primary successions, noting the common denomination that growth form and growth rate must match the capacity of the substratum to supply water and mineral nutrients.

Grubb (1986) also noted that the life-form of pioneers in primary succession recorded in different studies seemed to reflect physical differences in the nature of the substrata, with, for example, trees being common pioneers on rocky and bouldery substrata, iteroparous herbs on gravels and grasses on silty substrates. In an interesting case history Mazzoleni & Ricciardi describe how different plants colonizing the core of Vesuvius have selected different physical substrates.

However, any attempt to construct such general classifications is confounded by the tendency of plant species immigration to particular sites to be by particular mechanisms of dispersal. Thus it is not surprising that most species recorded on Krakatoa by Whittaker & Bush 100 years on in the succession appear to have arrived via the alimentary tracts of birds. It is similarly not surprising that del Moral found that most pioneers at Mount St Helens were either wind-dispersed or had survived the eruption. Yet the strandlines of the seashore and of gravel islands in fast-flowing rivers have colonists with a diversity of life histories; propagule mass is apparently irrelevant for dispersal by water.

There seem to be several reasons why trees and other woody shrubs are common pioneers in primary successions. Some genera such as *Salix*, *Populus* and *Betula* produce large numbers of small, light wind-dispersed propagules, which make them common pioneers in both primary and secondary successions. However, larger-seeded taxa, such as wind-dispersed *Pinus* and *Picea* spp. and bird-dispersed *Quercus* spp. are also common pioneers on landslips, mud flows and talus slopes. Grubb (1986) has suggested that many woody species of this type are good colonizers in primary succession because of their slower growth rates and hence lower nutrient requirements, while the relatively large seeds contain an appreciable initial source of nitrogen. Equally, their robust growth forms presumably allow them to survive in unstable rocky and bouldery areas where plants with softer tissues would be destroyed. Some trees also have remarkably extensive root systems; in stony substrates with intrinsically low water-holding capacities, it must be advantageous for a plant to have roots which can ramify down to the water-table or into a different substratum with a higher water content.

HABITAT

Initial characteristics

Grubb (1986), and in this volume del Moral and Mazzoleni & Ricciardi in particular, discuss how the physical characteristics of the substrate select for different groups of pioneers. Particularly coarse substrates develop a vegetation cover very slowly, perhaps mainly because of their very low water-holding capacity. Thus the west of Scotland has extensive raised beaches, 10 000 or more years old, which formed when the weight of the last glacial ice-cap had depressed the land relative to sea-level. Many of these beaches are of water-rounded boulders and are still largely devoid of vegetation, most notably on the Isle of Jura, apart from some epilithic lichens and mosses. Such substrates cannot of course weather into soil except on a geological time-scale, while the large spaces between boulders usually prevent the surface accumulation of appreciable amounts of airborne dust and organic matter.

A common characteristic of fine-grained primary substrates is their physical instability and hence proneness to movement by wind, water and frost action. Although records of the early stages of primary succession have usually only listed lichens, mosses and flowering plants, Wynn-Williams has clearly demonstrated the role of microbial crusts in stabilizing fine-textured primary substrates by binding mineral particles in a cyanobacterial and algal filament–mucigel matrix. Given the near ubiquity of cyanobacteria on primary sites (see Sprent), presumably because of their effective dispersal in the atmosphere, it seems highly probable that microbes may be the first, or among the first colonists, of all virgin substrates, and may initiate and facilitate plant succession.

Influence of autogenic factors

Walton and Wynn-Williams have both discussed the role of lower plants in primary successions. Almost all lower-plant taxa may have significant localized roles in accelerating the chemical weathering of minerals (so accelerating soil formation) and can be primary colonists of bare rocks and soils. Endolithic lichens can induce exfoliative weathering and hence promote the formation of fine-grained substrates, while terricolous lichens can stabilize the soil surface in areas prone to disruption by frost heaving. Mosses, like flowering plants, can scavenge and accumulate mineral particles, and so provide potential microsites and foci for establishment by flowering plants. The cryptoendolithic microbial communities described by Vestal may represent the self-renewing initial stages of primary successions which never proceed any further (unless the products of exfoliative weathering are taken into account) under existing climatic conditions.

The second stage at which autogenic factors seem to be crucial during primary successions is in establishing the store of organic nitrogen in the soil (and also of phosphorus in phosphorus-deficient substrates), which will allow sustained plant growth during succession. Most studies and all reviews have noted the importance of symbiotic nitrogen fixation in the early stages of primary succession. The presence of *Alnus* and *Dryas* spp. in succession behind retreating glaciers has long been known, and other examples were quoted at the symposium. Generally the colonist symbiotic nitrogen fixers are legumes, but these rarely appear in the earliest stages of primary succession, apparently because these large-seeded species generally lack effective long-distance dispersal mechanisms. A notable exception to this is the rapid dispersal of certain legumes, e.g. *Lupinus nootkatensis*, along the banks of Arctic rivers.

Although the role of nitrogen-fixing flowering plants in promoting succession was originally largely speculative, there is now increasing evidence, some experimental, that this is indeed the case. Beyond nitrogen fixation, many plants exert a 'nurse' effect, as well as acting as filters for trapping wind-blown material. As a site supports a higher plant biomass, a soil structure develops. However, although past studies have demonstrated favourable chemical trends during soil development in primary successions, studies of the development of a soil structure favourable to root growth and of the development of a soil ecosystem have been largely ignored.

Influence of allogenic factors

As soon as a virgin substrate is exposed, it is subject to weathering of the component minerals, leaching away of elements (including leaching of toxic elements in man-made wastes) and to inputs of materials from the air. Apart from del Moral's observations, little was heard at the symposium about allogenic factors. However, except on toxic wastes, it is difficult to separate the effects of allogenic and autogenic factors, because propagules of bacteria, fungi, algae, lichens, mosses and flowering plants, as well as invertebrates appear from the outset. Their growth immediately begins to influence the physical and chemical nature of the substrates, and so confounds the effects of purely allogenic factors.

Continued succession

Species interactions presumably occur from the outset, but those such as grazing of algae or bryophytes by invertebrates are not obvious and have rarely been studied in primary successions, although invertebrate faecal material plays an important role in soil development. Most studies of succession have been by plant ecologists, so most reports of species interactions are those between different species of flowering plant. Competition has been rarely studied, but simple inhibition — because the first-come species monopolizes space — does seem to be important. Equally, there are many records and inferences of particular species acting as 'nurse' plants (see del Moral). In particular, it seems clear that symbiotic nitrogen fixation by legumes and other nodulated taxa does facilitate vegetation development.

More intriguing are reports about iteroparous perennials such as those in this volume by Gray, by van der Putten, and by Thomson *et al.* The persistence of species such as *Ammophila arenaria* and *Spartina anglica* during succession is fascinating, and the reports here indicate habitat-correlated phenotypic plasticity, and also somatic variation related to clonal age. However, van der Putten reports reduced biomass production by and clonal degeneration of *Ammophila arenaria* in stable sand dunes, apparently caused by pathogenic soil organisms, probably both nematodes and fungi. In this latter case, competition between different flowering plant species is not needed for the pioneer to decline.

CONCLUSIONS

The symposium showed that there are few differences between primary and secondary successions, but that those that do exist are crucial. Arguably the most important is that newly exposed or deposited primary substrates almost always lack a reserve of nitrogen (only a very few rocks, all sedimentary, contain significant amount of this), while phosphorus and water availability also commonly limit plant growth. Stands of woody vegetation can only develop once adequate plant-available nutrient reserves have accumulated in the soil (Marrs & Bradshaw).

The two other crucial factors which clearly differentiate primary from secondary successions are substrate instability and the lack of any bank of organisms and their propagules in the soil. The exposed soil of a newly ploughed arable field is a complex mineral–organic matrix which is hundreds or thousands of years in time from a primary substrate. In contrast, a primary substrate is mineral only, and lacks the various mucigels and organic fractions which bind mineral particles together and stabilize the soil surface, so inhibiting erosion and facilitating colonizations by plants and animals. The lack of a propagule bank — a 'memory' of a previous vegetation cover — means that the development of species assemblages during primary succession depends on the probabilistic processes of species dispersal. It thus minimizes the chances of any orderliness or repeatability of successions on similar substrates but occurring at different places and times.

These differences suggest interestingly that the annual development of vegetation on coastal strandlines, which has traditionally been described as primary succession, is perhaps better described as secondary succession. Two key criteria — the lack of an adequate nitrogen supply and of a bank of propagules — are usually not met (Davy & Figueroa).

All other main features differentiating primary from secondary successions — rates and variability of ecosystem development, and variability in the characteristics of the pioneering flowering-plant species — follow from these crucial differences in the nature of the substrate.

The symposium highlighted our relative ignorance on how, how far and how quickly most species are dispersed, and on their requirements for establishing individuals and populations. This is perhaps especially the case for invertebrates; although a good deal is known about their role in soil development (e.g. Majer

1989), very little is known about the conditions for establishing populations and assemblages. Precise definitions of species niches — of 'microsites' for plants — are needed as much as ever. Grubb's (1986) hypotheses about the variability in life-form of flowering-plant pioneers of primary succession badly need experimental testing. Yet experimental studies such as that by Harper *et al.* (1965), which showed the very fine scale of discrimination of soil surface microtopography by establishing seedlings of different *Plantago* spp., remain rare.

Perhaps the best recent insights into primary succession have come from the careful and detailed studies following the 1980 volcanic eruption at Mount St Helens (del Moral), and from the analytical and experimental studies in Antarctica by the British Antarctic Survey (Smith; Walton; Wynn-Williams) and now extended by the French to their sub-Antarctic islands. The Mount St Helens work has clearly shown the intrinsically stochastic nature of the dispersal of macro-organisms, and hence of the course of succession. The studies in Antarctica have shown in particular just how important micro-organisms are in stabilizing the surfaces of newly exposed substrates, and hence in facilitating subsequent succession. The two contrasting areas also highlight the difference between the effective long-distance dispersal in the atmosphere of bryophyte spores and of micro-organisms (which can also be inferred from Sprent's note on the apparent ubiquity of cyanobacteria in primary successions), and of the slow and difficult-to-predict dispersal of most species of flowering plant. There remains a great deal to be learnt about primary succession.

REFERENCES

Clements, F.E. (1904). The development and structure of vegetation. *Botanical Survey of Nebraska*, **3**, 1–175.

Clements, F.E. (1916). Plant succession: an analysis of the development of vegetation. *Carnegie Institute of Washington Publication* No. 242.

Grubb, P.J. (1986). The ecology of establishment. *Ecology and Design in Landscape* (Ed. by A.D. Bradshaw, D.A. Goode & E. Thorp), pp. 83–97. Blackwell Scientific Publications, Oxford.

Harper, J.L., Williams, J.T. & Sagar, G.R. (1965). The behaviour of seeds in soil. I. The heterogeneity of soil surfaces and its role in determining the establishment of plants from seed. *Journal of Ecology*, **53**, 273–86.

Majer, J.D. (ed.) (1989). Animals in primary succession. *The Role of Fauna in Reclaimed Lands*. Cambridge University Press, Cambridge.

Index